What Is Mathematics, Really?

What Is Mathematics, Really?

Reuben Hersh

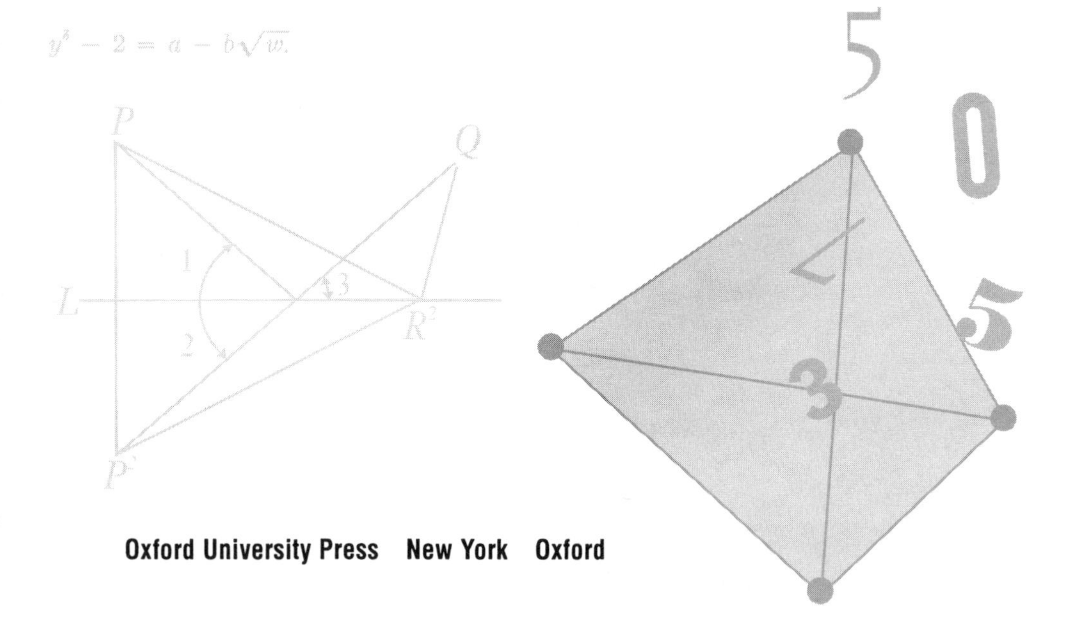

Oxford University Press New York Oxford

Oxford University Press

Oxford New York
Athens Auckland Bangkok Bogotá Buenos Aires Calcutta
Cape Town Chennai Dar es Salaam Delhi Florence Hong Kong
Istanbul Karachi Kuala Lumpur Madrid Melbourne Mexico City
Mumbai Nairobi Paris São Paolo Singapore Taipei Tokyo
Toronto Warsaw

and associated companies in

Berlin Ibadan

First published by Oxford University Press, Inc., 1997

First issued as an Oxford University Press paperback, 1999

Oxford is a registered trademark of Oxford University Press

Cataloging-in-Publication Data
Hersh, Reuben, 1927–
What is mathematics, really? / by Reuben Hersh.
p. cm. Includes bibliographical references and index.
ISBN 0-19-511368-3 (cloth) / 0-19-513087-1 (pbk.)
1. Mathematics—Philosophy. I. Title.
QA8.4.H47 1997 510'.1—dc20 96–38483

Illustration on dust jacket and p. vi "Arabesque XXIX" courtesy of Robert Longhurst. The sculpture depicts a "minimal surface" named after the German geometer A. Enneper. Longhurst made the sculpture from a photograph taken from a computer-generated movie produced by differential geometer David Hoffman and computer graphics virtuoso Jim Hoffman. Thanks to Nat Friedman for putting me in touch with Longhurst, and to Bob Osserman for mathematical instruction. The "Mathematical Notes and Comments" has a section with more information about minimal surfaces.

Figures 1 and 2 were derived from Ascher and Brooks, *Ethnomathematics*, Santa Rosa, CA.: Cole Publishing Co., 1991; and figures 6–17 from Davis, Hersh, and Marchisotto, *The Companion Guide to the Mathematical Experience*, Cambridge, Ma.: Birkhauser, 1995.

10 9 8 7 6 5 4 3

Printed in the United States of America
on acid free paper

to Veronka

Robert Longhurst, *Arabesque XXIX*

". . . , So long lives this, and this gives life to thee."

Shakespeare, *Sonnet 18*

Contents

Part Two

Summary and Recapitulation

Preface: Aims and Goals

Forty years ago, as a machinist's helper, with no thought that mathematics could become my life's work, I discovered the classic, *What Is Mathematics?* by Richard Courant and Herbert Robbins. They never answered their question; or rather, they answered it by *showing* what mathematics is, not by *telling* what it is. After devouring the book with wonder and delight, I was still left asking, "But what is mathematics, really?"

This book offers a radically different, unconventional answer to that question. Repudiating Platonism and formalism, while recognizing the reasons that make them (alternately) seem plausible, I show that *from the viewpoint of philosophy* mathematics must be understood as a human activity, a social phenomenon, part of human culture, historically evolved, and intelligible only in a social context. I call this viewpoint "humanist."

I use "humanism" to include all philosophies that see mathematics as a human activity, a product, and a characteristic of human culture and society. I use "social conceptualism" or "social-cultural-historic" or just "social-historic philosophy" for my specific views, as explained in this book.

This book is a subversive attack on traditional philosophies of mathematics. Its radicalism applies to philosophy of mathematics, not to mathematics itself. Mathematics comes first, then philosophizing about it, not the other way around. In attacking Platonism and formalism and neo-Fregeanism, I'm defending our right to do mathematics as we do. To be frank, this book is written out of love for mathematics and gratitude to its creators.

Of course it's obvious common knowledge that mathematics is a human activity carried out in society and developing historically. These simple observations are usually considered irrelevant to the philosophical question, what is mathematics? But without the social historical context, the problems of the philosophy of mathematics are intractable. In that context, they are subject to reasonable description and analysis.

The book has no mathematical or philosophical prerequisites. Formulas and calculations (mostly high-school algebra) are segregated into the final Mathematical Notes and Comments.

There's a suggestive parallel between philosophy of mathematics today and philosophy of science in the 1930s. Philosophy of science was then dominated by "logical empiricists" or "positivists" (Rudolf Carnap the most eminent). Positivists thought they had the proper methodology for all science to obey (see Chapter 10).

By the 1950s they noticed that scientists didn't obey their methodology. A few iconoclasts—Karl Popper, Thomas Kuhn, Imre Lakatos, Paul Feyerabend—proposed that philosophy of science look at what scientists actually do. They portrayed a science where change, growth, and controversy are fundamental. Philosophy of science was transformed.

This revolution left philosophy of mathematics unscratched. It's still dominated by its own dogmatism. "Neo-Fregeanism" is the name Philip Kitcher put on it. Neo-Fregeanism says set theory is the only part of mathematics that deserves philosophical consideration. It's a relic of the Frege-Russell-Brouwer-Hilbert foundationist philosophies that dominated philosophy of mathematics from about 1890 to about 1930. The search for indubitable foundations is forgotten, but it's still taken for granted that philosophy of mathematics is about—foundations!

Neo-Fregeanism is not based on views or practices of mathematicians. It's out of touch with mathematicians, users of mathematics, and teachers of mathematics. A few iconoclasts are working to bring in new ideas. P. J. Davis, J. Echeverria, P. Ernest, N. Goodman, P. Kitcher, S. Restivo, G.-C. Rota, B. Rotman, A. Sfard, M. Tiles, T. Tymoczko, H. Freudenthal, P. Henrici, R. Thomas, J. P. van Bendegem, and others. This book is a contribution to that effort.[1]

Mathematics Education

The United States suffers from "innumeracy" in its general population, "math avoidance" among high-school students, and 50 percent failure among college calculus students. Causes include starvation budgets in the schools, mental attrition by television, parents who don't like math.

There's another, unrecognized cause of failure: misconception of the nature of mathematics. This book doesn't report classroom experiments or make sug-

[1] This book is descended from my 1978 article, "Some Proposals for Reviving the Philosophy of Mathematics," published by Gian-Carlo Rota in *Advances in Mathematics*, and from Chapters 7 and 8 of *The Mathematical Experience*, co-authored with Philip J. Davis.

gestions for classroom practice. But it can assist educational reform, by helping mathematics teachers and educators understand what mathematics is.

There's discussion of teaching in Chapter 1, "The Plight of the Working Mathematician," and Chapter 13, "Teaching" and "Ideology."

Outline of Part 1

The book has two main parts. Part One, Chapters 1 through 5, is programmatic. Part Two, Chapters 6 through 12, is historical. The chapters are made of self-contained sections. Chapter 13 is a Summary and Recapitulation.

The book ends with Mathematical Notes and Comments. Often a mathematical concept mentioned in the main text receives more extended treatment in the Notes and Comments. *I signal this by a double asterisk* (**) *in the main text.*

Chapter 1 starts with a puzzle. How many parts has a four-dimensional cube? It's doubtful whether a four-dimensional cube exists. Yet as you read you'll figure out the number of its parts! After you've done so, the question returns. Does this thing exist? This is a paradigm for the main problem in philosophy of mathematics. In what sense do mathematical objects exist?

This beginning is followed by a quick overview of modern mathematics, and then a presentation of mathematical Platonism. Next comes the heart of the book: the social-historic philosophy of mathematics that I call humanism. Similar philosophies expounded by other recent authors are introduced in Chapters 11 and 12.

Chapter 2 evaluates criteria for evaluating a philosophy of mathematics. Some standard criteria are unimportant. Some neglected ones are essential. Later, in Chapter 13, I grade myself by these criteria. The first section of Chapter 3 exposes a scandal: Working mathematicians advocate two contradictory philosophies! The next section explains the front and the back of mathematics. Then we meet some mathematical myths, and knock them down with anecdotes from the back room. This chapter testifies that real-life mathematical experience supports humanism against Platonism or formalism.

Chapters 4 and 5 use the humanist point of view to reexamine familiar controversies:

proof
intuition
certainty
infinity
existence
meaning
object versus process
invention versus discovery

Why So Much History?

I advocate a historical understanding of mathematics. So it's natural to make an historical examination of different philosophies. We will find that foundationism and neo-Fregeanism are descendants of a centuries-old mating between mainstream philosophy of mathematics and religion/theology.

Raking up the past uncovers some surprises. René Descartes's famous Method is violated in the *Geometry* of René Descartes. Strange ideas about arithmetic and geometry were ardently held by George Berkeley and David Hume.

The history is told in two separate stories. First, starting with Pythagoras and Plato, we follow the idealists and absolutists, who see mathematics as superhuman or inhuman. I call them the Mainstream. Their story contains an unbroken thread of mutual support between idealist philosophy of mathematics and religion or theology.

Then, starting with Aristotle, we follow the thinkers who see mathematics as human activity. I call them "humanists and mavericks." ("Maverick" is taken from a fascinating article by Aspray and Kitcher.)

This unorthodox procedure finds some support in an interesting remark of Kurt Gödel: "I believe that the most fruitful principle for gaining an overall view of the possible world-views will be to divide them up according to the degree and the manner of their affinity to or, respectively, turning away from metaphysics (or religion.) In this way we immediately obtain a division into two groups, skepticism, materialism and positivism stand on one side, spiritualism, idealism and theology on the other. . . . Thus one would, for example, say that apriorism belongs in principle on the right and empiricism on the left side." (Gödel 1995, p. 375.)

"Dead White Males"

A glance at the index shows that nearly all the authors I cite are white males, many of them dead. Yet white males are a small fraction of the human race.

Why is this?

Art, music, poetry, botany, and architecture are available in some form to all peoples and both sexes. So are market-place arithmetic and architectural geometry. But disputing the meaning and nature of mathematics is ideological, not practical. Western society has been dominated by white males, and its ideologists have been white males.

Today this is no longer so true. I've been able to cite Juliet Floyd, Gila Hanna, Penelope Maddy, Anna Sfard, and Mary Tiles. I've been complimented in public for humanizing the male-hierarchical picture of mathematics.

Similar comments apply to the under-representation of persons of color.

In talking about "working mathematicians" or "academic philosophers" I lump the specimens I have met into some sort of statistic. Is it permissible to ignore differences? Not so many years ago national differences were thought interesting and important. Nowadays the differences thought significant are between males and females, and between advantaged and disadvantaged (white males and people of color).

Some people think female mathematicians and mathematicians of color see the nature of mathematics differently than do white male mathematicians. I'm not convinced such differences are present. If they are, I'm unqualified to write about them.

A defect of this book is neglect of non-Western mathematics. The crucial part of Arabic authors in restoring Greek science to Western Europe is well known. India and China sent important contributions to Europe. But compared with Greece, we hardly know the history of the philosophy of mathematics in Indo-America, Africa, or the Near and Far East. The literature on non-Western mathematics is valuable, but it's not philosophical. My report of Marcia Ascher's work in Chapter 12 goes beyond the Eurocentrism of the rest of the book. Sadly, I'm unprepared to translate archives in Mexico City or Beijing.

Did different religious philosophies in East and West result in different philosophies of mathematics? If so, did such differences affect mathematics itself? Future scholarship is sure to shed light on these fascinating questions.

Santa Fe, N.M.
December 1996

R. H.

Acknowledgments

Thanks to V. John-Steiner for wise advice, for numerous encouragements, and especially for lessons on socio-cultural theory and practice.

Hao Wang was a famous computer scientist, logician, and philosopher. His *Beyond Analytic Philosophy* contains careful, thorough critiques of Carnap and Quine. Wang favored "doing justice to what we know." He saw philosophy related to life, not just an abstract exercise.

He started a new field of research by asking: Is there a set of tiles that tile the plane nonperiodically, but not periodically? The answer was yes. Then crystallographers found such tilings in nature. They're "quasi-crystals," a potential source of new technology.

Wang wrote one of the first programs to prove theorems automatically. In a few minutes it proved the first 150 theorems in Russell and Whitehead's *Principia Mathematica.*

He was one of very few who had many conversations with Kurt Gödel. His *Reflections on Kurt Gödel* has unique historical importance.

I was looking forward to his criticism of these chapters when I heard the sad news of his death.

Thanks: for financial support, to Sam Goldberg and the Sloan Foundation. For use of facilities, to The Rockefeller University, the Courant Institute of New York University, Brown University, and the University of New Mexico. For valuable illustrations, to Caroline Smith.

For conversations and letters, to Jose-Luis Abreu, Archie Bahm, Mike Baron, Jon Barwise, Agnes Berger, Bill Beyer, Gus Blaisdell, Lenore Blum, Marcelo de Caravalho Borba, John Brockman, Felix Browder, Mario Bunge, Dorie Bunting,

Ida and Misha Burdzelan, John Busanich, Mutiara Buys, Bruce Chandler, Paul Cohen, Necia Cooper, Richard Courant, Chan Davis, Martin Davis, Hadassah and Phil Davis, Jim Donaldson, Burton Dreben, Mary and Jim Dudley, Freeman Dyson, Ann and Sterling Edwards, Ed Edwards, Peter Eggenberger, Jim Ellison, Bernie Epstein, Dick Epstein, Paul Erdös, Paul Ernest, Florence and Dave Fanshel, Sol Feferman, Dennis Flanagan, Lois Folsom, Marilyn Frankenstein, Hans Freudenthal, Tibor Gallai, Tony Gardiner, Tony Gieri, Jay Ginsburg, Sam Gitler, Nancy Gonzalez, Nick Goodman, Russell Goodman, Luis Gorostiza, Russell Goward, Jack Gray, Cindy Greenwood, Genara and Richard Griego, Liang-Shin Hahn, Gila Hanna, Leon Henkin, Malke, Daniel, Eva and Phyllis Hersh, Josie and Abe Hillman, Moe Hirsch, Doug Hofstadter, John Horvath, Takashi Hosoda, Ih-Ching Hsu, Kirk Jensen, Fritz John, Chris Jones, Maria del Carmen Jorge, Mark Kac, Ann and Judd Kahn, Evelyn Keller, Joe Keller, Philip Kitcher, Morris Kline, Vladimir Korolyuk, Martin Kruskal, Tom Kyner, Marian and Larry Kugler, George Lakoff, Anneli and Peter Lax, Uri Leron, Ina Lindemann, Lee Lorch, Ray Lorch, Wilhelm Magnus, Penelope Maddy, Elena Anne Marchisotto, Charlotte and Carl Marzani, Deena Mersky, Ray Mines, Merle Mitchell, Cathleen Morawetz, Don Morrison, Joseph Muccio, Gen Nakamura, Susan Nett, John Neu, Otto Neugebauer, Bob Osserman, George Papanicolaou, Alice and Klaus Peters, Stan Philips, Joanna and Mark Pinsky, George Pólya, Louise Raphael, Fred Richman, Steve Rosencrans, Gian-Carlo Rota, Muriel and Henry Roth, Brian Rotman, Paul Ryl, Sandro Salimbeni, Joe Schatz, Andy Schoene, Susan Schulte, Anna Sfard, Abe Shenitzer, David Sherry, Jill and Neal Singer, Melissa Smeltzer, Joel Smoller, Vera Sos, Ian Stewart, Gabe Stolzenberg, David Swift, Anatol Swishchuk, Béla Sz.-Nagy, Robert Thomas, William Thurston, Mary Tiles, Uri Treisman, Tim Trucano, Tom Tymoczko, Francoise and Stan Ulam, Istvan Vincze, Cotten and Larry Wallen, Solveig and Burt Wendroff, Myra and Alvin White, Raymond Wilder, Carla Wofsy, and Steve Wollman.

What Is Mathematics, Really?

Dialogue with Laura

I was pecking at my word processor when twelve-year-old Laura came over.

L: What are you doing?

R: It's philosophy of mathematics.

L: What's that about?

R: What's the biggest number?

L: There isn't any!

R: Why not?

L: There just isn't! How could there be?

R: Very good. Then how many numbers must there be?

L: Infinite many, I guess.

R: Yes. And where are they all?

L: Where?

R: That's right. Where?

L: I don't know. Nowhere. In people's heads, I guess.

R: How many numbers are in your head, do you suppose?

L: I think a few million billion trillion.

R: Then maybe everybody has a few million billion trillion or so?

L: Probably they do.

R: How many people could there be living on this planet right now?

L: Don't know. Probably billions.

R: Right. Less than ten billion, would you say?

L: Okay.

R: If each one has a million billion trillion numbers or less in her head, we can count up all their numbers by multiplying ten billion times a million billion trillion. Is that right?

L: Sounds right to me.

R: Would that number be infinite?

L: Would be pretty close.

R: Then it would be the largest number, wouldn't it?

L: Wait a minute. You just asked me that, and I said there couldn't be a largest number!

R: So there actually has to be a number bigger than the biggest number in anybody's head?

L: Right.

R: Where is that number, if not in anybody's head?

L: Maybe it's how many grains of sand in the whole universe.

R: No. The smallest things in the universe are supposed to be electrons. Much smaller than grains of sand. Cosmologists say the number of electrons in the universe is less than a 1 with 23* zeroes after it. Now, ten billion times a million billion trillion is a 1 with

$$1 + 9 + 6 + 9 + 12$$

zeroes after it. That's a 1 with 37 zeroes after it, which is a hundred trillion times as much as a one with 23 zeroes it, which is more than the number of elementary particles in the universe, according to cosmologists.

L: Cosmologists are people who figure out stuff about the cosmos?

R: Right.

L: Awesome!

R: So there are way more numbers than there are elementary particles in the whole cosmos.

L: Pretty weird!

R: Never mind "where." Let's talk about "when." How long do you suppose numbers have been around?

L: A real long time.

R: Have they told you in school about the Big Bang?

L: I heard about it. It was like fifteen billion years ago. When the cosmos began.

R: Do you think there were numbers at the time of the big bang?

* Friends tell me 23 is way, way, too small. My apologies to all, especially Laura.

L: Yes, I think so. Just to count what was going on, you know.

R: And before that? Were there any numbers before the Big Bang? Even little ones, like 1, 2, 3?

L: Numbers before there was a universe?

R: What do you think?

L: Seems like there couldn't be anything before there was anything, you know what I mean? Yet it seems like there should always be numbers, even if there isn't a universe.

R: Take that number you just came up with, 1 with 37 zeroes after it, and call it a name, any name.

L: How about 'gazillion'?

R: Good. Can you imagine a gazillion of anything?

L: Heck no.

R: Could you or anyone you know ever count that high?

L: No. I bet a computer could.

R: No. The earth and the sun will vanish before the fastest computer ever built could count that high.

L: Wow!

R: Now, what is a gazillion and a gazillion?

L: Two gazillion. How easy!

R: How do you know?

L: Because one anything and another anything is two anything, no matter what.

R: How about one little mousie and one fierce tomcat? Or one female rabbit and one male rabbit?

L: You're kidding! That's not math, that's biology.

R: You never saw a gazillion or anything near it. How do

you know gazillions aren't like rabbits?

L: Numbers can't be like rabbits.

R: If I take a gazillion and add one, what do I get?

L: A gazillion and one, just like a thousand and one or a million and one.

R: Could there be some other number between a gazillion and a gazillion and one?

L: No, because a gazillion and one is the next number after a gazillion.

R: But how do you know when you get up that high the numbers don't crowd together and sneak in between each other?

L: They can't, they've got to go in steps, one step at a time.

R: But how do you know what they do way far out where

you've never been?

L: Come on, you've got to be joking.

R: Maybe. What color is this pencil?

L: Blue.

R: Sure?

L: Sure I'm sure.

R: Maybe the light out here is peculiar and makes colors look wrong? Maybe in a different light you'd see a different color?

L: I don't think so.

R: No, you don't. But are you absolutely sure it's absolutely impossible?

L: No, not absolutely, I guess.

R: You've heard of being color blind, haven't you?

L: Yes, I have.

R: Could it be possible for a person to get some eye disease and become color blind without knowing it?

L: I don't know. Maybe it could be possible.

R: Could that person think this pencil was blue, when actually it's orange, because they had become color blind without knowing it?

L: Maybe they could. What of it? Who cares?

R: You see a blue pencil, but you aren't 100% sure it's really blue, only almost sure. Right?

L: Sure. Right.

R: Now, how about a gazillion and a gazillion equals two gazillion? Are you absolutely sure of that?

L: Yes I am.

R: No way that could be wrong?

L: No way.

R: You've never seen a gazillion. Yet you're more sure about gazillions than you are about pencils that you can see and touch and taste and smell. How do you get to know so much about gazillions?

L: Is that philosophy of mathematics?

R: That's the beginning of it.

Part One

one

Survey and Proposals

A Round Trip to the Fourth Dimension. Is There a 4-Cube?**

This section has two purposes. It's a worked exercise in Pólya's heuristic (see Chapter 11).

At the same time, it's an inquiry into mathematical existence. By guided induction and intelligent guessing, you'll count the parts of a 4-dimensional cube. Then you'll be asked, "Does your work make sense? What kind of sense does it make?"

You're familiar with two-dimensional cubes (squares) and three-dimensional cubes. Is there a four-dimensional cube?

To help you answer, here's a harder question:

"How many parts does a 4-cube have?"

I haven't explained the meaning of either "4-cube" or "part." What can you do? It's time for Pólya's problem-solving principle: *If you can't solve your problem, make up a related problem that you may be able to solve.*

In this case, what's an easier, related question? "How many parts has an *ordinary* cube, a 3-cube?"

What kinds of parts does it have?

A 3-cube has an interior (3-dimensional), some faces (2-dimensional), some edges (1-dimensional), and some vertices (0-dimensional). These are four different kinds of parts.

You can count each of these four kinds of parts.

Do it! (Hint: It has 12 edges.)

Write your four numbers in a row, from 0-dimensional to 3-dimensional. Add them up, and write the sum at the end of the row.

Answer: a 3-cube has 27 parts.

Ready for four dimensions?

Probably not yet.

What other related problem can you think of? Maybe a simpler one? (If you can't go up right away, try going down at first.)

How many kinds of parts has a *2-cube*, a square?

Count each kind. Write the numbers in a new row above the previous row, corresponding numbers above corresponding numbers. Add up the new numbers.

You get $4 + 4 + 1 = 9$. Write 9 above the 27 from the 3-cube.

You went down from 4 dimensions to 3, and from 3 to 2. What should you understand by a "*1-cube*"? How many kinds of parts does it have? How many of each kind? And what's the sum?

The answer is $2 + 1 = 3$.

Write these numbers above the corresponding numbers from the 2-cube and the 3-cube.

You have a table! It has three rows, one row for each "cube" from 1-dimensional to 3-dimensional.

The first row says 2, 1.

The middle row says 4, 4, 1.

The bottom row says 8, 12, 6, 1.

Your table has five columns. The first four columns give the number of parts of each dimension, from 0 to 3. The last column gives the sums.

We're trying to find the sum for the 4-cube. We have tabulated information for the 1-, 2-, and 3-cube, in three rows.

The 4-cube goes in the next row, the fourth! (It needs one additional space on the right, to count its four-dimensional part, the interior.)

You immediately see what to put first in row 4, below 2, 4, and 8. You've already found out how many vertices the 4-cube has! Stare at your table until you see what to put in the other places, for the one-, two- and three-dimensional parts of the 4-cube. Two of the diagonals follow obvious patterns. And there's a simple relation between every number in the table and a pair of numbers above it. Namely, each number equals the sum of the numbers diagonally above to the left plus double the number directly above. For instance, $6 = 4 + (2 \times 1)$ and $12 = 4 + (2 \times 4)$.

You've completed the fourth row! You know a 4-cube has 81 parts.

BUT!!

BACK TO PHILOSOPHY:

Does a 4-cube really exist?

If yes, where is it? How does it exist? In what sense?

How do you *know* it exists? Could you be mistaken?

If no, how could you find out so much about it? If there is no 4-cube, what's the meaning of the numbers you found? Should other readers of this exercise get the same numbers? *Why* should they get the same numbers, if there's no such thing as a 4-cube?

For that matter, is there even such a thing as a 3-cube? You've seen and touched physical objects called cubes. They aspired to approximate a cube. But they couldn't *be* cubes. No ice cube or ebony cube or brass cube has 12 edges all *exactly* the same length, 8 corners all *perfectly* square.

Only a mathematical 3-cube is a perfect cube. So a mathematical 3-cube is like a 4-cube, in not being a physical object! Then what is it? Where is it? Is there a big difference between asking, "Does a 3-cube exist?" and asking, "Does a 4-cube exist?"

Answering these questions is the point of this book.

For experts, here are some exercises in the philosophy of mathematics that anticipate much of what follows.

How would these questions about 3-cube and 4-cube be answered by Gödel or Thom? (See below, "Must We Be Platonists?")

By Frege or Russell in his logicist period?

By Brouwer or Bishop?

By Hilbert or Bourbaki (Chapter 8)?

By Wittgenstein (Chapter 11)?

By Quine? By Putnam in his phase I, II, or III (Chapter 9)?

Quick Overview

Even without three years of graduate school, you can get a rough notion of modern mathematics. Here's a mini-sketch of its method and matter.

The method of mathematics is "conjecture and proof." You come to an inherited network of concepts and facts, properties and connections, called a "theory." (For instance, classical solid geometry, including the 3-cube.) This presently existing theory is the result of a historic evolution. It is the cooperative and competitive work of generations of mathematicians, associated by friendship and rivalry, by mutual criticism and correction, as leaders and followers, mentors and protégés.

Starting with the theory as you find it, you fill in gaps, connect to other theories, and spin out enlargements and continuations—like going up one dimension to dream of a "hypercube."

You just solved the hypercube problem. But you didn't solve it in isolation. You were handed the problem in the first place. Then you got helpful hints and encouragement as you went along. When you finally got the answer, you received confirmation that your answer was right.

Believe it or not, a mathematician has needs similar to yours. He/she needs to discover a problem connected to the existing mathematical culture. Then she needs reassurance and encouragement as she struggles with it. And in the end when she proposes a solution she needs agreement or criticism. No matter how isolated and self-sufficient a mathematician may be, the source and verification of his work goes back to the community of mathematicians.

Sometimes new theories seem to spin out of your head and the heads of your predecessors. Sometimes they're suggested by real-world subjects, like physics. Today the infinite-dimensional spaces of higher geometry are models for the elementary particles of physics.

Mathematical discovery rests on a validation called "proof," the analogue of experiment in physical science. A proof is a conclusive argument that a proposed result follows from accepted theory. "Follows" means the argument convinces qualified, skeptical mathematicians. Here I am giving an overtly social definition of "proof." Such a definition is unconventional, yet it is plainly true to life.

In logic texts and modern philosophy, "follows" is often given a much stricter sense, the sense of mechanical computation. No one says the proofs that mathematicians write actually *are* checkable by machine. But it's conventional to insist that there be *no doubt* they *could* be checked that way.

Such lofty rigor isn't found in all mathematics. From one specialty to another, from one mathematician to another, there's variation in strictness of proof and applicability of results. Mathematics that stresses results above proof is often called "applied mathematics." Mathematics that stresses proof above results is sometimes called "pure mathematics," more often just "mathematics." (Outsiders sometimes say "theoretical mathematics.")

A naive non-mathematician—perhaps a neo-Fregean analytic philosopher—looks into Euclid, or a more modern math text of formalist stripe, and observes that axioms come first. They're right on page one. He or she understandably concludes that in mathematics, axioms come first. First your assumptions, then your conclusions, no?

But anyone who has done mathematics knows what comes first—a problem. Mathematics is a vast network of interconnected problems and solutions. Sometimes a problem is called "a conjecture."

Sometimes a solution is a set of axioms!

I explain.

When a piece of mathematics gets big and complicated, we may want to systematize and organize it, for esthetics and for convenience. The way we do that is to axiomatize it. Thus a new type of problem (or "meta-problem") arises:

"Given some specific mathematical subject, to find an attractive set of axioms from which the facts of the subject can conveniently be derived."

Any proposed axiom set is a proposed solution to this problem. The solution will not be unique. There's a history of re-axiomatizations of Euclidean geometry, from Hilbert to Veblen to Birkhoff the Elder.

In developing and understanding a subject, axioms come late. Then in the formal presentations, they come early.

Sometimes someone tries to invent a new branch of mathematics by making up some axioms and going from there. Such efforts rarely achieve recognition or permanence. Examples, problems, and solutions come first. Later come axiom sets on which the already existing theory can be "based."

The view that mathematics is in essence derivations from axioms is backward. In fact, it's wrong.

An indispensable partner to proof is mathematical intuition. This tells us what to try to prove. We relied heavily on intuition in our hypercube exercise. It often gives true theorems, even with gappy proofs. We return to intuition and proof in Chapter 4.

So far I've described mathematics by its methods. What about its content? The dictionary says math is the science of number and figure ("figure" meaning the shapes or figures of geometry.) This definition might have been O.K. 200 years ago. Today, however, math includes the groups, rings, and fields of abstract algebra, the convergence structures of point-set topology, the random variables and martingales of probability and mathematical statistics, and much, much more. *Mathematical Reviews* lists 3,400 subfields of mathematics! No one could attempt even a brief presentation of all 3,400, let alone a philosophical investigation of them all. To identify a branch of study as part of mathematics, one is guided by its method more than its content.

Formalism: A First Look

Two principal views of the nature of mathematics are prevalent among mathematicians—Platonism and formalism. Platonism is dominant, but it's hard to talk about it in public. Formalism feels more respectable philosophically, but it's almost impossible for a working mathematician to really believe it.

The next section is about Platonism. Here I take a quick glance at formalism. I return to it in the section of Chapter 9 on David Hilbert. The third major school, intuitionism or constructivism, is also discussed in Chapter 9, in the sections on foundations, on L. E. J. Brouwer, and on Errett Bishop.

The formalist philosophy of mathematics is often condensed to a short slogan: "Mathematics is a meaningless game." ("Meaningless" and "game" remain undefined. Wittgenstein showed that games have no strict definition, only a family resemblance.)

What do formalists mean by "game" when they call mathematics a game? Perhaps they use "game" to mean something "played by the rules." (Now "play" and "rule" are undefined!)

For a game in that sense, two things are needed:

(2) people to play by the rules.
(1) rules.

Rule-making can be deliberate, as in Monopoly or Scrabble—or spontaneous, as in natural languages or elementary arithmetic.

In either case, the *making* of rules doesn't follow rules!

Wittgenstein and some others seem to think that since the making of rules doesn't follow rules, then the rules are arbitrary. They could just as well be any way at all. This is a gross error.

The rules of language and of mathematics are historically determined by the workings of society that evolve under pressure of the inner workings and interactions of social groups, and the physical and biological environment of earth. They are also simultaneously determined by the biological properties, especially the nervous systems, of individual humans. Those biological properties and nervous systems have permitted us to evolve and survive on earth, so of course they reflect somehow the physical and biological properties of this planet. Complicated, certainly. Mysterious, no doubt. Arbitrary, no.

People often make rules deliberately. Not only for games, but also for computer languages, for parliamentary procedure, for stopping at STOP signs, and for Orthodox weddings. These rule-making tasks don't follow rules. But that doesn't make them arbitrary. Rules are made for a purpose. To be played or accepted or performed by people, they have to be playable or acceptable by people. Tradition, taste, judgment, and consensus matter. Eccentricities of individual rule-makers matter. The resultant of such social and personal factors is what makes us make the rules we make. The outcome of rule-making isn't arbitrary. Neither is it rule governed.

Some details of a rule system may seem arbitrary or optional. In chess, for instance, the rule for castling might be varied without ruining the game.

Is there a sharp separation between playing by the rules and making the rules?

Some formalists in philosophy of mathematics say discovery is lawless—has no logic—while proof or justification is nothing but logic. If such a philosopher notices that real mathematical life isn't that way, the discrepancy seems like a scandal that must be kept out of the newspapers, or a crime calling for correction by Georg Kreisel's "logical hygiene."

In real life, in all games including mathematics (supposing for the moment that it is a game), the separation between playing the game and making the rules is imperfect, partial, incomplete.

Chess players don't change the rules of chess as they go along. Not in tournament chess, at any rate. Disputes are settled according to written procedures. But these procedures aren't *rules*. Settling disputes comes down to judgments and opinions. Big league baseball has plenty of rules. But the game would be impossible without umpires who use their judgment. In street stick ball, first base is supposed to be the left front fender on the closest car parked on the right side of the street. If no car is there, we improvise.

In real life there are no totally rule-governed activities. Only more or less rule-governed ones, with more or less definite procedures for disputes. The rules and procedures evolve, sometimes formally like amending the U.S. Constitution, sometimes informally, as street games evolve with time and mixing of cultures. Is there totally unruly or ruleless behavior? Perhaps not. Mathematics is in part a rule-governed game. But one can't overlook how the rules are made, how they evolve, and how disputes are resolved. That isn't rule governed, and can't be.

Computer proof is changing the way the game of mathematics is played. Wolfgang Haken thinks computer proof is permitted under the rules. Paul Halmos thinks it ought to be against the rules. Tom Tymoczko thinks it amounts to changing the rules. In the long run, what mathematicians publish, cite, and especially teach, will decide the rules. We have no French Academy to set rules, no cabal of team owners to say how to play our game. Our rules are set by our consensus, influenced and led by our most powerful or prestigious members (of course).

These considerations on games and rules in general show that one can't understand mathematics (or any other nontrivial human activity) by simply finding rules that it follows or ought to follow. Even if that could be done, it would lead to more interesting questions: Why and whence those rules?

The notion of strictly following rules without any need for judgment is a fiction. It has its use and interest. It's misleading to apply it literally to real life.

Must We Be Platonists?

Platonism, or realism as it's been called, is the most pervasive philosophy of mathematics. It has various variations. The standard version says mathematical entities exist outside space and time, outside thought and matter, in an abstract realm independent of any consciousness, individual or social. Today's mathematical Platonisms descend in a clear line from the doctrine of Ideas in Plato (see "Plato" in Chapter 6). Plato's philosophy of mathematics came from the Pythagoreans, so mathematical "Platonism" ought to be "Pythago-Platonism." I defer to custom and say "Platonism." (This debt of Plato is discussed by John Dewey in his 1929 Gifford lectures and by Bertrand Russell in Chapter 9.)

There are Platonisms of mathematicians and Platonisms of philosophers. I quote half a dozen eminent Platonists of past and present, mostly mathematicians. (Somerville and Everett are copied from Leslie White's article in *The World of Mathematics.*)

Edward Everett (1794–1865), the first American to receive a doctorate at Göttingen, an orator who shared the platform with Abraham Lincoln at Gettysburg, wrote: "In the pure mathematics we contemplate absolute truths which existed in the divine mind before the morning stars sang together, and which will continue to exist there when the last of their radiant host shall have fallen from heaven."

The scholar and mathematician Mary Somerville (1780–1872): "Nothing has afforded me so convincing a proof of the unity of the Deity as these purely mental conceptions of numerical and mathematical science which have been by slow degrees vouchsafed to man, and are still granted in these latter times by the Differential Calculus, now superseded by the Higher Algebra, all of which must have existed in that sublimely omniscient Mind from eternity."

G. H. Hardy, the leading English mathematician of the 1920s: "I have myself always thought of a mathematician as in the first instance an observer, who gazes at a distant range of mountains and notes down his observations. His object is simply to distinguish clearly and notify to others as many different peaks as he can. There are some peaks which he can distinguish easily, while others are less clear. He sees A sharply, while of B he can obtain only transitory glimpses. At last he makes out a ridge which leads from A and, following it to its end, he discovers that it culminates in B. B is now fixed in his vision, and from this point he can proceed to further discoveries. In other cases perhaps he can distinguish a ridge which vanishes in the distance, and conjectures that it leads to a peak in the clouds or below the horizon. But when he sees a peak, he believes that it is there simply because he sees it. If he wishes someone else to see it, he *points to it*, either directly or through the chain of summits which led him to recognize it himself. When his pupil also sees it, the research, the argument, the *proof* is finished" (1929, p. 18). Here the "chain of summits" is the chain of statements in a proof, connecting known facts (peaks) to new ones. Hardy uses a chain of summits to find a new peak. Once he sees the new peak, he believes in it because he sees it, no longer needing any chain.

The preeminent logician, Kurt Gödel: "Despite their remoteness from sense experience, we do have something like a perception also of the objects of set theory, as is seen from the fact that the axioms force themselves upon us as being true. I don't see any reason why we should have less confidence in this kind of perception, i.e., in mathematical intuition, than in sense perception. . . . This, too, may represent an aspect of objective reality."

The French geometer and Fields Medalist René Thom, father of catastrophe theory: "Mathematicians should have the courage of their most profound convictions and thus affirm that mathematical forms indeed have an existence that is independent of the mind considering them. . . . Yet, at any given moment, mathematicians have only an incomplete and fragmentary view of this world of ideas."

Thom's world of ideas is geometric; Gödel's is set-theoretic. They believe in an independent world of ideas—but not the same world!

Paul Erdös was a famous Hungarian mathematician who talked about "The Book." "The Book" contains all the most elegant mathematical proofs, the known and especially the unknown. It belongs to "the S. F."—"the Supreme Fascist"—Erdös's pet name for the Almighty. Occasionally the S. F. permits someone a quick glimpse into the Book.

The Book is a perfect metaphor for Platonism. But Erdös said he's not interested in philosophy. The Book and the S. F. are "only a joke."

However, in a film about Erdös (*N is a Number*, produced by Paul Csicsery) his friend and collaborator Fam Chung, says, "In Paul's mind there is only one reality, and that's mathematics."

Ron Graham, a well-known combinatorialist, collaborator friend of Erdös and husband of Chung, goes even further: "I personally feel that mathematics is the essence of what's driving the universe."

Another Erdös collaborator, Joel Spencer: "Where else do you have absolute truth? You have it in mathematics and you have it in religion, at least for some people. But in mathematics you can really argue that this is as close to absolute truth as you can get. When Euclid showed that there were an infinite number of primes, *that's it!*. There are an infinite number of primes, no ifs, ands, or buts! That's as close to absolute truth as I can see getting."

(As a small point of historical fidelity, Euclid never could have said there was an infinite number of anything. Proposition 20, Book IX, says, in Heath's translation, "Prime numbers are more than any assigned multitude of prime numbers"—there is no greatest prime. Heath immediately paraphrases this as "the important proposition that the number of prime numbers is infinite." Heath's and Spencer's formulation is natural in today's context of infinite sets. Not in Euclid's context.)

Why do mathematicians believe something so unscientific, so far-fetched as an independent immaterial timeless world of mathematical truth?

The mystery of mathematics is its objectivity, its seeming certainty or near-certainty, and its near-independence of persons, cultures, and historical epochs (see the section on Change in Chapter 5).

Platonism says mathematical objects are real and independent of our knowledge. Space-filling curves, uncountably infinite sets, infinite-dimensional manifolds—all the members of the mathematical zoo—are definite objects, with definite properties, known or unknown. These objects exist outside physical space and time. They were never created. They never change. By logic's law of the excluded middle, a meaningful question about any of them has an answer, whether we know it or not. According to Platonism a mathematician is an empirical scientist, like a botanist. He can't invent, because everything is already there. He can only discover. Our mathematical knowledge is objective and unchanging because it's knowledge of objects external to us, independent of us, which are indeed changeless.

An inarticulate, half-conscious Platonism is nearly universal among mathematicians. Research or problem-solving, even at the elementary level, generates a naive, uncritical Platonism. In math class, everybody has to get the same answer. Except for a few laggards, they *do* all get the same answer! That's what's special about math. *There are right answers.* Not right because that's what Teacher wants us to believe. Right because *they are* right.

That universality, that independence of individuals, makes mathematics seem immaterial, inhuman. Platonism of the ordinary mathematician or student is a recognition that the facts of mathematics are independent of her or his wishes. This is the quality that makes mathematics exceptional.

Yet most of this Platonism is half-hearted, shamefaced. We don't ask, How does this immaterial realm relate to material reality? How does it make contact with flesh and blood mathematicians? We refuse to face this embarrassment:

Ideal entities independent of human consciousness violate the empiricism of modern science. For Plato the Ideals, including numbers, are visible or tangible in Heaven, which we had to leave in order to be born. For Leibniz and Berkeley, abstractions like numbers are thoughts in the mind of God. That Divine Mind is still real for Somerville and Everett.

Heaven and the Mind of God are no longer heard of in academic discourse. Yet most mathematicians and philosophers of mathematics continue to believe in an independent, immaterial abstract world—a remnant of Plato's Heaven, attenuated, purified, bleached, with all entities but the mathematical expelled.

Platonism without God is like the grin on Lewis Carroll's Cheshire cat. The cat had a grin. Gradually the cat disappeared, until all was gone—except the grin. The grin remained without the cat.

MacLane is unusual in his unequivocal rejection of Platonism, without turning to formalism. "The platonic notion that there is somewhere the ideal realm of sets, not yet fully described, is a glorious illusion" (p. 385). He thinks there's no need to consider the question of existence of mathematical entities.

The Platonisms of philosophers are more sophisticated than those of mathematicians. One of them is logicism, once preached by Gottlob Frege and Bertrand Russell. Today's "most influential philosopher," W. V. O. Quine, has his own pragmatic-type Platonism (see Chapter 9). Here we talk mainly about "garden variety" or "generic" Platonism, Platonism among the broad mathematical masses.

The objections to Platonism are never answered: the strange parallel existence of two realities—physical and mathematical; and the impossibility of contact between the flesh-and-blood mathematician and the immaterial mathematical object. Platonism shares the fatal flaw of Cartesian dualism. To explain the existence and properties of mind and matter, Descartes postulated a different "substance" for each. But he couldn't plausibly explain how the two substances interact, as mind and body do interact. In similar fashion, Platonists explain mathematics by a separate universe of abstract objects, independent of the material universe. But how do the abstract and material universes interact? How do flesh-and-blood mathematicians acquire the knowledge of number?

To answer, you have to forget Platonism, and look in the socio-cultural past and present, in the history of mathematics, including the tragic life of Georg Cantor.

The set-theoretic universe constructed by Cantor and generally adopted by Platonists is believed to include all mathematics, past, present, and future. In it, the uncountable set of real numbers is just the beginning of uncountable chains of uncountables. The cardinality of this set universe is unspeakably greater than that of the material world. It dwarfs the material universe to a tiny speck. And it was all there before there was an earth, a moon, or a sun, even before the Big Bang. Yet this tremendous reality is unnoticed! Humanity dreams on, totally unaware of it—*except for us mathematicians.* We alone notice it. But only since Cantor revealed it in 1890. Is this plausible? Is this credible? Roger Penrose declares himself a Platonist, but draws the line at swallowing the whole set-theoretic hierarchy.

Platonists don't acknowledge the arguments against Platonism. They just re-avow Platonism.

Frege's point of view persists today among set-theoretic Platonists. It goes something like this:

1. Surely the empty set exists—we all have encountered it!
2. Starting from the empty set, perform a few natural operations, like forming the set of all subsets. Before long you have a magnificent structure in which you can embed the real numbers, complex numbers, quaternions, Hilbert spaces, infinite-dimensional differentiable manifolds, and anything else you like.
3. Therefore it's vain to talk of inventing or creating mathematics. In this all-encompassing, set-theoretic structure, everything we could ever want or dream of is already present.

Yet most advances in mainstream mathematics are made without reference to any set-theoretic embedding. Saying Hilbert space was already there in the set universe is like telling Rodin, "*The Thinker* is a nice piece of work, but all you did was get rid of the extra marble. The statue was there inside the marble quarry before you were born."

Rodin made *The Thinker* by removing marble. Hilbert, von Neumann, and the rest made the theory of Hilbert space by analyzing, generalizing, and rearranging mathematical ideas that were present in the mathematical atmosphere of their time.

A Way Out

What's the nature of mathematical objects?

The question is made difficult by a centuries-old assumption of Western philosophy: "There are two kinds of things in the world. What isn't physical is mental; what isn't mental is physical."

Mental is individual consciousness. It includes private thoughts—mathematical and philosophical, for example—before they're communicated to the world and become social—and also perception, fear, desire, despair, hope, and so on.

Physical is taking up space—having weight or energy. It's flesh and bones, sound waves, X-rays, galaxies.

Frege showed that mathematical objects are neither physical nor mental. He labeled them "abstract objects." What did he tell us about abstract objects? Only this: They're neither physical nor mental.

Are there other things besides numbers that aren't mental or physical?

Yes! Sonatas. Prices. Eviction notices. Declarations of war.

Not mental or physical, but not abstract either!

The U.S. Supreme Court exists. It can condemn you to death!

Is the Court physical? If the Court building were blown up and the justices moved to the Pentagon, the Court would go on. Is it mental? If all nine justices expired in a suicide cult, they'd be replaced. The Court would go on.

The Court isn't the stones of its building, nor is it anyone's minds and bodies. Physical and mental embodiment are necessary to it, but they're not *it. It's a social institution.* Mental and physical categories are insufficient to understand it. It's comprehensible only in the context of American society.

What matters to people nowadays?

Marriage, divorce, child care.
Advertising and shopping.
Jobs, salaries, money.
The news, and other television entertainment.
War and peace.

All these entities have mental and physical aspects, but none is a mental or a physical entity. Every one is a social entity.

Social reality distinct from physical and mental reality was explained by Émile Durkheim a century ago. These quotations are taken from an essay by L. White.

"Collective ways of acting and thinking have a reality outside the individuals who, at every moment of time, conform to it. These ways of thinking and acting exist in their own right. The individual finds them already formed, and he cannot act as if they did not exist or were different from how they are. . . . Of course, the individual plays a role in their genesis. But for a social fact to exist, several individuals, at the very least, must have contributed their action; and it is this combined action which has created a new product. Since this synthesis takes place outside each one of us (for a plurality of consciousness enters into it), its necessary effect is to fix, to institute outside us, certain ways of acting and certain judgments which do not depend on each particular will taken separately" (1938, p. 56).

"There are two classes of states of consciousness that differ from each other in origin and nature, and in the end toward which they aim. One class merely expresses our organisms and the object to which they are most directly related. Strictly individual, the states of consciousness of this class connect us only with ourselves, and we can no more detach them from us than we can detach ourselves from our bodies. The states of consciousness of the other class, on the contrary, come to us from society; they transfer society into us and connect us with something that surpasses us. Being collective, they are impersonal; they turn us toward ends that we hold in common with other men; it is through them and them alone that we can communicate with others. . . . In brief, this duality corresponds to the double existence that we lead concurrently: the one purely individual and rooted in our organism, the other social and nothing but an extension of society" (1964, p. 337).

Concepts have their own life, said Durkheim. "When once born they obey laws all their own. They attract each other, repel each other, unite, divide themselves and multiply" (1976, p. 424).

Mathematics consists of concepts. Not pencil or chalk marks, not physical triangles or physical sets, but concepts, which may be suggested or represented by physical objects.

In reviewing *The Mathematical Experience*, the mathematical expositor and journalist Martin Gardner made this objection: When two dinosaurs wandered to the water hole in the Jurassic era and met another pair of dinosaurs happily sloshing, there were four dinosaurs at the water hole, even though no human was present to think, "$2 + 2 = 4$." This shows, says Gardner, that $2 + 2$ really is 4 in reality, not just in some cultural consciousness. $2 + 2 = 4$ is a law of nature, he says, independent of human thought.

To untangle this knot, we must see that "2" plays two linguistic roles. Sometimes it's an adjective; sometimes it's a noun.

In "two dinosaurs," "two" is a *collective adjective*. "Two dinosaurs plus two dinosaurs equals four dinosaurs" is telling about dinosaurs. If I say "Two discrete, reasonably permanent, noninteracting objects collected with two others makes four such objects," I'm telling part of what's meant by discrete, reasonably permanent noninteracting objects. That is a statement in elementary physics.

John Stuart Mill pointed out that with regard to discrete, reasonably permanent non-interacting objects, experience tells us

$$2 + 2 = 4.$$

In contrast, "Two is prime but four is composite" is a statement about the pure numbers of elementary arithmetic. Now "two" and "four" are *nouns*, not adjectives. They stand for pure numbers, which are concepts and objects. They are *conceptual objects*, shared by everyone who knows elementary arithmetic, described by familiar axioms and theorems.

The collective adjectives or "counting numbers" are finite. There's a limit to how high anyone will ever count. Yet there isn't any last counting number. If you counted up to, say, a billion, then you could count to a billion and one. In pure arithmetic, these two properties—finiteness, and not having a last—are contradictory. This shows that the counting numbers aren't the pure numbers.

Consider the pure number $10^{(10^{10})}$. We easily ascertain some of its properties, such as: "The only prime factors of $10^{(10^{10})}$ are 2 and 5." But we can't count that high. In that sense, there's no counting number equal to $10^{(10^{10})}$.

Körner made the same distinction, using uppercase for Counting Numbers (adjectives) and lowercase for "pure" natural numbers (nouns). Jacob Klein wrote that a related distinction was made by the Greeks, using their words "arithmos" and "logistiké."

So "two" and "four" have double meanings: as Counting Numbers or as pure numbers. The formula

$$2 + 2 = 4$$

has a double meaning. It's about counting—about how discrete, reasonably permanent, noninteracting objects behave. And it's a theorem in pure arithmetic (Peano arithmetic if you like). This linguistic ambiguity blurs the difference between Counting Numbers and pure natural numbers. But it's convenient. It's comparable to the ambiguity of nonmathematical words, such as "art" or "America."

The pure numbers rise out of the Counting Numbers. In a process related to Aristotle's abstraction, they disconnect from "real" objects, to exist as shared concepts in the mind/brains of people who know elementary arithmetic. In that realm of shared concepts, $2 + 2 = 4$ is a different fact, with a different meaning. And we can now show that it follows logically from other shared concepts, which we usually call axioms.

Platonist philosophy masks this social mode of existence with a myth of "abstract concepts."

From living experience we know two facts:

Fact 1: Mathematical objects are created by humans. Not arbitrarily, but from activity with existing mathematical objects, and from the needs of science and daily life.

Fact 2: Once created, mathematical objects can have properties that are difficult for us to discover. This is just saying there are mathematical problems which are difficult to solve. Example: Define x as the 200th digit in the decimal expansion of $23^{(45^{6789})}$. x is thereby determined. Yet I have no effective way to find it.

These two facts aren't theses waiting to be established! They're experiences needing to be understood. We need to "unpack" their philosophical consequences and their paradoxes.

Once created and communicated, mathematical objects are *there*. They detach from their originator and become part of human culture. We learn of them as external objects, with known properties and unknown properties. Of the unknown properties, there are some we are able to discover. Some we can't discover, even though they are our own creations. Does this sound paradoxical? If so, it's because of thinking that recognizes only two realities: the individual subject (the isolated interior life), and the exterior physical world. The existence of mathematics shows the inadequacy of those two categories. The customs, traditions, and institutions of our society are real, yet they are neither in the private inner nor the nonhuman outer world. They're a different reality, a social-cultural-historical reality. Mathematics is that third kind of reality—"inner" with respect to society at large, "outer" with respect to you or me individually.

To say mathematical objects are invented or created by humans makes them different from natural objects—rocks, X-rays, dinosaurs. Some philosophers (Stephen Körner, Hilary Putnam) argue that the subject matter of pure mathematics is the physical world—not its actualities, but its potentialities. "To exist in mathematics," they think, means "to exist potentially in the physical world." This interpretation is attractive, because it lets mathematics be meaningful. But it's unacceptable, because it tries to explain the clear by the obscure.

Consider this famous theorem of Georg Cantor: "If C is the set of points on the real line, and P is the set of all subsets of C, then it's impossible to put the points of C into 1–1 correspondence with the subsets of C—the elements of P." P can be regarded as the set of all functions of a real variable taking on the values 0 or 1. Nearly all these functions are nowhere continuous and nowhere measurable. We have no way to interpret them as physical possibilities.

The common sense of the working mathematician says this theorem is just a theorem of pure mathematics, not part of any physical interpretation. It's a human idea, recently invented. It wasn't timelessly or tenselessly existing, either as a Platonic idea or as a latent physical potentiality.

Why do these objects, our own creations, so often become useful in describing nature? To answer this in detail is a major task for the history of mathematics, and for a psychology of mathematical cognition that may be coming to birth in Piaget and Vygotsky. To answer it in general, however, is easy. Mathematics is part of human culture and history, which are rooted in our biological nature and our physical and biological surroundings. Our mathematical ideas in general match our world for the same reason that our lungs match earth's atmosphere.

Mathematical objects can have well-determined properties because mathematical problems can have well-determined answers. To explain this requires investigation, not speculation. The rough outline is visible to anyone who studies or teaches mathematics. To acquire the idea of counting, we handle coins or beans or pebbles. To acquire the idea of an angle, we draw lines that cross. In higher grades, mental pictures or simple calculations are *reified* (term of Anna Sfard) and become concrete bases for higher concepts. These shared activities—first physical manipulations, then paper and pencil calculations—have a common product—shared concepts.

Not everyone achieves the desired result. The student who doesn't catch on doesn't pass the course.Why can we converse about polynomials? We've been trained to, by a training evolved for that purpose. We do it without a definition of "polynomial." Even without a definition, polynomial is a shared notion of middle-school students and teachers. And polynomials are objective: They have certain properties, whether we know them or not. These are implicit in our common notion, "polynomial."

To unravel in detail how we attain this common, objective notion is a deep problem, comparable to the problem of language acquisition. No one understands clearly how children acquire rules of English or Navajo, which they follow without being able to state them. These implicit rules don't grow spontaneously in the brain. They come from the shared language-use of the community of speakers. The properties of mathematical objects, like the properties of English sentences, are properties of shared ideas.

The observable reality of mathematics is this: an evolving network of shared ideas with objective properties. These properties may be ascertained by many kinds of reasoning and argument. These valid reasonings are called "proofs." They differ from one epoch to another, and from one branch of mathematics to another.

Looking at this fact of experience, we find questions. How are mathematical objects invented? What's the interplay of mathematics with the ideas and needs of science? How does proof become refined as errors are uncovered? Does the network of mathematical reasoning have an integrity stronger than any link, so that the fracture of any link affects only the closest parts?

These questions can be studied by historians of mathematics. Thomas Kuhn showed the insight that the history of science can give to the philosophy of science. Such work is beginning in the history and philosophy of mathematics.

Generally speaking, before an answer is interesting or even makes sense, there has to be a question. This trivial remark applies to mathematics as well as to anything else. Mathematical statements, mathematical theorems, are answers to questions. Modern mathematics has been sarcastically described as "answers to questions that nobody asked." This is unfair. Most likely the mathematician who found the answer did first ask the question. And very likely he'll publish the answer without mentioning the question. To an unwary reader it can then look like a self-subsisting, self-justifying piece of information, a question-less answer.

The mystery of how mathematics grows is in part caused by looking at mathematics as answers without questions. That mistake is made only by people who have had no contact with mathematical life. It's the questions that drive mathematics. Solving problems and making up new ones is the essence of mathematical life. If mathematics is conceived apart from mathematical life, of course it seems—dead.

To learn how mathematics grows, study how mathematical problems are recognized, how they're attractive. It has to be both something somebody would *like* to do and something somebody might be *able* to do.

An adequate description of today's mathematics (or any other period's) has to include some problems that are considered interesting. That's one reason a formal axiomatic description is incomplete and misleading.

This is recognized by Kitcher in his *Nature of Mathematical Knowledge*. It's implicit in Lakatos's *Proofs and Refutations*. It's fatally absent in Frege, Russell, and their epigones.

Psychological and historical studies won't make mathematical truth indubitable. But why expect mathematical truth to be indubitable? Correcting errors by confronting them with experience is the essence of science. What's needed is explication of what mathematicians do—as part of general human culture, as well as in mathematical terms. The result will be a description of mathematics that mathematicians recognize—the kind of truth that's obvious once said.

Certain kinds of ideas (concepts, notions, conceptions, and so forth) have science-like quality. They have the rigidity, the reproducibility, of physical science. They yield reproducible results, independent of particular investigators. Such kinds of ideas are important enough to have a name.

Study of the lawful, predictable parts of the physical world has a name: "physics." Study of the lawful, predictable, parts of the social-conceptual world also has a name: "mathematics."

A world of ideas exists, created by human beings, existing in their shared consciousness. These ideas have objective properties, in the same sense that material objects have objective properties. The construction of proof and counterexample is the method of discovering the properties of these ideas. This branch of knowledge is called mathematics.

An Objection

There's a logical difficulty we have to look at.

I say the 3-cube or the 4-cube—any mathematical object you like—exists at the social-cultural-historic level, in the shared consciousness of people (including retrievable stored consciousness in writing). In an oversimplified formulation, "mathematical objects are a kind of shared thought or idea."

A mathematical 3-cube is just an idea we share.

This statement is open to an objection. If you turn it around, as by ordinary logic it seems you have a right to do, you get "A certain idea we share is a mathematical 3-cube."

That is, an idea has volume, and vertices, edges, and faces—all of which is nonsense. Probe my mind-brain anyway you like; you won't find inside it a cube or a hyper-cube.

What are we trying to say?

Things become clear if we turn to familiar material objects. We have an idea of a chair, but our idea of a chair isn't a chair. It's our mind-brain's representation of a chair, analogous to a photograph of a chair or to the definition of "chair" in Webster. We know little about the construction or functioning of ideas in the mind-brain. But there's no logical confusion between a chair and the idea of a chair.

Between a 4-cube and the idea of such, there is a confusion. Why? Because we have nowhere to point, to show a "real" 4-cube as distinct from the idea of a 4-cube.

There are two ways to go from here. One well-worn path is the Platonist way. "There *is* a real 4-cube. It's a transcendental immaterial inhuman abstraction. Our idea of a cube is a representation of this transcendental thing, parallel to our idea of chair being a representation of real chairs."

The other way is fictionalism. There is no more a "real" 4-cube than a "real" Mickey Mouse. Oedipus and Mickey Mouse exemplify shared ideas that don't represent anything real. They show that there can be representation without a represented.

Our mental picture of a 4-cube is only a picture, not a 4-cube. It doesn't have vertices or edges, but it does have representations of vertices and edges. It's different from a 4-cube, because it does exist (on the social-cultural-historic level) while the 4-cube, itself doesn't exist. Or, as I prefer to say, it exists only in its social and mental representations.

A 4-cube has 16 vertices. At each vertex, 4 edges meet at right angles. But there is no 4-cube! So *nothing* has 16 vertices at which 4 edges meet at right angles—except as we have a shared idea of such a thing, an idea so consistent, rigid, and reliable that we share each other's reasonings, and come to the same conclusions.

This may sound paradoxical. It's an honest account of the actual state of affairs.

It's a Futile Question

Some questions, which at first seem meaningful, are *futile*—to answer them is neither possible nor necessary.

Why are there rigid, reproducible concepts such as number or circle?

Why is there consciousness?

Why is there a cosmos?

We need not answer Kant's question, "How is mathematics possible?" any more than we need answer Heidegger's question, "Why should anything exist?"

I haven't heard about progress on either problem.

People who think up such questions may get compliments for asking amusing questions. But no physicist and few philosophers feel obliged to answer Heidegger's question. The existence of a world is the starting point from which we go forward.

Once upon a time an important question was, "How can the world be so simple, complicated, and beautiful unless Someone made it?" Now many would say that's a futile question.

Some of today's questions about cosmology, ethics, determinism, or cognition may be futile.

Kant answered his question, "How is mathematics possible?" If not because of the existence of external mathematical objects, then, he thought, our minds ("intuitions") must impose arithmetic and geometry universally.

Ethnology, comparative history, developmental psychology, the development of non-Euclidean geometry, and general relativity, all show that Euclidean geometry is not built into everyone's mind/brain. We think about space in more than one way. We reject Kant's answer. Must we still accept his question?

I counter Kant's question with a counter-question: "Why should your question have an answer?"

This much is clear: Mathematics *is* possible. It's the old saying, "What *is* happening *can* happen."

How does mathematics come about, in a daily, down-to-earth sense? That question belongs to psychology, to the history of thought, and to other disciplines of empirical science. It can't be answered by philosophy. Accept the *possibility* of mathematics as a fact of experience.

Major empirical discoveries about it are coming. Neuro-scientists are hunting for the brain structures we use in counting and spatial thinking. George Lakoff, George Johnson, Terry Regier, and others, using work of Antonio Damasio, Gerald Edelman, and others, may be approaching that goal.

When such discoveries come they'll have tremendous importance, both scientific and practical. But they won't decide philosophical controversies.

To see why not, consider a comparable question. Is what our eyes see really there? That is, is matter an illusion, as many brilliant idealists have said? Or, as Kant taught, is it impossible for us to know whether it's an illusion?

These questions have been of the highest concern to great philosophers.

Today, we realize that those philosophers had limited understanding of the workings of the eye and brain. We do know something about those workings. Maybe some day we'll understand them completely, for practical purposes. Would that understanding tell us whether the visible is real? No. Idealists and skeptics could find new distinctions, and go on being idealists or skeptics as long as they wished.

The reasons that apply to visual reality apply to mathematical reality. The philosophical issues around it will be influenced by empirical discovery, but not settled.

We can study how mathematics develops, in history, in society, and in the individual. We can study how mathematical theories give rise to one another. We can study how mathematics springs from and goes back to physics and other sciences. But the question, "How is mathematics possible?" tries to push mathematics into a pigeonhole: physical, mental, transcendental. None fits. I reject the question and its old alternatives.

Since Dedekind and Frege in the 1870s and 1880s, philosophy of mathematics has been stuck on a single problem—find a solid foundation to which all mathematics can be reduced, a *foundation* to make mathematics indubitable, free of uncertainty, free of any possible contradiction (see section in Chapter 8).

That goal is now admitted to be unattainable. Yet, with the exception of a few mavericks, philosophers continue to see "foundation" as the main interesting problem in philosophy of mathematics.

The key assumption in all three foundationist viewpoints is mathematics as a source of indubitable truth. Yet daily experience finds mathematical truth to be fallible and corrigible, like other kinds of truth.

None of the three can account for the existence of its rivals. If Platonism is right, the existence of formalism and constructivism is incomprehensible. If constructivism is right, the existence of Platonism and formalism is incomprehensible. If formalism is right, the existence of Platonism and constructivism is incomprehensible.

Humanism sees that constructivism, formalism, and Platonism each fetishizes one aspect of mathematics, insists that one limited aspect *is* mathematics.

This account of mathematics looks at what mathematicians do. The novelty is conscious effort to avoid falsifying or idealizing.

If we give up the obligation of mathematics to be a source of indubitable truths, we can accept it as a human activity. We give up age-old hopes, but gain a clearer idea of what we are doing, and why.

1. Mathematics is human. It's part of and fits into human culture.
2. Mathematical knowledge isn't infallible. Like science, mathematics can advance by making mistakes, correcting and recorrecting them. (This fallibilism is brilliantly argued in Lakatos's Proofs and Refutations.)
3. There are different versions of proof or rigor, depending on time, place, and other things. The use of computers in proofs is a nontraditional rigor. Empirical evidence, numerical experimentation, probabilistic proof all help us decide what to believe in mathematics. Aristotelian logic isn't always the only way to decide.
4. Mathematical objects are a distinct variety of social-historic objects. They're a special part of culture. Literature, religion, and banking are also special parts of culture. Each is radically different from the others.

Music is an instructive example. It isn't a biological or physical entity. Yet it can't exist apart from some biological or physical realization—a tune in your head, a page of sheet music, a high C produced by a soprano, a recording, or a radio broadcast. Music exists by some biological or physical manifestation, but it makes sense only as a mental and cultural entity.

What confusion would exist if philosophers could conceive only two possibilities for music—either a thought in the mind of an Ideal Musician, or a noise like the roar of a vacuum cleaner.

I have two concluding points.

Point 1 is that mathematics is a social-historic reality. This is not controversial. All that Platonists, formalists, intuitionists, and others can say against it is that it's irrelevant to their concept of philosophy.

Point 2 *is* controversial: There's no need to look for a hidden meaning or definition of mathematics beyond its social-historic-cultural meaning. Social-historic is all it needs to be. Forget foundations, forget immaterial, inhuman "reality."

two

Criteria for a Philosophy of Mathematics

Taking the Test

One way to test a philosophy of mathematics is to confront it with test questions:

What makes mathematics different?

What is mathematics about?

Why does mathematics achieve near-universal consensus?

How do we acquire knowledge of mathematics, apart from proof?

Why are mathematical results independent of time, place, race, nationality, and gender, in spite of the social nature of mathematics?

Does the infinite exist? How?

The social-historical approach gives better answers than the neo-Fregean, the intuitionist-constructivist, or any other proposed philosophy I know of.

Regarding the infinite, if mathematical objects are like stories we make up, we can make up fantastic weird stories if we want to, just so they fit together with other stories. N, an infinite set, is no harder to accept than $10^{(10^{10})}$. Both are socially validated inventions, real, not as physical objects, but as other socially validated inventions are real. Though they are our inventions, their properties are not arbitrary. They're forced to be what they are, by the purposes for which we invented them.

Guiding Principles

Evaluate a body of thought according to its own goals and presuppositions. Understand it historically, in the sense of history of ideas. Pay attention to its consequences, theoretical and practical. Beneficial consequences don't verify a doctrine. Harmful consequences don't falsify it. But consequences are as important as plausibility, consistency, or explanatory power.

I now list criteria for a philosophy of mathematics. No philosophy, including our own, is satisfactory by all criteria. The list is a vantage point from which to evaluate theories, including our own. We do so in Chapter 13.

1. Breadth

An adequate philosophy of mathematics would be aware at least of *some* active field of mainstream mathematical research (dynamical systems, say, or stochastic processes, or algebraic/differential geometry/topology). It would look at how mathematics is being used somewhere, whether in hydrodynamics, or meteorology, or geophysics. It would notice that theoretical physicists do mathematics differently from either "pure" or "applied" mathematics. It wouldn't ignore computing in mathematics today—real computing with real machines, not just idealized theory of ideal machines. It would be compatible with the history of mathematics and with how people learn mathematics.

Who dares to write philosophy of science without some acquaintance with quantum mechanics and relativity? But you must search long and far to find a philosopher of mathematics who claims a nodding acquaintance with functional analysis or differential topology.

This complaint can be parried by claiming that all mathematics "can be got down" to set theory and arithmetic (quote from Quine in Chapter 9). Such a claim depends on what you mean by "got down." It has a self-serving flavor. The argument, "What I don't know, *ipso facto* doesn't matter" isn't new. Age hasn't made it palatable.

2. Links to Epistemology and Philosophy of Science

Philosophy of mathematics should articulate with epistemology and philosophy of science. But virtually all writers on philosophy of mathematics treat it as an encapsulated entity, isolated, timeless, ahistorical, inhuman, connected to nothing else in the intellectual or material realms. Philosophy of mathematics is routinely done without reference to mind, science, or society. (We'll see exceptions in due course.)

Your view of mathematics should fit your view of physical science. Your view of mathematical knowledge should fit your view of knowledge in general. If you write philosophy of mathematics, you aren't expected simultaneously to write philosophy of science and general epistemology. To write on philosophy of mathematics alone is daunting enough. But to be adequate, it needs a connection with epistemology and with philosophy of science.

3. Valid against Practice

Philosophy of mathematics should be tested against five kinds of mathematical practice: research, application, teaching, history, computing. In all areas of mathematical practice, an essential role is played, one way or another, by something called mathematical intuition. There is an extended discussion of intuition in Chapter 4. Here I want to make clear that an adequate philosophy of mathematics must recognize and deal with mathematical intuition.

The need to check philosophy of mathematics against mathematical *research* doesn't require explication. Many important philosophers of mathematics were mathematical researchers: Pascal, Descartes, Leibniz, d'Alembert, Hilbert, Brouwer, Poincaré, Rényi, and Bishop come to mind.

Applied mathematics isn't illegitimate or marginal. Advances in mathematics for science and technology often are inseparable from advances in pure mathematics. Examples: Newton on universal gravitation and the infinitesimal calculus; Gauss on electromagnetism, astronomy, and geodesy (the last inspired that beautiful pure subject—differential geometry); Poincare on celestial mechanics; and von Neumann on quantum mechanics, fluid dynamics, computer design, numerical analysis, and nuclear explosions.

Not only did the same great mathematicians do both pure and applied mathematics, their pure and applied work often fertilized each other. This was explicit in Gauss and Poincaré. Nearer our time is Norbert Wiener. He was generally known for cybernetics, but his life work was mathematical analysis. His study of infinite-dimensional stochastic processes was guided by their physical interpretation as Brownian motion, illuminated by experiments of the French physicist, Jean Perrin (see Wiener, 1948). The standard mathematical model for Brownian motion is the Wiener process. His stochastic processes were useful in controlling anti-aircraft fire. When he renounced military research, he took up prosthetics for the blind—more applied work!

G. H. Hardy "famously" boasted: "I have never done anything 'useful'. No discovery of mine has made, or is likely to make, directly or indirectly, for good or ill, the least difference to the amenity of the world." Nevertheless, the Hardy-Weinberg law of genetics is better known than his profound contributions to analytic number theory. What's worse, cryptology is making number theory applicable. Hardy's contribution to that pure field may yet be useful.

Twenty years after the war, mathematical purism was revived, influenced by the famous French group "Bourbaki." That period is over. Today it's difficult to find a mathematician who'll say an unkind word about applied math.

Instead, we see the spectacular merger of elementary-particle physics and high-dimensional differential geometry. One famous practitioner, Ed Witten, is a physicist whose physical insight permits him unexpected discoveries in mathematics.

A philosophy of mathematics that ignores applied mathematics, or treats it as an afterthought, is out of date. The relationship between pure and applied mathematics is a central philosophical question. A philosophy of mathematics blind to this challenge is inadequate.

An interesting fact about mathematics is that it's taught and learned. (Even more interesting—sometimes taught but *not* learned!) A credible philosophy of mathematics must accord with the experience of *teaching and learning* mathematics. To a formalist or Platonist who presents an inhuman picture of mathematics, I ask, "If this were so, how could anyone learn it?"

We know a little about how mathematics is learned. It's not done by memorizing the times table or Peano's axioms. Mathematics is learned by computing, by solving problems, and by conversing, more than by reading and listening.

An account of the nature of mathematics incompatible with these facts is wrong. Piaget was one of the few writers on philosophy of mathematics to take teaching seriously. Today Thomas Tymoczko and Paul Ernest are doing so.

There is also *the historical test.* Mathematics has seen enormous changes. Its story reaches from Babylonia to Maya, pre-colonial Africa, India, China, Japan. A philosophy about today's mathematics that leaves inexplicable the mathematics of 500 years ago is inadequate.

André Weil, a leading mathematician of the postwar period and now a historian of mathematics, wrote that there could hardly be two disciplines further apart than history of mathematics and philosophy of mathematics. Perhaps in this statement he was identifying philosophy of mathematics with foundations of mathematics.

One philosophical issue in which history is relevant is the reduction of all mathematics to set theory,** as in foundations textbooks, and as mentioned previously in point 1, breadth.

On the basis of this reduction, philosophers of mathematics generally limit their attention to set theory, logic, and arithmetic.

What does this assumption, that all mathematics is fundamentally set theory, do to Euclid, Archimedes, Newton, Leibniz, and Euler? No one dares to say they were thinking in terms of sets, hundreds of years before the set-theoretic reduction was invented. The only way out (implicit, never explicit) is that their own understanding of what they did must be ignored! We know better than they how to explicate their work!

That claim obscures history, and obscures the present, which is rooted in history.

An adequate philosophy of mathematics *must* be compatible with the history of mathematics. It *should* be capable of shedding light on that history. Why did the Greeks fail to develop mechanics, along the lines that they developed geometry? Why did mathematics lapse in Italy after Galileo, to leap ahead in England, France, and Germany? Why was non-Euclidean geometry not conceived until the nineteenth century, and then independently rediscovered three times? The philosopher of mathematics who is historically conscious can offer such questions to the historian. But if his philosophy makes these questions invisible, then instead of stimulating the history of mathematics, he stultifies it.

Computing is a major part of mathematical practice. The use of computing machines in mathematical proof is controversial. An adequate philosophy of mathematics should shed some light on this controversy.

Formalist and logicist philosophies, each in its own way, picture mathematics as essentially calculation. For the logicist, mathematical theorems are true tautologically, as logical identities. For the formalist, the undefined terms of mathematics are meaningless, so mathematical theorems are meaningless. Both say the essence of mathematics is proof in the sense of formal logic. Proof as a formal calculation in a formal language according to formal rules.

When these conceptions were developed early in the twentieth century, the possibility of realizing them was fantasy. The most famous attempt to formalize statements and proofs of mathematical theorems was the *Principia Mathematica* of Bertrand Russell and Alfred Whitehead. I'm told that finally on page 180 or so they prove 1 is different from 0.

Applied mathematicians have long used computers as a matter of course. Today computers are used more and more in pure mathematics research. Number theorists and algebraists use them to test and make up conjectures. With a computer you sometimes can finish a proof by a calculation impossible by hand. The four-color conjecture** is a famous example (see "Proof" in Chapter 5).

There are differences of opinion about computers. Clifford Truesdell, a leading authority on continuum mechanics, calls them a menace and an abomination. Some say the proof of the four-color theorem violates the traditional idea of proof, since it requires believing that computers work, which is not a mathematical belief. Such objections are ironic. For decades philosophers said valid mathematical proofs should be checkable by machine. Now, when part of a proof *is* done on a machine, some say, "That's not a proof!"

So far our criteria for a philosophy of mathematics have been external—how the philosophy relates to mathematics. There are criteria within the philosophical doctrine itself: consistency, elegance, economy, simplicity, comprehensibility, and precision.

4. Elegance

Elegance is more common in mathematical theories than in philosophical ones. Paul Cohen remarked that no one could accuse philosophy of being beautiful. He had in mind foundationist philosophy. Beauty or elegance is desirable, but not prerequisite. Otherwise, Aristotle and Kant would go out the window.

5. Economy

Good old Ockham's razor: Use what you need, nothing more. This principle justifies the set-theoretic reduction of mathematics. It's claimed that set membership suffices to define number systems, spaces, geometric figures, and all operations and operands in mathematics past, present, and future. That being so, there should be nothing in mathematics but sets, since there needn't be. But economy can conflict with other criteria such as comprehensibility and applicability. It may be fun to find minimal generators of a theory, even if they're neither unique nor convenient. Economy is like elegance: desirable, but optional.

6. *Comprehensibility*

Comprehensibility is valued by readers, not by all writers.

Philosophy students think that among professional philosophers, incomprehensibility gets "Brownie points," and comprehensibility gets demerits.

Unworthy suspicions aside, it's a question of comprehensibility *to whom.* What's impenetrable to you may be crystalline to the Heidegger expert.

This book aims to be easily comprehensible to anyone. If some allusion is obscure, skip it. It's inessential.

7. *Precision*

Should the philosophy of mathematics be precise? Analytic philosophers sometimes use pseudo-mathematical notation. Call a claim "Claim A" or "Hypothesis B," and obscure a conversation that would have gone better in some natural language, English, for example. If you notice that no other branch of philosophy even hopes for precision, the dispensability of precision in philosophy of mathematics becomes apparent.

Mathematics is precise; philosophy cannot be. Expecting philosophy of mathematics to be a branch of mathematics, with definitions and proofs, is like thinking philosophy of art can be a branch of art, with landscapes and still lives.

Art can be beautiful; philosophy of art cannot be beautiful.

Philosophy of politics cannot win at the ballot. Philosophy of law cannot fill the wallets of lawyers. Philosophical specialties, including the philosophy of mathematics, should be evaluated by philosophical standards, not the standards of the field they critique.

Is philosophy of mathematics, part of mathematics or part of philosophy? It's *about* mathematics, but it's *part* of philosophy. It happens that the creators of foundationist philosophy of mathematics were mathematicians (Hilbert, Brouwer) or mathematically trained (Husserl, Frege, Russell). This training may explain their bias. They sought to turn philosophical problems into mathematical problems, *to make them precise.* This bias was fruitful mathematically. Some of today's mathematical logic descended from the search for mathematical solutions to philosophical problems. But, even though mathematically fruitful, it was philosophically misguided. Today we can turn away from the philosophical failure of the foundationist schools. We can think of philosophy of mathematics, not as a branch of mathematics, but as a philosophical enterprise based on mathematical experience. Give up the illusion of mathematical precision. Aim for insight, enlightenment.

8. *Simplicity*

Both simplicity and precision are desirable, in science and in philosophy of science, especially in philosophy of mathematics.

But in science, philosophy of science, and philosophy of mathematics, there's another desideratum—truthfulness, faithfulness to the facts, simple honesty.

One wants all three: truthfulness, precision, and simplicity. But one can't usually maximize at once goal A and goal B. If you're not willing to pick one goal and ignore the others (maximum cash flow, for instance—reputation and legality be damned!) then you have to do some balancing or juggling. (Work on cash flow, but don't actually go to jail.)

Precision is easier to achieve in a simple situation than in a complicated one. Some phenomena are inherently imprecise.

Precision in philosophy of mathematics is sought by trying to mathematize it. Axiomatic set theory is a branch of mathematics. If the philosophy of mathematics were no more than a style of doing axiomatic set theory, it would attain mathematical precision. But it's impossible to talk about all interesting aspects of mathematics in the language of axiomatic set theory. And it's not necessary to do so! The notion that philosophy of mathematics is a branch of axiomatic set theory is no divine ordinance or self-enforcing decree. It's just one school, one trend, that hopes to borrow the prestige of mathematics by doing its nonmathematical thing *more mathematico*. In this it's reminiscent of certain "mathematical" specialties in economics, sociology, and psychology.

Simplicity goes with single-mindedness. Where several factors interact to give a complex result, simplicity can be created by ignoring all factors but one. Different scholars may single out different factors. This kind of simplicity leads to fruitless controversy, like between Red Sox fans and White Sox fans. For example, both formalization and construction are central features of mathematics. But the *philosophies* of formalism and constructivism are long-standing rival schools. It would be more productive to see how formalization and construction interact than to choose one and reject the other.

The notions I advocate are less precise and less simple than familiar philosophies. This permits better faithfulness to experience. Some may think no loss of simplicity or precision is acceptable.

Putting simplicity and precision ahead of truthfulness is treating philosophy as a game, *art pour l'art*, like an exotic branch of algebra. Philosophy can be serious—no less than how and why to live. Being serious means putting truthfulness first. First get it right, then go for precision and simplicity.

My first assumption about mathematics is: It's something people do. An account of mathematics is unacceptable unless it's compatible with what people do, especially what mathematicians do.

9. Consistency

This is highly valued by logicians and logic-minded philosophers. Others downplay it. Ralph Waldo Emerson wrote, "A foolish consistency is the hobgoblin of little minds, adored by little statesmen and philosophers and divines."

Walt Whitman wrote:

> Do I contradict myself?
> Very well then I contradict myself.
> (I am large, I contain multitudes.)

Bourbaki wrote:

> Historically speaking, it is of course quite untrue that mathematics is free from contradiction. Non-contradiction appears as a goal to be achieved, not as a God-given quality that has been granted us once for all. There is no sharply drawn line between the contradictions which occur in the daily work of every mathematician, beginner or master of his craft, as the result of more or less easily detected mistakes, and the major paradoxes which provide food for logical thought for decades and sometimes centuries. (N. Bourbaki, 1949)

In a first course in logic, the teacher shows how, from any proposition A together with its negation you can deduce any other proposition B.**

From *any* contradiction, *all* propositions (and their negations) follow! Everything's both true and false! The theory collapses in ruins!

Outside of logic class, it isn't that way. The U.S. Constitution contains contradictions. Any courtroom prosecution or defense probably contains contradictions.

Quantum electrodynamics gives the most precise predictions of any physical theory. Yet physicists have known from its birth that it's self-contradictory. They make *ad hoc* rules for handling the inconsistency. Divergent series of divergent terms are manipulated and massaged. In a *Festschrift* for the famous physicist John Wheeler I found this praise: "He's never stopped by a formal contradiction" (i. e., a mathematical contradiction).

Players in this game know contradiction is seldom fatal. Richard Rorty quotes Aquinas: "Coherence is a matter of avoiding contradictions, and St. Thomas' advice, 'When you meet a contradiction, make a distinction,' makes that pretty easy. As far as I could see, philosophical talent was largely a matter of proliferating as many distinctions as were needed to wriggle out of a dialectical corner."

If A is B and also not B, make a distinction, A_1 and A_2. Example: "Mathematics is precise and imprecise." Whoops, contradiction! What to do? Distinguish between formal and informal mathematics. Formal is precise, informal is imprecise. Contradiction gone!

That contradiction is even useful! It calls attention to an interesting distinction.

Classical logic says all the consequences of a set of axioms exist—are derived—instantly, as soon as the axiom set is laid down. There are infinitely many consequences. This whole infinite theory, created instantly! Consistency holds or it's violated immediately, instantly.

In practice, consequences are derived step by step. At any time only a finite number have been derived. If a contradiction appears, some device is brought in to wall it off, to keep the rest of the theory from infection.

I once wrote that mathematicians hate contradiction. That's not accurate. We love it—like a duck hunter loves ducks. Nothing draws us to the chase like a contradiction in a famous theory.

In evaluating a mathematical theory, consistency is important. But it's less important than fruitfulness (inside and outside of mathematics), imaginative appeal, and linking new mathematical devices to old, respected problems. A contradiction can generally be fixed up one way or another.

As Bourbaki explained, "freedom from contradiction is attained in the process, not guaranteed in advance." They didn't notice that this fact discredits all standard foundationist or logicist theories about mathematics. In practice, we can't always prove in advance the consistency of all possible deductions. Instead, we develop a technique for preserving partial consistency—absence of contradiction up to the latest set of results. In that way we continue to forestall contradiction each time it raises its ugly head.

Frege announced that his building had collapsed. Then after its collapse he tried to patch it up, just as an ordinary, nonfoundationist mathematician would do. Mathematical buildings collapse—lose interest, are forgotten—not because of contradictions, but because their questions are no longer interesting, or because another theory answers them better.

10. Originality, Novelty

This quality is external; it relates to other philosophers and philosophies. It's rare for philosophical writing to be entirely novel. The basic ideas are in Plato and Aristotle. Nothing here is without some antecedent. Presentation, examples, arrangement, and flavor are mine. I hope to be provocative, even convincing to some, but not to all.

11. Certitude and Indubitability

Today the goals of certitude and indubitability are abandoned. The foundationist project has lost its philosophical rationale. "Foundations" (axiomatic set theory and related topics) is just one mathematical specialty of many. It compares axiomatic setups with respect to convenience, effectiveness, or strength. It doesn't pretend to give certainty or indubitability, the founding motive for "foundations."

An adequate philosophy of mathematics must account for the special role of proof in mathematics. Mark Kac, a famous probabilist, asked why mathematicians are obsessed with a need to prove everything. Many people, physicists among them, are willing to believe without proof. Yet mathematicians want proof. They even say, "Without a proof it's nothing." Is this the very nature of mathematics? Or is it one aspect among several of equal importance?

12. *Applicability*

This does not refer to mathematical applications, but to philosophical ones. Your philosophy may increase your feeling of being at home in the universe, or your ability to sleep with a clear conscience. But it should also be helpful in analyzing philosophical problems, perhaps even in solving one or two. If it's useless, who needs it? Chapters 4 and 5 below offer applications of humanist philosophy of mathematics.

13. Acceptability

This criterion is never explicitly demanded. Yet in practice it's the most important. It's why Cartesians prospered while Spinozists were damned (Chapters 6 and 7).

Mathematical theories "ahead of their time" have been ignored for decades, even centuries (Desargues, Grassmann.) In every field of learning, theory can't prosper if it too grossly violates current acceptability. In Europe until the seventeenth century, acceptability was conformity to the Holy Roman Catholic Church. In the Soviet Union, it was Marxism-Leninism-Stalinism, in genetics, linguistics, literature, and music. Here and now, no philosophy penetrates *philosophia academica* without bowing to that establishment's sine qua non.

The acceptability of this book—to mathematicians, philosophers, and the general public—remains to be seen.

Summing Up

Not all of these thirteen criteria are essential.

The first three are essential:

1. Recognize the scope and variety of mathematics.
2. Fit into general epistemology and philosophy of science.
3. Be compatible with mathematical practice—research, application, teaching, history, calculation, and mathematical intuition.

The next five are desirable:

4. Elegance
5. Economy
6. Comprehensibility
7. Precision
8. Simplicity

The next, 9. Consistency, is essential, but not as hard to attain as 1–3.
We reject 10. Novelty, Originality, as inessential and unattainable.
We reject 11. Certainty, Indubitability, as false and misleading.

We recognize 12. Applicability, as essential in practice. If you can't do anything with it, what good is it?

13. Acceptability can't be a goal, yet it can't be evaded.

Notice that in this list moral, ethical, and political considerations are excluded. I don't ask whether a theory is beneficial or harmful, progressive or reactionary, humane or inhumane. The preceding thirteen criteria suffice.

We return to these criteria in Chapter 13, and consider there the linkage between philosophical opinions and political opinions.

three

Myths/Mistakes/ Misunderstandings

This chapter is an impressionistic report of how mathematics looks to mathematicians today. I emphasize, but do not exaggerate, aspects ignored in conventional accounts. This picture of mathematical life is evidence that a humanist or social-historical account is truer to real life than traditional foundationist or neo-Fregean accounts.

Mathematics Has a Front and a Back

In *The Presentation of Self in Everyday Life* by the U.S. sociologist Erving Goffman, there's a chapter called "Regions and Region Behavior." There Goffman develops the concepts of "front" and "back." In a restaurant the dining area is the front; the kitchen is the back. In a theater, front stage is for the audience; backstage is for actors and stagehands. In front, actors (waiters) wear costumes (uniforms); in back, they change clothes or wear casual dress. In front the public is served; in back, professionals prepare to serve them. Front is where the public is admitted; back is where it's excluded.

Goffman's contribution is extending "front" and "back" from restaurants and theaters to all our institutions. In the university, classrooms and libraries are the front, where the public (students) is served. The chairman's and dean's offices are the back, where the products (classes and courses) are prepared.

This separation is a necessity. Goffman gives an example of distress from mixing front and back: a gas station where customers walk into the toolroom and help themselves to wrenches.

He quotes George Orwell: "It is an instructive sight to see a waiter going into a hotel dining room. As he passes the door a sudden change comes over him. The set of his shoulders alters; all the dirt and hurry and irritation have dropped off in an instant. He glides over the carpet, with a solemn, priest-like air . . . he entered the dining room and sailed across it, dish in hand, graceful as a swan" (p. 121).

The purpose of separating front from back isn't only to keep customers from interfering with the cooking. It's also to keep them from knowing too much about the cooking.

Everybody down front knows the leading lady wears powder and blusher. They don't know exactly how she looks without them.

Diners know what's supposed to go into the ragout. They don't know for sure what *does* go into it.

Traditional philosophy recognizes only the front of mathematics. But it's impossible to understand the front while ignoring the back.

The front and back of mathematics aren't physical locations like dining room and kitchen. They're its public and private aspects. The front is open to outsiders; the back is restricted to insiders. The front is mathematics in finished form—lectures, textbooks, journals. The back is mathematics among working mathematicians, told in offices or at cafe tables.

The front divides into subregions—first, second, third class. A restaurant may include banquet hall and snack bar. Theaters can have box, orchestra, and balcony. The mathematical public includes professionals, graduate students, and undergraduates.

The back also divides into subregions. In a restaurant, there are domains of salad chef, pastry chef, and dishwasher. Mathematicians divide into finite group theorists, numerical linear algebraists, nonstandard differential topologists, and so on and on.

Front mathematics is formal, precise, ordered, and abstract. It's broken into definitions, theorems, and remarks. Every question either is answered or is labeled: "open question." At the beginning of each chapter, a goal is stated. At the end of the chapter, it's attained.

Mathematics in back is fragmentary, informal, intuitive, tentative. We try this or that. We say "maybe," or "it looks like."

This distinction is described by George Pólya in his preface to *Mathematics and Plausible Reasoning*. "Finished mathematics presented in a finished form appears as purely demonstrative, consisting of proofs only. Yet mathematics in the making resembles any other human knowledge in the making. You have to guess a mathematical theorem before you prove it; you have to have the idea of the proof before you carry through the details. You have to combine observations and follow analogies; you have to try and try again."

Mainstream philosophy doesn't know that mathematics has a back. Finished, published mathematics—the front—is taken as a self-subsistent entity. Worse, mainstream philosophy doesn't look at the real front—real articles and treatises. It contemplates idealized texts as logicians would have them, not mathematical texts as they are and must be.

It would make as much sense for a restaurant critic not to know there are kitchens, or a theater critic not to know there is backstage.

Myths

The front/back separation makes possible a myth about the seasoning of the ragout or about the leading lady's complexion. This kind of myth takes the performance in front at face value, innocently unaware that it was concocted behind the scenes.

A myth isn't bad *per se*. It has allegorical or metaphorical power. It may increase the customer's enjoyment. It may be essential for the performance.

The myth of the divine right of kings was useful. So are the myths of Christmas, Easter, and those of other religions.

There's an unwritten criterion separating the professional from the amateur, the insider from the outsider: The outsider is taken in by the myths. The insider is not.

I will describe a few myths of mathematics. I limit myself to a few provocative comments. To present them in detail and refute them would fill a volume. The reader can dig deeper or extend the list, with help from the references.

The myth of Euclid: "Euclid's *Elements* contains truths about the universe which are clear and indubitable." Today advanced students of geometry know Euclid's proofs are incomplete and his axioms are unintelligible. Nevertheless, in watered-down versions that ignore his impressive solid geometry, Euclid's *Elements* is still upheld as a model of rigorous proof.

The plaster Newton fabricated in the eighteenth century is intact. Alexander Pope offered an epitaph:

> Nature and Nature's Laws lay hid in Night;
> GOD said, *Let Newton be!* and all was Light.

The strange, complex historical Newton is almost unknown, even among the mathematically literate.

The myths of Russell, Brouwer, and Bourbaki—of logicism, intuitionism, and formalism—are treated in *The Mathematical Experience*. I return to them in Part 2, in articles on foundationism, the foundationist philosophers, and Lakatos, their critic.

Now four more general myths.

1. *Unity*: There's only one mathematics, indivisible now and forever. Mathematics is a single inseparable whole.
2. *Universality*: The mathematics we know is the only mathematics there can be. If little green critters from Quasar X9 showed us their textbooks, we'd find again $A = \pi r^2$.
3. *Certainty*: Mathematics has a method, "rigorous proof," which yields absolutely certain conclusions, given truth of premises.

4. *Objectivity*: Mathematical truth is the same for everyone. It doesn't matter who discovers it. It's true whether or not anybody discovers it.

It would be easy to give citations to show that these four myths are generally accepted. Documentation would be pedantic.

Being a myth doesn't entail its truth or falsity. Myths validate and support institutions; their truth may not be determinable. Who can say the divine right of kings is false? Without a channel to the mind of God, this can't be proved or disproved. In its day it was credible and useful.

Myths 1–4 are almost universally accepted, but they aren't self-evident or self-proving. It's possible to question, doubt, or reject them. Some people do reject them. Standard and official though they are, they aren't taken literally or naively by backstage people.

Part of preparing mathematics for public presentation—in print or in person—is tying up loose ends. If there's disagreement whether a theorem has been proved, it's left out of the text. The standard exposition purges mathematics of the personal, the controversial, the tentative, leaving little trace of humanity in the creator or the consumer. This style is the "front" of mathematics.

Without it, the myths would lose their aura. If mathematics were presented in the style in which it's created, few would believe its universality, unity, certainty, or objectivity. These myths support the institution of mathematics. For mathematics is not only an art and a science, but also an institution, with budgets, administrations, rank, status, awards, and grants.

Let's examine 1–4 from backstage.

Myth 1 is unity. But at meetings of the American Mathematical Society, any contributed talk is understandable to only a small fraction of those present.

Sometimes pure and applied mathematicians interact. Most of the time they're oblivious to each other, working to different audiences, different standards, and different criteria. The purists sometimes even declare applied mathematics isn't mathematics. "Where are the definitions? Where are the theorems?"

The "unity" claimed in principle doesn't exist in practice

Myth 2 is universality. If there's intelligent life on Quasar X9, it may be blobs of plasma we can't recognize as life. What would it mean to talk about their literature, art, or mathematics? To ask if their mathematics is the same as ours requires a possibility of comparing. Comparing demands communication. The possibility of comparing isn't universal. It's conditional on their being enough like us to communicate with us.

Myth 3 is certainty. We're certain

$$2 + 2 = 4,$$

though we don't all mean the same thing by that equation. It's another matter to claim certainty for the theorems of contemporary mathematics. Many of these theorems have proofs that fill dozens of pages. They're usually built on top of other contemporary theorems, whose proofs weren't checked in detail by the mathematician who quotes them. The proofs of these theorems replace boring details with "it is easily seen" and "a standard argument yields" and "a calculation gives." Many papers have several coauthors, no one of whom thoroughly checked the whole paper. They may use machine calculations that none of the authors completely understands.

A mathematician's confidence in some theorem doesn't necessarily means she knows every step from the axioms of set theory up to the theorem she's interested in. It may include confidence in the word of fellow researchers, journals, and referees.

Certainty, like unity, can be claimed in principle—not in practice.

Myth 4 is objectivity. This myth is more plausible than the first three. Yes! There's amazing consensus in mathematics as to what's correct or accepted.

But just as important is what's interesting, important, deep, or elegant. Unlike correctness, these criteria vary from person to person, specialty to specialty, decade to decade. They're no more objective than esthetic judgments in art or music.

Mathematicians want to believe in unity, universality, certainty, and objectivity, as Americans want to believe in the Constitution and free enterprise, or other nations in their Gracious Queen or their Glorious Revolution. But while they believe, they know better.

To become a professional, you must move from front to back. You get a more sophisticated attitude to myth. Backstage, the leading lady washes off powder and blusher. She's seen with her everyday face.

The front-back codependence makes it hopeless to understand the front while ignoring the back. That's what Mainstream philosophy of mathematics tries to do. You can't understand a restaurant meal if you're unaware of the kitchen. Yet you *can* present yourself as a philosopher of mathematics, and be aware only of publications washed and ironed for public consumption.

Of the three historic schools, only intuitionism pays attention to the producer of mathematics. Formalists, logicists, and Platonists sit at a table in the dining room, discussing their ragout as a self-created, autonomous entity.

The Mathematician's Philosophical Dilemma

What is the mathematicians's dilemma?

How did it come about?

Can he escape?

Writers agree: The working mathematician is a Platonist on weekdays, a formalist on weekends. On weekdays, when doing mathematics, he's a Platonist, convinced he's dealing with an objective reality whose properties he's trying to determine. On weekends, if challenged to give a philosophical account of this reality, it's easiest to pretend he doesn't believe in it. He plays formalist, and pretends mathematics is a meaningless game.

I quote two famous mathematicians:

Jean Dieudonné: "We believe in the reality of mathematics but of course when philosophers attack us with their paradoxes we rush to hide behind formalism and say, 'Mathematics is just a combination of meaningless symbols,' and then we bring out Chapters 1 and 2 on set theory. Finally we are left in peace to go back to our mathematics and do it as we have always done, with the feeling each mathematician has that he is working with something real. This sensation is probably an illusion, but is very convenient. That is Bourbaki's attitude toward foundations."

(Bourbaki is a Parisian mathematical clique in which Dieudonné was a captain.)

Paul Cohen: "To the average mathematician who merely wants to know his work is securely based, the most appealing choice is to avoid difficulties by means of Hilbert's program. Here one regards mathematics as a formal game and one is only concerned with the question of consistency. . . . The Realist [Platonist] position is probably the one which most mathematicians would prefer to take. It is not until he becomes aware of some of the difficulties in set theory that he would even begin to question it. If these difficulties particularly upset him, he will rush to the shelter of Formalism, while his normal position will be somewhere between the two, trying to enjoy the best of two worlds."

(Cohen transformed logic and set theory in 1963 when he proved that the continuum hypothesis and the axiom of choice are independent of the other axioms of Zermelo-Frankel.)

This is shocking news! Most mathematicians hold contradictory views on the nature of their work.

Does it matter? Yes. Truth and meaning aren't recondite technical terms. They concern anyone who uses or teaches mathematics. Ignoring them leaves you captive to unexamined philosophical preconceptions. This has practical consequences.

Consequence 1: "What's interesting in mathematics?" is an urgent question for anyone doing research, or hiring or promoting researchers. There's no public discussion of this question. No vehicle for public discussion of it. No language or viewpoint that could be used for such a discussion.

Not to say there should be agreed-on standards of what's interesting. Precisely because tastes differ, we need discussion on taste. We have some common standards. That's proved by our identity as one profession, and our agreement that certain feats in mathematics deserve the highest rewards. Bringing out those standards for analysis and controversy would be important philosophical work. Our inability to sustain public discussion on values betrays philosophical unawareness and incompetence.

Consequence 2: In the middle of the twentieth century formalism was the philosophy of mathematics most advocated in public. (Recall the section introducing formalism in Chapter 1.) In that period, the style in mathematical journals, texts, and treatises was to insist on details of definitions and proofs and tell little or nothing of why a problem is interesting or why a method of proof is used that changed drastically in the last quarter century.

It would be difficult to document a connection between formalism in expository style and formalism in philosophical attitude. Still, ideas have consequences. What I think mathematics *is* affects how I present it.

Consequence 3: The unfortunate importation into primary and secondary schools, during the 1960s, of set-theoretic notation and axiomatics. This wasn't an inexplicable aberration. It was a predictable consequence of a philosophical doctrine: Mathematics *is* axiomatic systems expressed in set-theoretic language.

Critics of formalism in high school say "This is the wrong thing to teach, and the wrong way to teach." Such criticism leaves unchallenged the dogma that real mathematics *is* formal derivations from formally stated axioms. If this dogma rules, the critic of formalism is seen as asking for lower quality, to give students something "watered down" instead of the "real thing."

The fundamental question is, "What is mathematics?" Controversy about high-school teaching can't be resolved without controversy about the nature of mathematics. To discredit formalism in teaching, discredit the formalist picture of the nature of mathematics. The critic of formalism has to give a more satisfactory account of the nature of mathematics.

Mathematicians don't usually discuss philosophical issues. We think somebody else has taken that over—"the professionals." But with few exceptions, philosophers have little to say to us. Nearly all of them have only a remote notion of what we do. This isn't discreditable, in view of the technical prerequisites for understanding our work. But a mathematician who seeks enlightenment in philosophical books and journals is disappointed. Some philosophers are unfamiliar with anything beyond arithmetic and elementary geometry. Others are competent in logic or axiomatic set theory, but are as narrow technical specialists as we are.

There are a few philosophical comments by a few mathematicians whose interests go beyond set theory and logic. But philosophical discourse isn't well developed today among mathematicians. Philosophical issues as much as mathematical ones deserve careful argument, developed analysis, and consideration of objections. A bald statement of one's opinion is insufficient, even in philosophy.

The philosophical questions in the university math curriculum were raised by foundationists 50 years ago. "None of the three foundationisms succeeded," we tell students, "and there's no prospect of a 'solid' foundation. There's only one philosophical problem of interest—the foundation of the real number system—and that's intractable."

We can understand the working mathematician's oscillation between formalism and Platonism, if we look at her work experience and at the philosophical dogmas she inherited—Platonism and formalism. Both dogmas say mathematical truth must possess absolute certainty. Her experience in mathematics, on the other hand, offers plenty of uncertainty.

Emilio D. Roxin described Platonism: "Most mathematicians feel like a hunter in a jungle, where the theorems sit in the trees or fly around, where the definitions are like convenient ladders awaiting to be used in order to trap the theorems and corollaries which are there, even if nobody finds them." Pick a familiar theorem: Cauchy's integral formula, or Cantor's theorem on the uncountability of the continuum. Is it a true statement about the world? Does one discover such a theorem? Does such a discovery increase our knowledge? If our mathematician says yes, he's a Platonist for the moment.

The basis for Platonism is awareness that the problems and concepts of mathematics are independent of him as an individual. The roots of a polynomial are where they are, regardless of what he thinks or knows. It's easy to imagine that this objectivity is outside human consciousness as a whole. The working mathematician's Platonism expresses this aspect of his daily working experience. Yet, I repeat, it's a halfhearted, shamefaced Platonism, because it's incompatible with his general philosophy or worldview.

Platonism in the strong sense—belief in the existence of ideal entities, independent of or prior to human consciousness—was tenable with belief in a Divine Mind. Once mysticism is left behind, once scientific skepticism is focused on it, Platonism is hard to maintain,

The next question is: "To what objects or features of the world do such statements refer?" You don't meet complex integrals or uncountable sets walking down the street or flying in outer space. Outside our thoughts, where are complex integrals or uncountable sets? Perhaps such things aren't real!

The available alternative to Platonism is formalism. Instead of saying theorems are truths about eternal extra-human ideals, it says they're just transformations of symbols. If she retreats to this cautious disclaimer, she's a formalist for the moment. For the moment, she thinks mathematics doesn't mean anything, so she needn't worry about its meaning. "I cannot imagine that I shall ever return to the creed of the true Platonist, who sees the world of the actual infinite spread out before him and believes that he can comprehend the incomprehensible," said the American logician Abraham Robinson in declaring himself a formalist.

But formalism needs its own act of faith. How do we know that our latest theorem about diffusion on manifolds is formally deducible from Zermelo-Frankel set theory? No such formal deduction will ever be written down. If it were, the likelihood of error would be greater than in the usual informal or semi-formal mathematical proof.

Now another question: "How come these examples were known before their axioms were known? If a theorem is only a conclusion from axioms, then do you say Cauchy didn't know Cauchy's integral formula? Cantor didn't know Cantor's theorem?"

Formalism doesn't work! Back to Platonism.

We don't quit mathematics, of course. Just quit thinking about it. Just do it. That's roughly the situation of the mathematician's philosophy of mathematics.

Mathematicians mostly don't want to bother about philosophy. A bee gathers honey without wondering why. A salmon climbs up the river without wondering why. Mathematicians make conjectures and try to prove them, without wondering why. "Just to have fun and make a living," they say. No need to have a clue of what it's all about, any more than a bee or a salmon.

Yet more is possible for us. "The unexamined life isn't worth living," said Socrates. At least, the unexamined life is less worth living than the examined. If a mathematician thinks examining his life is worthy, he must examine the meaning of his obsession—mathematics.

Platonism and formalism, each in its own way, falsify part of daily experience. We talk formalism when compelled to face the mystical, antiscientific essence of Platonic idealism; we fall back to Platonism when we realize that the formalist description of mathematics has only a distant resemblance to our actual knowledge of mathematics. To abandon both, we must abandon absolute certainty, and develop a philosophy faithful to mathematical experience.

Mistakes

How is it possible that mistakes occur in mathematics?

René Descartes's Method was so clear, he said, a mistake could happen only by inadvertence. Yet, as we see in Chapter 5, his *Géométrie* contains conceptual mistakes about three-dimensional space.

Henri Poincaré said it was strange that mistakes happen in mathematics, since mathematics is just sound reasoning, such as anyone in his right mind follows. His explanation was memory lapse—there are only so many things we can keep in mind at once.

Wittgenstein said that mathematics could be characterized as the subject where it's possible to make mistakes. (Actually, it's not just possible, it's inevitable.) The very notion of a mistake presupposes that there is right and wrong independent of what we think, which is what makes mathematics mathematics. We mathematicians make mistakes, even important ones, even in famous papers that have been around for years.

Philip Davis displays an imposing collection of errors, with some famous names. His article shows that mistakes aren't uncommon. It shows that mathematical knowledge is fallible, like other knowledge.

Dreben, Andrews, and Anderaa show that lemmas in Jacques Herbrand's 1929 thesis are false. Herbrand used false lemmas to "prove" a theorem that has been influential for fifty years. Dreben, Andrews, and Anderaa prove his theorem, replacing false lemmas with correct ones. (Rohit Parikh informs me that for years Herbrand's thesis was physically inaccessible. Its errors could have been found sooner in normal circumstances.)

Some mistakes come from keeping old assumptions in a new context.

Infinite dimensional space is just like finite dimensional space—except for one or two properties, which are entirely different.

Kummer tried to prove Fermat's last theorem. He assumed falsely that every integral domain of algebraic numbers has the unique factorization property. To correct his blunder, he founded a new branch of algebraic number theory.

Riemann stated and used what he called "Dirichlet's principle" incorrectly.

Julius Koenig and David Hilbert each thought he had proved the continuum hypothesis. (Decades later, it was proved undecidable by Kurt Gödel and Paul Cohen.)

Sometimes mathematicians try to give a *complete classification* of an object of interest. It's a mistake to claim a complete classification while leaving out several cases. That's what happened, first to Descartes, then to Newton, in their attempts to classify cubic curves (Boyer).

Is a gap in a proof a mistake? Newton found the speed of a falling stone by dividing 0/0. Berkeley called him to account for bad algebra, but admitted Newton had the right answer (see Chapter 6). Mistake or not?

Euler had an art of working with divergent series. His answer was *usually* correct (see Pólya and Putnam in Chapters 9 and 10). Later a rigorous theory of divergent series showed in what sense Euler was right.

"The mistakes of a great mathematician are worth more than the correctness of a mediocrity." I've heard those words more than once. Explicating this thought would tell something about the nature of mathematics. For most academic philosophers of mathematics, this remark has nothing to do with mathematics or the philosophy of mathematics. Mathematics for them is indubitable—rigorous deduction from premises. If you made a mistake, your deduction wasn't rigorous. By definition, then, it wasn't mathematics!

So the brilliant, fruitful mistakes of Newton, Euler, and Riemann, weren't mathematics, and needn't be considered by the philosopher of mathematics.

Riemann's incorrect statement of Dirichlet's principle was corrected, implemented, and flowered into the calculus of variations. On the other hand, thousands of *correct* theorems are published every week. Most lead nowhere.

A famous oversight of Euclid and his students (don't call it a mistake) was neglecting the relation of "between-ness"** of points on a line. This relation was used *implicitly* by Euclid in 300 B.C. It was recognized *explicitly* by Moritz Pasch over 2,000 years later, in 1882. *For two millennia, mathematicians and philosophers accepted reasoning that they later rejected.*

Can we be sure that we, unlike our predecessors, are not overlooking big gaps? We can't. Our mathematics can't be certain.

Gossip

A friend of mine was interested in some exciting work on "translation representations of linear operators" by our Professor Q. At another school, Professor Z worked on the same problem. When Z was in our neighborhood, my friend talked to him. She confided, "Neither one of them understands the other."

Can this be? Mathematics is straightforward. No secrets, just "If A then B." Yet here are top experts, neither understanding the other, *neither admitting he didn't understand the other.*

Maybe their use of different methods makes it hard for them to understand each other. When Q meets a difficulty, he turns to his usual device E. He may have trouble forgetting E to think about Z's favorite device F.

In a philosophy journal it's no surprise if author X says author Y doesn't understand him. The charge of not understanding is popular with philosophers. Popper is one of the clearest philosophical writers, yet he said he'd been understood only 2 or 3 times. Russell was Wittgenstein's mentor, and the world champion philosopher of science. But Wittgenstein said Russell didn't understand his *Tractatus.*

Mathematicians are different. You never see mathematicians accusing each other in print of not understanding, even when they *don't* understand, and it's apparent that they don't. Mathematics changes. The whole point of view can change in a few years. In the early part of this century mathematics was "set-theorized"—reformulated in Cantor's set theory. In the 1930s, led by Bourbaki, "structure" was the word; examples were buried in the exercises. The algebraization of geometry and topology is a third example.

When these transformations occur, a few outstanding mathematicians lead the revolution. Then more and more fresh PhDs are trained to think the new way. Meanwhile, some of the older generation continue in the old style. Among them are brilliant veterans of the previous revolution. If they don't master the new methods, that says something about mathematics. If it were simply correct reasoning from arbitrary premises, good mathematicians couldn't fail to understand good mathematics.

Solomon Lefschetz was a gray eminence of modern topology. According to Gian-Carlo Rota, Lefschetz's theorems were always correct, his proofs never correct.

My friend W. collaborated with B. (not Lefschetz.). "B is the most amazing mathematician I ever knew," W said. "His theorems are always right, his proofs are always wrong."

Professors D and K are probabilists, creators and shapers of their field. D uses measure-theoretic arguments. K works "analytically" or "classically." D doesn't use K's methods. K doesn't use D's methods. S, a student of both D and K, combined their methods to go further than either had gone separately.

In several branches of mathematics you hear: "Our deepest theorems were found by Professor Z. No one but he ever understood his methods. By now we can get his results with methods anyone can follow. With one or two exceptions, all his formulas and theorems were correct." The same story is told by probabilists, partial differential equators, algebraists, and topologists. The name of the hero changes.

Professor Z's knowledge without proof is inexplicable in the formalist account of mathematics.

At an International Congress of Mathematicians, a famous analyst reports his new results. They're not quite certain, he says, because there hasn't been time for other famous analysts to check them. *Until you check with other people, you don't know if you've overlooked something*, he tells us. A well-known example of this was Wiles's original "proof" of Fermat's last theorem

What about Ramanujan, the brilliant self-taught Indian mathematician? Dozens of amazing results that he did not see the need to prove, or didn't know how to. American mathematicians Ed Witten and Bill Thurston today are honored for fascinating results that, it is said, they left in part for others to prove.

I can't omit this reminiscence of J. J. Sylvester, one of the supreme algebraists of the nineteenth century. "Sylvester's *Methods*! He had none. Statements like the following were not unfrequent in his lectures: 'I haven't proved this, but I am as sure as I can be of anything that it must be so. From this it will follow, etc.' At the next lecture it turned out that what he was so sure of was false. Never mind, he kept on forever guessing and trying, and presently a wonderful discovery followed, then another and another. Afterward he would go back and work it all over again, and surprise us with all sorts of side lights. He then made another leap in the dark, more treasures were discovered, and so on forever" (E. W. Davis, 1890).

Sylvester's partner, Arthur Cayley, had the opposite style. Always careful and responsible, never saying more than he knew.

One of my honored teachers (a world class mathematician) astonished me by saying he didn't read mathematical papers. When something interesting happens, somebody tells him.

The same was reported of David Hilbert, in his day the world champion of mathematics. Didn't read.

Every young math PhD has had this experience: You've done something good. You talk about it. Your seminar or colloquium has an audience of a dozen or two. You feel successful if two or three show they understood.

Facing these ugly facts, the easy choice is hypocrisy. *Pretend you don't notice the gap between preaching and practice.* If you don't choose hypocrisy, you must give up either myth or reality. You can hold onto the myth, by saying mathematics as practiced by mathematicians isn't what it ought to be. You can hold onto reality, by admitting that mathematical proof isn't a mechanical procedure, even "in principle."

Isn't there a middle way? How about, "We're not as careful as we should be, but that doesn't detract from the ideal." Yes, let's try not to make mistakes!

But if you think mathematics would be infallible if we wrote our proofs to be checkable by computer, you're mistaken. As you already know, if you've done much debugging. A proof doesn't become certain by being formalized. To the doubtfulness of the proof are added the doubtfulness of the coding, the programming, the logical design of the machine, and the physical functioning of its components. In real computers and real computations, the correctness of a formal proof (a code, in computer lingo) has to be verified by a human being. Her *understanding* of the meaning and purpose of the program permits her to check its formal correctness. Programs don't work until they're checked and corrected by people. And even then, if they're big enough, they're still "buggy." We have no formal definition of "understanding." Nevertheless, understanding is required to verify formal computations, as well as the other way around.

Sometimes mathematicians make mistakes. Sometimes we disagree whether a proof is correct. There isn't absolute certainty in mathematics. There's *virtual* certainty, as in other areas of life.

four

Intuition/Proof/Certainty

There's an old joke about a theory so perfectly general it had no possible application. Humanist philosophy is applicable.

From the humanist point of view, how would one investigate such knotty problems of the philosophy of mathematics as mathematical proof, mathematical intuition, mathematical certainty? It would be good to compare the humanist method with methods suggested by a Platonist or formalist philosophy, but I am really not aware how either of those views would lead to a method of philosophical investigation. A humanist sees mathematics as a social-cultural-historic activity. In that case it's clear that one can actually look, go to mathematical life and see how proof and intuition and certainty are seen or not seen there.

What is proof? What should it be? Old and difficult questions.

Rather than, in the time-honored way of philosophers, choose a priori the right definitions and axioms for proof, I attempt to *look carefully, with an open mind.* What do mathematicians (including myself) do that we call proving? And what do we mean when we talk about proving? Immediately I observe that "proof" has different meanings in mathematical practice and in logico-philosophical analysis. What's worse, this discrepancy is not acknowledged, especially not in teaching or in textbooks. How can this be? What does it mean? How are the two "proofs" related, in theory and in practice?

The articles on intuition and certainty, and half a dozen more in the next chapter, are in the same spirit.

"Call this philosophy? Isn't it sociology?"

The sociologist studying mathematics comes as an outsider—presumably an objective one. I rely, not on the outsider's objectivity, but on my insider's experience and know-how, as well as on what my fellow mathematicians do, say, and write. I watch what's going on and I understand it as a participant.

This is not sociology. Neither is it introspection, that long discredited way to find truth by looking inside your own head. My experience and know-how aren't isolated or solipsistic. They're part of the web of mutual understanding of the mathematical community.

There used to be a kind of sociology called "Verstehen." That was perhaps close to what I've been talking about. But my looking and listening with understanding isn't an end in itself. It's a preparation for analysis, criticism, and connection with the rest of mathematics, philosophy, and science. That's why I call it philosophy.

Proof

The old, colloquial meaning of "prove" is: *Test, try out, determine the true state of affairs* (as in Aberdeen *Proving* Ground, galley *proof*, "the *proof* of the pudding," "the exception that proves the rule," and so forth.) How is the mathematical "prove" related to the old, colloquial "prove"?

We accuse students of the high crime of "not even knowing what a proof is." Yet we, the math teachers, don't know it either, if "know" means give a coherent, factual explanation. (Of course, we know how to "give a proof" in our own specialty.)

The trouble is, "mathematical proof" has two meanings. In practice, it's one thing. In principle, it's another. We *show* students what proof is in practice. We *tell* them what it is in principle. The two meanings aren't identical. That's O.K. But *we never acknowledge the discrepancy.* How can that be O.K.?

Meaning number 1, the *practical* meaning, is informal, imprecise. *Practical mathematical proof is what we do to make each other believe our theorems.* It's argument that convinces the qualified, skeptical expert. It's done in Euclid and in *The International Archive Journal of Absolutely Pure Homology*. But what is it, *exactly*? No one can say.

Meaning number 2, theoretical mathematical proof, is formal. Aristotle helped make it. So did Boole, Peirce, Frege, Russell, Hilbert, and Gödel. It's transformation of certain symbol sequences (formal sentences) according to certain rules of logic (modus ponens, etc). A sequence of steps, each a strict logical deduction, or readily expanded to a strict logical deduction. This is supposed to be a "formalization, idealization, rational reconstruction of the idea of proof" (P. Ernest, private communication).

Problem A: *What does meaning number 1 have to do with meaning number 2?*
Problem B: *How come so few notice Problem A?* Is it uninteresting? Embarrassing?
Problem C: *Does it matter?*

Problem C is easier than A and B. It matters, morally, psychologically, and philosophically.

When you're a student, professors and books claim to prove things. But they don't say what's meant by "prove." You have to catch on. Watch what the professor does, then do the same thing.

Then you become a professor, and pass on the same "know-how" without "knowing what" that your professor taught you.

There is an official, standard viewpoint. It makes two interesting assertions.

Assertion 1. Logicians don't tell mathematicians what to do. They make a theory out of what mathematicians actually do. Logicians supposedly study us the way fluid dynamicists study water waves. Fluid dynamicists don't tell water how to wave. They just make a mathematical model of it.

A fluid dynamicist studying water waves usually does two things. In advance, seek to derive a model from known principles. After the fact, check the behavior of the model against real-world data.

Is there a published analysis of a sample of practical proofs that derives "rigorous" proof as a model from the properties of that sample? Are there case studies of practical proof in comparison with theoretical proof? I haven't heard of them. The claim that theoretical proof models practical proof is an assertion of belief. It rests on intuition. Gut feeling.

You might say there's no need for such testing of the logic model against practice. If a mathematician doesn't follow the rules of logic, her reasoning is simply wrong. I take the opposite standpoint: What mathematicians at large sanction and accept *is* correct mathematics. Their work is the touchstone of mathematical proof, not *vice versa*. However that may be, anyone who takes a look must acknowledge that formal logic has little visible resemblance to what's done day to day in mathematics.

It commonly happens in mathematics that we believe something, even without possessing a complete proof. It also sometimes happens that we *don't* believe, even in the *presence* of complete proof. Edmund Landau, one of the most powerful number theorists, discovered a remarkable fact about analytic functions called the two constants theorem.** He proved it, but couldn't believe it. He hid it in his desk drawer for years, until his intuition was able to accept it (Epstein and Hahn, p. 396n).

There's a famous result of Banach and Tarski,** which very few can believe, though all agree, it has been proved.

Professor Robert Osserman of Stanford University once startled me by reporting that he had heard a mathematician comment, "After all, Riemann never proved anything." Bernhard Riemann, German (1826—1866), is universally admired as one of the greatest mathematicians. His influence is deeply felt to this day in many parts of mathematics. I asked Osserman to explain. He did:

"Clearly, it's not literally true that Riemann never proved anything, but what is true is that his fame derives less from things that he proved than from his other contributions. I would classify those in a number of categories. The first consists of conjectures, the most famous being the Riemann Hypothesis about the zeros of the zeta function. The second is a number of definitions, such as the Riemann integral, the definition of an analytic function in terms of the Cauchv-Riemann equations (rather than via an 'analytic' expression), the Riemann curvature tensor and sectional curvature in Riemannian geometry. The third, related to the second, but somewhat different, consists of new concepts, such as a Riemann surface and a Riemannian manifold, which involve radical rethinking of fundamental notions, as well as the extended complex plane, or Riemann sphere, and the fundamental notions of algebraic curve. The fourth is the construction of important new examples, such as Riemann's complete periodic embedded minimal surface and what we now call hyperbolic space, via the metric now referred to as the 'Poincaré metric', which Riemann wrote down explicitly. And finally, there is the Riemann mapping theorem, whose generality is simply breathtaking, and again required a depth of insight, but whose proof, as given by Riemann, was defective, and was not fully established until many years later. Perhaps one could combine them by saying that Riemann was a deep mathematical thinker whose vision had a profound impact on the future of mathematics."

Many readers have not heard of all these contributions of Riemann. But this example is enough to refute the catch-phrases. "A mathematician is someone who proves theorems" and "Mathematics is nothing without proof."

As a matter of principle there can't be a strict demonstration that Meaning 2, formal proof, is what mathematicians do in practice. This is a universal limitation of mathematical models. In principle it's impossible to give a mathematical proof that a mathematical model is faithful to reality. All you can do is test the model against experience.

The second official assertion about proof is:

Assertion 2. Any correct practical proof can be filled in to be a correct theoretical proof.

"If you can do it, then do it!"

"It would take too long. And then it would be so deadly boring, no one would read it."

Assertion B is commonly accepted. Yet I've seen no practical or theoretical argument for it, other than absence of counter-examples. It may be true. It's a matter of faith.

Take a mathematically accepted proof and undertake to fill in the gaps, turn it into a formal proof. If you meet no obstacles, very well! And if you do meet an obstacle? That is, a mathematically accepted step that you can't break down into successive *modus ponens*? You've discovered an implicit assumption, a hidden lemma! Join it to the hypotheses of the theorem and go merrily forward.

If this is what's meant by formalization of proof, it can always be done. There's a preassigned way around any obstacle! There remains one little worry. Is your enlarged set of assumptions consistent? This can be hard to answer. You may say, if there's doubt about their consistency, it was wrong to claim the original proof was correct. On the other hand, we don't even know if the hypotheses of our number system are consistent. We presume it's O.K. on pragmatic grounds.

No one has proved that definition 2, the logic model of proof, is wrong. If a counter-example were found, logic would adapt to accommodate it.

By "proof" we mean correct proof, complete proof. By the standards of formal logic, ordinary mathematical proofs are incomplete. When an ordinary mathematical proof is offered to a referee, she may say, "More detail is needed." So more detail is supplied, and then the proof is accepted. But this acceptable version is still incomplete as a formal proof! The original proof was mathematically incomplete, the final proof is mathematically complete, but *both are formally incomplete.* The formal concept of proof is irrelevant at just the point of concern to the mathematician!

We prefer a beautiful proof with a serious gap over a boring hyper-correct one. If the *idea* is beautiful, we think it will attain valid mathematical expression. We even change the meaning of a concept to get a more beautiful theory. Projective geometry is a classic example. Euclidean geometry uses the familiar axiom, "Two points determine a line, and two lines determine a point, *unless the lines are parallel.*" Projective geometry brings in ideal points at infinity—one point for each family of parallel lines. The axiom becomes: "Two points determine a line, and two lines determine a point." This is "right." The Euclidean axiom by comparison is awkward and clumsy.

When a mathematician submits work to the critical eyes of her colleagues, it's being tested, or "proved" in the old sense. With few exceptions, mathematicians have no other way to test or "prove" their work—invite whoever's interested to have a shot at it. Then the mathematical "Meaning number 1" of "proof" agrees with the old meaning. The *proof* of the pudding is in the eating. The *proof* of the theorem (in that sense of being tested) is in the refereeing.

Philosophers call this social validation "warranting."

Computers and Proof

In the late 1970s or early 1980s, while visiting a respected engineering school, I was told that the dean was vexed with his mathematicians. Other professors used his computer center; mathematicians didn't. Today no one's complaining that mathematicians don't compute. We were just 10 or 15 years late.

The issue about proof today is the impact of computers on mathematical proof. The effects are complex and manifold. I describe 2 trends, 3 examples, 2 critics, and 5 reasons.

First, two trends: (Trend A) Computers are being used as aids in proving theorems; and (Trend B) computers are encroaching on the central role we give proof in mathematics.

Three examples of trend A:

Example 1. In 1933, before general-purpose computers were known, Derrick Henry Lehmer built a computer to study prime numbers. It collected number-theoretic data and examples, from which he formulated conjectures. This was a mechanization of Lagrange and Gauss, who conjectured the prime number theorem from tables of prime numbers.

Example 2. Wolfgang Haken and Kenneth Appel's proof that a plane map needs at most four colors. The conjecture had been in place since 1852. Appel and Haken turned it into a huge computation. Then their computer did the computation. Thus they proved the conjecture.

Example 3. In the Feigenbaum-Lanford discovery of a universal critical point for doubling of bifurcations, computers played two distinct essential roles. Mitchell Feigenbaum discovered the universal critical point by computer exploration, comparable to Lehmer's number-theoretic explorations. Later, rigorous proofs were given by Oscar Lanford III (9, 10, 11) and by M. Campanino, H. Epstein, and D. Ruelle. At that stage the computer played a role like its role in the four-color problem: assistant in completing the proof.

Now the proof wasn't algebra but analysis. A certain derivative had to be estimated. This was accomplished by the computer, using approximation methods from numerical analysis, with a rigorous error estimate.

Example 1, Lehmer's exploratory use, didn't challenge the standard notion of proof. That notion doesn't care how a conjecture is made. I heard a distinguished professor tell his class, "It doesn't matter if you find the answer lying in a mud puddle. If you prove it's the answer, it makes no difference how you found it."

Examples 2 and 3 are different. Now the machine computation is *part of the proof.* For some people, such a proof violates time-honored doctrines about the difference between empirical knowledge and mathematical knowledge. According to Plato, Kant, and many others, common knowledge and scientific knowledge are a posteriori. They come from observation of the material world. Mathematical knowledge is a priori—independent of contingent facts about the material world. Daily experience and physical experiment could conceivably be other than what they are, but mathematical truths will hold in every possible world, or so thought Plato and Kant.

The operation of computers depends on properties of copper and silicon, on electrodynamics and quantum mechanics. Confidence in computers comes from confidence in physical facts and theories. These are not a priori. We learn the laws of physics and the electrical properties of silicon and copper from experience. It seems, then, that while old-fashioned theorems proved "by hand" are a priori, computer-assisted theorems are a posteriori! (Philip Kitcher thoroughly goes into the issue of a priori knowledge in mathematics.)

But despite the illusions of idealist philosophy, old-fashioned person-made proofs are also a posteriori. They depend on credence in the world of experience, the material world. We believe our scribbled notes don't change from hour to hour, that our thinking apparatus—our brain—is reliable more often than not, that our books and journals are what they pretend to be. We sometimes believe a proof without checking every line, every reference, and every line of every reference. Why? We depend on the integrity and competence of certain human beings—our colleagues. But human beings are *less* reliable at long computations than a Cray or a Sun! And still less in collaborations by large groups, as in the classification theorem of simple finite groups. Recognizing this demolishes the dream of a priori knowledge in advanced mathematics, and in general.

Most mathematicians see the difference between computer-assisted proofs and traditional proofs as a difference of degree, not of kind. But Paul Halmos does object to the Haken-Appel proof. "I do not find it easy to say what we learned from all that. We are still far from having a good proof of the Four Color Theorem. I hope as an article of faith that the computer missed the right concept and the right approach. 100 years from now the map theorem will be, I think, an exercise in a first-year graduate course, provable in a couple of pages by means of the appropriate concepts, which will be completely familiar by then. The present proof relies in effect on an Oracle, and I say down with Oracles! They are not mathematics."

Why does Halmos call the computer an Oracle? He can't know every step in the calculation. Indeed, the physical processes that make computers work aren't fully understood. Believing a computer is like believing a successful, well-reputed fortune teller. Should we regard the four-color conjecture as true? Should our confidence in it be increased by the Appel-Haken proof? Halmos doesn't say.

Halmos dislikes the Appel-Haken proof because it uses an Oracle, and because, he thinks, we can't learn anything from it. His criticism isn't for logical defects—incompleteness, incorrectness, or inaccuracy. It's esthetic and epistemological. This is normal in real-life mathematics. It would be senseless in formalized or logicized mathematics.

Views like Halmos were also vented by Daniel Cohen. "Our pursuit is not the accumulation of facts about the world or even facts about mathematical objects. The mission of mathematics is understanding. The Appel and Haken work on the Four Color Problem amounts to a confirmation that a map-maker with only four paint pots will not be driven out of business. This is not really what mathematicians were worried about in the first place. Admitting the computer shenanigans of Appel and Haken to the ranks of mathematics would only leave us intellectually unfulfilled."

The eloquence of Halmos and Cohen won't deter mathematicians who hope a computer will help on their problem. Computers in pure mathematics will increase for at least five reasons.

Reason 1. Access to more powerful computers spreads ever more widely.
Reason 2. As oldsters are replaced by youngsters schooled after the computer revolution, the proportion of us at home with computers rises.
Reason 3. Despite the scolding of Halmos, Cohen, and sympathizers, success like that of Haken-Appel and Lanford inspires emulation.
Reason 4. Scientists and engineers were computerized long ago. You can't imagine a mathematician interacting with a scientist or engineer today without a computer.
Reason 5. Reasons 1, 2, 3, and 4 stimulate those branches of mathematics that use computation, leaving the others at a disadvantage. This reinforces Reasons 1, 2, 3, and 4.

Some will still reject computer proofs. They'll have as much effect as old King Canute. (By royal order, he commanded the tide to turn back.)

In fields like chaos, dynamical systems, and high Reynolds number fluid dynamics, we come to trend B. Here, as in number theory, the computer is an explorer, a scout. It goes much deeper into unknown regions than rigorous analysis can. In these fields we no longer think of computer-obtained knowledge as tentative—pending proper proof. We prove what we can, compute what we can. This part of mathematical reality rests on computation and analysis closely yoked together. The computer finds "such and such." We believe it—not as indubitable, but as believable. Proof of "such and such" then would help us see *why* it's so, though not perhaps increase our conviction it *is* so.

This effect of computers, Trend B, is profound. It erodes the time-honored understanding of what mathematics is. Machine computation as part of proof is radical, but still more radical is machine computation accepted as *empirical evidence of mathematical truth, virtually a weak form of proof.* Such acceptance contradicts the official line that mathematics *is* deductions from axioms. It makes mathematics more like an empirical science.

An interesting proposal to control this disturbing new tendency was made in the *Bulletin of the American Mathematical Society* by the distinguished mathematical physicist Arthur Jaffe and the distinguished topologist Frank Quinn. They want to save rigorous proof, the chief distinction of modern mathematics, from promiscuous contamination with mere machine calculations. Yet they acknowledge that "empirical" or "experimental" computer mathematics will not go away, and may have a place in the world. Their solution is—compulsory rigid distinction between the two. Genuine mathematics stays as it should be. Experimental or numerical or empirical mathematics is labeled appropriately. Analogous to kosher chickens versus non-kosher. I don't have the impression that this proposal met with general acclaim.

A different departure from traditional proof was found by Miller (1976), Rabin (1976), Davis (1977), and Schwartz (1980). There's now a way to say of an integer n whose primality or compositeness is unknown, "On the basis of available information, the probability that n is prime is p." If n is really prime, you can make p arbitrarily close to 1. Yet the primality of a given n is *not* a random variable. n either is prime or it isn't. But if n is large, finding its primality by a deterministic method is so laborious that random errors must be expected in the computation. Rabin showed that if n is very large, the probability of error in the deterministic calculation is greater than the probability p in his fast probabilistic method!

How will we adjust to such variation in proof? We can think of proofs as having variable quality. Instead of "proved," label them either "proved by hand" or "proved by machine." Even provide an estimate for the reliability of the machine calculation used in a proof (Swart, 1980).

There are two precedents for this situation. Between the world wars, some mathematicians made it a practice to state explicitly where they used the axiom of choice. And in 1972 Errett Bishop said the clash between constructivists and classicists would end if classicists stated explicitly where they used the law of the excluded middle (L. E. M.). (Constructivists reject the L. E. M. with respect to infinite sets.) These issues didn't involve computing machines. They involved disagreement about proof.

If experience with the axiom of choice and the L. E. M. is indicative, Swart's proposal is unpromising. Nobody worries any more about the axiom of choice, and few worry about the L. E. M. Perhaps we don't care much about distinctions of quality or certainty in proof. Perhaps few care today if proofs are handmade or computer-made.

There's a separate trend in computer proof, that so far has had more resonance in computer science than in main-line mathematics. In this trend, the idea is not to use the computer to supply a missing piece of a mathematician's proof. Rather, the computer is a logic assistant. Give it some axioms, and send it out to find interesting theorems. Of course, you have to give it a way to measure "interesting." Or tell it what you want proved—a conjecture you haven't proved yourself, or a proved theorem, to see what different proof it might find.

Since the computer is expected to follow the rules of logic, the relation to definition 2 is apparent. Since it really does find proofs that convince mathematicians, it conforms to definition 1. People who think mathematicians will become obsolete have this kind of thing in mind—computer proofs of real theorems.

Professor Wos writes movingly, "We have beaten the odds, done the impossible, automated reasoning so effectively that our programs have even answered open questions. Yet skeptics still exist, funding is not abundant, and recognition of our achievements is inappropriately small and not sufficiently widespread."

This work can be regarded as practical justification of the formal logic notion of proof. To the extent that computers following only the rules of formal logic do reproduce discoveries of live mathematicians, they show that formal logic is an adequate model of real live proof. But "technical limitations" restrict automated proof to relatively simple theorems.

There are centers of this research in Austin, Texas, and at Argonne National Laboratory. There is a *Journal of Automated Reasoning*. Bledsoe and Loveland is an instructive and readable review.

I've said nothing about applied mathematics. In applied mathematics, infiltration and domination by computers has long been a fait accompli. Under the influence of computers, pure mathematics is becoming more like applied mathematics.

Fallibility

Philosophical discussions of mathematical proof usually talk about it only as it's seen in journals and textbooks. There, proof functions as the last judgment, the final word before a problem is put to bed. But the essential mathematical activity is *finding* the proof, not checking after the fact that indeed it is a proof.

How does formal proof differ from real live proof?

Real-life proof is informal, in whole or in part. A piece of formal argument—a calculation—is meaningful only to complete or verify some informal reasoning. The formal-logic picture of proof is a topic for study in logic rather than a truthful picture of real-life mathematics.

Formal proof exists only in a formalized theory, cleaned and purged of all associations and connotations. It uses a formal vocabulary, formal axioms, and formal inference rules.

The passage from informal to formalized theory must entail loss of meaning or change of meaning. The informal has connotations and alternative interpretations not in the formalized theory. Consequently, anything proved formally can be challenged: "How faithful are this statement and proof to the informal concept we're actually interested in?"

For some investigations, formalization and complete formal proof would take time and persistence beyond human capability, beyond any foreseeable computer.

Mathematicians say that any "Theorem and Proof" in a pure mathematics journal must be formalizable *in principle*. A glance into any mathematical *Archive* or *Bulletin* or *Journale* or *Zeitschrift* shows a great proportion of the text in natural language. Even pages of solid calculation turn out, on inspection, not to be formalized. In presenting calculations for publication, we include steps that we consider nonroutine, which should be explained to fellow specialists. And in every calculation there are routine steps that needn't be explained to fellow experts. These, naturally, we leave out. (If we include them, the editor throws them out.)

So the published proof is incomplete. The reader accepts the result on faith, or fills in the steps herself.

Even in a graduate math class proof isn't completely formalized. The professor leaves out as a matter of course what she considers routine or trivial. A nonroutine step may be assigned as homework.

Practically none of the mathematical literature is formalized (except for computer programs, which are another story.) Yet these far-from-formalized proofs are accepted by mathematicians as "formalizable in principle."

Why are they accepted? Because they're convincing to the experts, who'd be only too happy to find a serious error or gap.

We accept incomplete, natural-language proofs on the basis of our experience, our know-how in looking for the weakest link. And also on the author's reputation. A known bungler, or just an unknown, or an authority of proved accomplishment?

Peano hoped that his formal language would guarantee proofs to be correct. In today's language, that proofs could be checked by computer. But trying to check proofs by computer may introduce new errors. There is random error by physical fluctuations of the machine, and there is human error in designing and producing hardware and software (logic and programming.) For most nontrivial mathematics, the vision of formal proof is still visionary. And when a machine does part of a proof, as in the four-color theorem of Appel and Haken, some mathematicians reject it because the details of machine computation are inevitably hidden. More than *whether* a conjecture is correct, we ask *why* it is correct. We want to understand the proof, not just be told it exists.

The issue of machine error is more than just electrical engineering. It's the difference between computation in principle (infallible) and computation in practice (fallible). A simple calculation shows that no matter how small the chance of error in one step, if the calculation or formal proof is long enough, it's almost sure to contain errors. But we can have other grounds for believing the conclusion of a proof, apart from the claimed certainty of step-by-step reasoning. Examples and special cases, analogy with other results, expected symmetry, unexpected elegance, even an inexplicable feeling of rightness. These illogical logics may say, "It's true!" If you have something you hope is a proof, such nonrigorous reasons can make you sure of the conclusion, even while you know the "proof" is sure to contain uncorrected errors. Such intuition is fallible in principle. Attempted rigorous proof is fallible in practice.

Beyond the certainty of error in long calculations, you face the fact that calculation is finite, mathematics infinite. There's a limit to the biggest computer, how tightly it can be packed, how fast it can run, how long it will operate. The life of the human race is a limit. Put these limits together, and you have a bound on how much anyone will ever compute. If you can't *know* anything in mathematics except by formal proof (a particular kind of computation), you've set a bound on how much mathematics you'll ever know. The physical bound on computation implies a bound on the number and length of theorems that will ever be proved. The only way out would be a faster, less certain way to mathematical knowledge (Knuth, 1976).

Meyer (1974, p. 481) quotes a theorem, proved with L. J. Stockmeyer: "If we choose sentences of length 616 in the decidability theory of WSIS (weak monadic second-order theory of the successor function on the nonnegative integers) and code these sentences into $6 \times 616 = 3{,}696$ binary digits, then any logical network with 3,696 inputs which decides truth of these sentences contains 10^{123} operations." (WSIS is much weaker than ordinary arithmetic. The conclusion applies a fortiori to sentences longer than 616 digits.)

"We remind the reader," Meyer writes, "that the radius of a proton is approximately 10^{-13} cm., and the radius of the known universe is approximately 10^{28} cm. Thus for sentences of length 616, a network whose atomic operations were performed by transistors the size of a proton connected by infinitely thin wires would densely fill the entire universe." No decision procedure for sentences of length 616 in WSIS can be physically realized. Yet WSIS is "decidable": One says that a decision procedure "exists" for sentences of any length.

Our notion of rigorous proof isn't carved in granite. We'll modify it. We'll allow machine computation, numerical evidence, probabilistic algorithms, if we find them advantageous. We mislead our pupils if we make "rigorous proof" a shibboleth in class.

In Class

The role of proof in class isn't the same as in research. In research, it's to *convince*. In class, students are all too easily convinced! Two special cases do it. In a first course in abstract algebra, proof of the fundamental theorem of algebra is often omitted. Students believe it anyway.

The student needs proof to *explain*, to give insight why a theorem's true. Not proof in the sense of formal logic. As the graduate student said to the Ideal Mathematician in *The Mathematical Experience*, she never saw such a proof in class. In class informal or semiformal proofs are presented in natural language. They include calculations, which are formal subproofs inside the overall informal proof.

Some instructors think, "If it's a math class, you prove. If you don't prove anything, it isn't math." That makes a kind of sense. If proof is math and math is proof, then in math class you're duty bound to prove. The more you prove, the more honest and rigorous you feel your class is.

Exposure to proof can be more emotional than intellectual. If the instructor gives no better reason for proof than "That's math!," the student knows she saw a proof, but not wherefore, except: "That's math!"

I call this view "absolutist," despite that word's unfortunate associations (absolute monarchy, absolute zero, etc.). If mathematics is a system of absolute truths, independent of human construction or knowledge—then mathematical proofs are external and eternal. They're to admire. The absolutist teacher wants to tell only what he intends to prove (or order the students to prove). He'll usually try for the shortest proof or the most general one. The main purpose of proof isn't explanation. The purpose is certification: admission into the catalog of absolute truths.

The view I favor is humanism. To the humanist, mathematics is *ours*—our tool, our plaything.

Proof is complete explanation. Give it when complete explanation is appropriate, rather than incomplete explanation or no explanation.

The humanist math teacher looks for enlightening proofs, not necessarily the most general or the shortest. Some proofs don't explain much. They're called "tricky," "pulling a rabbit out of a hat." Give that kind of proof when you want your students to see a rabbit pulled out of a hat. But in general, give proofs that explain. And if the only proof you can find is unmotivated and tricky, if your students won't learn much from it, must you do it "to stay honest"? That "honesty" is a figment, a self-imposed burden. Better try to be clear, well-motivated, even inspiring.

This attitude disturbs people who think proof is the be-all and end-all of mathematics—who say "a mathematician is someone who proves theorems" and "without proof, there's no mathematics." From that viewpoint, a mathematics in which proof is less than absolute is heresy.

For the humanist, the purpose of proof, as of all teaching, is understanding. Whether to give a proof as is, elaborate it, or abbreviate it, depends on what he thinks will increase the student's understanding of concepts, methods, and applications.

This policy uses the notion of "understanding," which isn't precise or likely to be made precise. Do we understand what it means "to understand"? No. Can we teach to foster understanding? Yes. We recognize understanding, though we can't say precisely what it is.

In a stimulating article, Uri Leron (1983) borrowed an idea from computing—"structured proof." A structured proof is like a structured program. Instead of starting with little lemmas whose significance appears at the end, start by breaking the proof-task into chunks. Then break each chunk into subchunks. The little lemmas come at the end, where you see why you need them.

In the general classroom, the motto is: "Proof is a tool in service of teacher and class, not a shackle to restrain them."

In teaching future mathematicians, "Proof is a tool in service of research, not a shackle on the mathematician's imagination."

Proof can convince, and it can explain. In research, convincing is primary. In high-school or undergraduate class, explaining is primary.

Intuition

If we look at mathematical practice, the intuitive is everywhere. We consider intuition in the mathematical literature and in mathematical discovery.

A famous example was the letter from Ramanujan to Hardy, containing astonishing formulas for infinite sums, products, fractions, and roots. The letter had gone to Baker and to Hobson. They ignored it. Hardy didn't ignore it.

Ramanujan's formulas prove there is mathematical intuition, for they're correct, even though Ramanujan didn't prove them, and in some cases had hardly an idea what a proof would be. But what about Hardy? He also made a correct judgment without proof—the judgment that Ramanujan's formulas were true, and that Ramanujan was a genius. How did he do that? Not by checking his formulas with complete proofs. By some mental faculty associated with mathematics, Hardy made a sound judgment of Ramanujan's formulas, without proofs. Was Hardy's judgment of Ramanujan's letter a mathematical judgment? Of course it was, in any reasonable understanding of the word mathematics. It was an exceptional event, yet not essentially different from mathematical judgments made every day by reviewers and referees, by teachers and paper-graders, by search committees and admission committees. The faculty called on in these judgments is mathematical intuition. It's reliable mathematical belief without the slightest dream of being formalized.

Since intuition is an essential part of mathematics, no adequate philosophy of mathematics can ignore intuition.

The word intuition, as mathematicians use it, carries a heavy load of mystery and ambiguity. Sometimes it's a dangerous, illegitimate substitute for rigorous proof. Sometimes it's a flash of insight that tells the happy few what others learn with great effort. As a first step to explore this slippery concept, consider this list of the meanings and uses we give this word.

1. Intuitive is the opposite of rigorous. This usage is not completely clear, for the meaning of "rigorous" is never given precisely. We might say that in this usage intuitive means lacking in rigor, yet the concept of rigor is defined intuitively, not rigorously.

2. Intuitive means visual. Intuitive topology or geometry differs from rigorous topology or geometry in two ways. On one hand, the intuitive version has a meaning, a referent in the domain of visualized curves and surfaces, which is absent from the rigorous formal or abstract version. In this the intuitive is superior; it has a valuable quality the rigorous version lacks. On the other hand, visualization may mislead us to think obvious or self-evident statements that are

dubious or false. The article by Hahn, "The Crises in Intuition" is a beautiful collection of such statements.

3. Intuitive means plausible, or convincing in the absence of proof. A related meaning is, "what you might expect to be true in this kind of situation, on the basis of experience with similar situations." "Intuitively plausible" means reasonable as a conjecture, i.e., as a candidate for proof.

4. Intuitive means incomplete. If you take a limit under the integral sign without using Lebesgue's theorem, if you expand a function in a power series without checking that it's analytic, you acknowledge the logical gap by calling the argument intuitive.

5. Intuitive means based on a physical model or on some special examples. This is close to "heuristic."

6. Intuitive means holistic or integrative as opposed to detailed or analytic. When we think of a theory in the large, when we're sure of something because it fits everything else we know, we're thinking intuitively. Rigor requires a chain of reasoning where the first step is known and the last step is the conjecture. If the chain is very long, rigorous proof may leave doubt and misgiving. It may actually be less convincing than an intuitive argument that you grasp as a whole, which uses your faith that mathematics is coherent.

In all these usages intuition is vague. It changes from one usage to another. One author takes pride in avoiding the "merely" intuitive—the use of figures and diagrams as aids to proof. Another takes pride in emphasizing the intuitive—showing visual and physical significance of a theory, or giving heuristic derivations, not just formal post hoc verification.

With any of these interpretations, the intuitive is to some degree extraneous. It has desirable and undesirable aspects. It's optional, like seasoning on a salad. It's possible to teach mathematics or to write papers without thinking about intuition.

However, if you're not doing mathematics, but watching people do mathematics and trying to understand what they're doing, dealing with intuition becomes unavoidable.

I maintain that:

1. All the standard philosophical viewpoints rely on some notion of intuition.
2. None of them explain the nature of the intuition that they postulate.
3. Consideration of intuition as actually experienced leads to a notion that is difficult and complex, but not inexplicable.
4. A realistic analysis of mathematical intuition should be a central goal of the philosophy of mathematics.

Let's elaborate these points. By the main philosophies, I mean as usual constructivism, Platonism, and formalism. For the present we don't need refined distinctions among versions of the three. It's sufficient to characterize each crudely with one sentence.

The constructivist regards the natural numbers as the fundamental datum of mathematics, which neither requires nor is capable of reduction to a more basic notion, and from which all meaningful mathematics must be constructed.

The Platonist regards mathematical objects as already existing, once and for all, in some ideal and timeless (or tenseless) sense. We don't create, we discover what's already there, including infinites of a complexity yet to be conceived by mind of mathematician.

The formalist rejects both the restrictions of the constructivist and the theology of the Platonist. All that matters are inference rules by which he transforms one formula to another. Any meaning such formulas have is nonmathematical and beside the point.

What does each of these three philosophies need from the intuition? The most obvious difficulty is that besetting the Platonist. If mathematical objects constitute an ideal nonmaterial world, how does the human mind/brain establish contact with this world? Consider the continuum hypothesis. Gödel and Cohen proved that it can neither be proved nor disproved from the set axioms of contemporary mathematics. The Platonist believes this is a sign of ignorance. The continuum is a definite thing, independent of the human mind. It either does or doesn't contain an infinite subset equivalent neither to the set of integers, nor to the set of real numbers. Our *intuition* must be developed to tell us which is the case. The Platonist needs intuition to connect human awareness and mathematical reality. But his intuition is elusive. He doesn't describe it, let alone analyze it. How is it acquired? It varies from person to person, from one mathematical genius to another mathematical genius. It has to be developed and refined. By whom, by what criteria, does one develop it? Does it directly perceive an ideal reality, as our eyes perceive visible reality? Then intuition would be a second ideal entity, the subjective counterpart of Platonic mathematical reality. We have traded one mystery for two: first, the mysterious relation between timeless, immaterial ideas and the mundane reality of change and flux; and second, the mysterious relation between the flesh and blood mathematician and his intuition, which directly perceives the timeless and eternal. These difficulties make Platonism hard for a scientifically oriented person to defend.

Mathematical Platonists simply disregard them. For them, intuition is something unanalyzable but indispensable. Like the soul in modern Protestantism, the intuition is there but no questions can be asked about it.

The constructivist, as a conscious descendant of Kant, knows he relies on intuition. The natural numbers are given intuitively. This doesn't seem problematical. Yet Brouwer's followers have disagreed on how to be a constructivist. Of course, every philosophical school has that experience. But it creates a difficulty for a school that claims to base itself on a universal intuition.

The dogma that the intuition of the natural numbers is universal violates historical, pedagogical, and anthropological experience. The natural number system seems an innate intuition to mathematicians so sophisticated they can't remember or imagine before they acquired it; and so isolated they never meet people who haven't internalized arithmetic and made it intuitive (the majority of the human race!).

What about the formalist? Does intuition vanish along with meaning and truth? You can avoid intuition as long as you consider mathematics to be no more than formal deductions from formal axioms. A. Lichnerowicz wrote, "Our demands on ourselves have become infinitely larger; the demonstrations of our predecessors no longer satisfy us but the mathematical facts that they discovered remain and we prove them by methods that are infinitely more rigorous and precise, methods from which geometric intuition with its character of badly analyzed evidence has been totally banned."

Geometry was pronounced dead as an autonomous subject; it was no more than the study of certain particular algebraic-topological structures. The formalist eliminates intuition by concentrating on refinement of proof and dreaming of an irrefutable final presentation. To the natural question, Why should we be interested in these superprecise, superreliable theorems?, formalism turns a deaf ear. Obviously, their interest derives from their meaning. But the all-out formalist throws out meaning as nonmathematical. Then how did our predecessors find correct theorems by incorrect reasoning? He has no answer but "Intuition."

Surely Cauchy knew Cauchy's integral theorem, even though (in the formalist's sense of knowing the formal set-theoretic definition) he didn't know the meaning of any term in the theorem. He didn't know what is a complex number, what is an integral, what is a curve; yet he found the complex number represented by the integral over this curve! How can this be? Cauchy had great intuition.

"But during my last night, the 22–23 of March, 1882—which I spent sitting on the sofa because of asthma—at about 3:30 there suddenly arose before me the Central Theorem, as it has been prefigured by me through the figure of the 14-gon in (Ges. Abh., vol. 3, p. 126). The next afternoon, in the mail-coach (which then ran from Norden to Emden) I thought through what I had found, in all its details. Then I knew I had a great theorem. . . . The proof was in fact very difficult. I never doubted that the method of proof was correct, but everywhere I ran into gaps in my knowledge of function theory or in function theory itself. I could only postulate the resolution of these difficulties, which were in fact completely resolved only 30 years later (in 1921) by Koebe" (F. Klein).

But what *is* this intuition? The Platonist believes in real objects (ideal, to be sure), which we "intuit." The formalist believes no such things exist. So what is there to intuit? The only answer is, unconscious formalizing. Cauchy subconsciously knew a correct proof of his theorem, which means knowing the correct definitions of all the terms in the theorem.

This answer is interesting to the many mathematicians who've made correct conjectures they couldn't prove. If their intuitive conjecture was the result of unconscious reasoning, then: (a) Either the unconscious has a secret method of reasoning that is better than any known method; or (b) the proof is there in my head, I just can't get it out!

Formalists willing to consider the problem of discovery and the historical development of mathematics need intuition to account for the gap between their account of mathematics (a game played by the rules) and the real experience of mathematics, where more is sometimes accomplished by breaking rules than obeying them.

Accounting for intuitive "knowledge" in mathematics is the basic problem of mathematical epistemology. What do we believe, and why do we believe it? To answer this question we ask another question: what do we teach, and how do we teach it? Or what do we try to teach, and how do we find it necessary to teach it? We try to teach mathematical concepts, not formally (memorizing definitions) but intuitively—by examples, problems, developing an ability to think, which is the expression of having successfully internalized something. What? An intuitive mathematical idea. The fundamental intuition of the natural numbers is a shared concept, an idea held in common after manipulating coins, bricks, buttons, pebbles. We can tell by the student's answers to our questions that he gets the idea of a huge bin of buttons that never runs out.

Intuition isn't direct perception of something external. It's the effect in the mind/brain of manipulating concrete objects—at a later stage, of making marks on paper, and still later, manipulating mental images. This experience leaves a trace, an effect, in the mind/brain. That trace of manipulative experience is your representation of the natural numbers. Your representation is equivalent to mine in the sense that we both give the same answer to any question you ask. Or if we get different answers, we compare notes and figure out who's right. We can do this, not because we have been explicitly taught a set of algebraic rules, but because our mental pictures match. If they don't, since I'm the teacher and my mental picture matches the one all the other teachers have, you get a bad mark.

We have intuition because we have mental representations of mathematical objects. We acquire these representations, not mainly by memorizing formulas, but by repeated experiences (on the elementary level, experience of manipulating physical objects; on the advanced level, experiences of doing problems and discovering things for ourselves). These mental representations are checked for veracity by our teachers and fellow students. If we don't get the right answer, we flunk the course. Different people's representations are always being rubbed against each other to make sure they're congruent. We don't know how these representations are held in the mind/brain. We don't know how *any* thought or knowledge is held in the mind/brain. The point is that as shared concepts, as mutually congruent mental representations, they're real objects whose existence is just as "objective" as mother love and race prejudice, as the price of tea or the fear of God.

How do we distinguish mathematics from other humanistic studies? There's a fundamental difference between mathematics and literary criticism. While mathematics is a humanistic study with respect to its subject matter—human ideas—it's science-like in its objectivity. Those results about the physical world that are reproducible—which come out the same way every time you ask—are called scientific. Those subjects that have reproducible results are called natural sciences. In the realm of ideas, of mental objects, those ideas whose properties are reproducible are called mathematical objects, and *the study of mental objects with reproducible properties is called mathematics.* Intuition is the faculty by which we consider or examine these internal, mental objects.

There's always some discrepancy between my intuition and yours. Mutual adjustment to keep agreement is going on all the time. As new questions are asked, new parts of the structure come into sight. Sometimes a question has no answer. The continuum hypothesis doesn't have to be true or false.

We know that with physical objects we may ask questions that are inappropriate, which have no answer. What are the exact velocity and position of an electron? How many trees are there growing at this moment in Minnesota? For mental objects as for physical ones, what seems at first an appropriate question is sometimes discovered, perhaps with great difficulty, to be inappropriate. This doesn't contradict the existence of the particular mental or physical object. There are questions that *are* appropriate, to which reliable answers can be given.

The difficulty in seeing what intuition is arises because of the expectation that mathematics is infallible. Both formalism and Platonism want a superhuman mathematics. To get it, each of them falsifies the nature of mathematics in human life and in history, creating needless confusion and mystery.

Certainty

Even if it's granted that the need for certainty is inherited from the ancient past, and is religiously motivated, its validity is independent of its history and its motivation. The question remains: is mathematical knowledge indubitable?

Set aside history and motivation. Look at samples of mathematical knowledge and ask: Is this indubitable?

We take three examples. First, good old

$2 + 2 = 4$.**

Second, familiar to all former high-school students,

"The angle sum of any triangle equals two right angles." Finally, a more sophisticated example: a convergent infinite series.

Label the first example

Formula A: $2 + 2 = 4$.

Everyone knows Formula A is a mathematical truth. Everyone knows it's indubitable. *Ergo*, at least one mathematical truth is indubitable.

There's Russell's cavil: indubitable, but no great truth.

> 2 + 2 by definition means (1 + 1) + (1 + 1)
> 4 by definition means 1 + (1 + (1 + 1))

By the associative law of addition, Formula A then is:

$$1 + 1 + 1 + 1 = 1 + 1 + 1 + 1.$$

Indubitable but unimpressive! As Frege said, it is an analytic a priori truth, not a synthetic one. (See Chapters 7 and 8 about Kant and Frege.)

Bertrand Russell thought *every* mathematical truth is a tautology like 2 + 2 = 4 —trivially indubitable. He said that a mathematical theorem says no more than "the great truth that there are three feet in a yard." This view is regarded by mathematicians as absurd and without merit.

To probe into Formula A, we must ask, what is 2? What is 4? What is +? What is =? Trivial as these questions may seem, they serve to distinguish the different schools (logicist, formalist, intuitionist, empiricist, conventionalist).

The most elementary answer is the empiricist one. "2 + 2 = 4" means "Put two buttons in a jar, put in two more, and you have four buttons in the jar." John Stuart Mill is the classic advocate of this interpretation. For him, formula A is *not* indubitable. It's about buttons. Buttons are material objects, which never can be known with certainty (Heraclitus et al., as expounded, for instance, in Russell's *History of Western Philosophy*, 1945). Who knows if some exotic chemical reaction might give

> two buttons + two buttons = zero buttons

or

> two buttons + two buttons = five buttons.

For indubitability, forget buttons.

Another answer, along formalist or logicist lines, might be given by a graduate student of mathematics. "1, +, = are symbols defined by the Dedekind-Peano axioms.**

> "2 is short for 1 + 1,
> "3 is short for 2 + 1,
> "4 is short for 3 + 1.
> "Now Formula A can be proved."

(Our reduction above of Formula A to

$$1 + 1 + 1 + 1 = 1 + 1 + 1 + 1$$

is a sketch of the formal proof in the Mathematical Notes and Comments.)

So then is Formula A doubtable?

Before I worry about doubtability, how sure am I that the proof is even correct? It *seems* to me that it is correct. Could I be mistaken? Overlooked something staring me in the face? I make mistakes in math. I think I'm sure the proof is O.K., but am I *really* sure that I'm *totally* certain?

A second worry is more substantial. How do I know Peano's axioms produce the same number system 1,2,3, . . . that I had in mind (or Dedekind had in mind) in the first place? They seem to work. How certain can I be that they'll always work? The numbers 1,2. . . . with which we start (before anybody gives us axioms) are an informal, "intuitively given " system. For that reason, it's impossible to prove formally that they correspond to any formal model. Such a proof is possible only between two formal models. Does the formal model correspond to the original intuitive idea? That question can never by answered by a rigorous, formal proof! It must rest on informal, intuitive reasoning that has no claim to be rigorous, let alone indubitable.

I have no doubt that Peano's axioms actually do describe the intuitive natural numbers 1,2,. . . . But I can't claim this belief as indubitable. Consider my knowledge that I'm now writing on a yellow pad with a blue pen sitting at a round table, and so forth. I can barely conceive that this might be false. But by the standards of Heraclitus, Plato, and others, such knowledge is doubtable or dubitable. If so, I can more readily believe Peano's seemingly convincing axioms might be doubtable. If they are, then every theorem in Peano arithmetic is doubtable. Even Formula A,

$$2 + 2 = 4 \text{ !!}$$

Another often-cited distinction between sensory knowledge and mathematical knowledge is that sensory knowledge *could conceivably be* other than it is. It's *conceivable* that my blue pen is really yellow and my yellow pad really blue.

Mathematical knowledge, on the other hand, such as Formula A, is supposed to be not only indubitably true, it's supposed to be inconceivable that it could be false. We supposedly can't imagine

$$2 + 2 = 3$$

as we supposedly can imagine a blue pad that looks yellow. This observation has been held to demonstrate that Formula A is indubitable.

But whether the denial of Formula A is inconceivable can't be judged until the import and significance of Formula A is explicated. It has one meaning as a *report of a property of buttons*, coins, or other such discrete objects. It has a second meaning as *a theorem in an axiom system* with +, =, and 1 as undefined terms. But we've just seen that as a formula about buttons, Formula A is doubtable. And as a theorem in an abstract axiom system, it's doubtable still. Whether in terms of buttons or of Peano arithmetic, Formula A is doubtable. Its negation is conceivable.

Let's consider the angle-sum theorem in Euclidean plane geometry.** It was Baruch Spinoza's favorite example of an absolutely certain statement. We'll follow proper terminology and call it

Formula B: Angle A + Angle B + Angle C = 2 right angles.

The angles A, B, and C on the left side of the equation are the internal angles of a triangle, any triangle at all.

Trite as this example seems to us today, it was already the twentieth century when Gottlob Frege berated David Hilbert for not understanding that there can be only one real, true geometry (Euclid's, of course).

We have theorem and proof. They have withstood scrutiny for 2,000 years. Isn't this indubitable?

Consider what is meant by the term "angle." Some say the theorem tells what will happen if you find a triangle lying somewhere and measure its angles. The sum should be 180 degrees.

If you actually do it, you find the sum not exactly 180 degrees. You retreat. It *would* be 180 degrees if you could draw perfect straight lines and measure perfectly. But you can't. You claim only that the sum will be *close* to 180 degrees, and even closer if you remeasure more accurately. That's *not* what Euclid said. It's a modern reinterpretation based on our post-Newtonian idea of "limit" or "approximation." Euclid's theorem is about the angles themselves, not about measurements or approximations.

A more sophisticated interpretation says Euclid's talking about ideal triangles, not triangles you actually draw or measure—an idealization that our mind/brain creates or discovers after seeing lots of real triangles. The indubitability of the theorem flows from the indubitability of the argument and the indubitability of the axioms with respect to the ideal points and lines.

The most important of the axioms about lines is the parallel postulate, "Euclid's Fifth." It can be stated in many equivalent forms. Most popular is Playfair's: "Through any given point, parallel to any given line, can be drawn exactly one line." For thousands of years the fifth axiom was accepted as intuitively obvious. Using it, we obtain Euclidean geometry, including the "indubitable" theorem about the angle sum of a triangle. But it was too complicated for an axiom, some thought. It ought to be a theorem.

In the nineteenth century Carl Friedrich Gauss, Janos Bolyai, a Hungarian army officer, and Nicolai Ivanovich Lobachevsky, professor and later rector at the University of Kazan, all without knowing of each other's work, had the same idea: Suppose the fifth postulate is *false*, and see what happens. Others had made the same supposition, in search of a contradiction that would prove the fifth postulate. Gauss, Bolyai and Lobatchevsky recognized that they got, not a contradiction, but a new geometry!

There is more information about non-Euclidean geometry in the article about Kant and in the mathematical notes and comments. Here we see how this discovery affects the certitude of the angle sum theorem. Suppose we replace Euclid's fifth by the non-Euclidean "anti-Playfair" postulate: "Through any point there pass more than one line parallel to any given line." From this strange axiom it follows that the angle sum of any triangle is *less* than 180 degrees. And if we try a different non-Euclidean fifth postulate—assume there are *no* parallel lines—then the angle sums of triangles are *greater* than 180 degrees.

We can't claim the angle-sum theorem is indubitable unless Euclid's parallel postulate is indubitable, for if we alter that postulate, the angle sum theorem is falsified. It's tempting just to declare that obviously or intuitively Euclid is correct. This wasn't believed by Gauss. There's a story that he tried to settle the question by measuring angles of a triangle whose vertices were three mountain tops. The larger the triangle, the likelier would be a measurable deviation from Euclideanness. According to the legend, the measurement was indecisive.

Some hundred years later, Einstein's general relativity depended on Riemannian geometry, a far-reaching generalization of non-Euclidean geometry. In relativity texts non-Euclidean geometry represents relativistic velocity vectors. So physics gives no license to favor Euclid over non-Euclid. Our prescientific, intuitive notions of space are learned on a small scale, relative to the universe at large. Locally, "in the small," Euclidean and non-Euclidean geometries are indistinguishable. Local intuition can't tell which is true in the large. Belief in the Euclidean angle sum theorem as indubitable was based on belief in an infallible spatial intuition. That belief has been refuted by non-Euclidean geometry and the following development of Riemannian geometry with its application in relativity.

Then where are we? Surely the angle-sum theorem is true in some sense? Of course. It's true as a theorem in Euclidean geometry. It follows from the axioms—not from Euclid's axioms alone, however. They need correction and completion. Hilbert took care of that, so we finally have a correct proof of the angle sum theorem, from a corrected and completed Euclidean axiom set.

We're in the same position as in the interpretation of

$$2 + 2 = 4$$

by Peano's axioms. We can't be completely sure which axioms describe the triangle we had in mind. By the nature of the case, such a thing can't be proved. Geometry is worse off than arithmetic, for its axioms are more subtle and elaborate. No one says Hilbert's axioms for Euclidean plane geometry are self-evident or indubitable.

Formula C is our last:

$$1 + \frac{1}{4} + \frac{1}{9} + \frac{1}{16} + \frac{1}{25} + \ldots = \frac{\pi^2}{6}$$

This formula was first written down by Euler. On the left is an infinite series. It says, "Add all terms of the form $1/n^2$, where n is any positive whole number." It asserts that the result of such an addition is the same as if you take π (the ratio of the circumference of a circle to its radius), square it, and then divide by 6.

The novice should be staggered. How can he ever add infinitely many numbers? And why would the answer have anything to do with a circle? The proof is easy when you know how to expand functions in series of sines and cosines (Fourier series).**

Formula C is abstruse and remote compared to Formulas A an B. You check

$$2 + 2 = 4$$

on your fingers. You check the angle sum theorem by measuring angles. What can you do here? Well, if you have a computer, or a calculator plus patience, you add up the series—as much as you have patience for. Then press the π button on your calculator, square and divide by 6.

A reader in Johannesburg sent these numbers (found after 30 hours on a hand calculator):

$$\sum \frac{1}{n^2} = 1.644914943$$

$$\frac{\pi^2}{6} = 1.644934067$$

What should you conclude from this?

If you're an optimist, and the error after 30 hours of calculation is only 0.000019124, you expect it would be even less after 40 hours. The formula is probably right. This is hope, not certainty.

No amount of addition could yield certainty, because the formula is about the *limit*—infinitely many terms! Any computation is finite.

If you want indubitable truth for this formula, you need *proof*. Fortunately, from the calculation of Euler or Fourier, which yields Formula C, it takes only a few more steps to such a proof.

The proof relies on the real number system. That has served long and well, but Gödel told us we can't prove it's consistent. Intuitionist and constructivist mathematicians don't rely on it without radical trimming. For practical purposes, we confidently write

$$\sum \frac{1}{n^2} = \frac{\pi^2}{6}$$

At the same time, if the angle-sum theorem and the formula "2 + 2 = 4" are subject to possible doubt, then this formula, involving limits, irrationals, and infinites, is more doubtable.

five

Five Classical Puzzles

"Vacancy in a schedule" uses a doctor's appointment to drive home the real existence of nonphysical, nonmental, nontranscendental entities. "Creating/discovering" is a new solution to the old dilemma, based on real math-talk. The other sections develop related ramifications of the humanist philosophy of mathematics.

Vacancy in a Schedule

I phone Dr. Caldwell for an appointment. He's booked way ahead, but his receptionist finds me an opening.

It's fortunate that the opening in Caldwell's schedule exists. That existence may mean he'll see me in time to detect a life-threatening illness.

The opening in his schedule can't be weighed, or measured, or analyzed chemically. It's not a physical object. The suggestion is absurd.

Is it a mental object? A thought in the receptionist's mind/brain? No. If she overlooked it, the opening would still be "there." A thought in Dr. Caldwell's mind/brain? No. When you ask about his schedule, he asks the receptionist.

The opening in Dr. Caldwell's schedule really exists. It's definite; it's recognizable. It has the right to be called "an object." But neither physical, mental, nor transcendental object.

It exists in a web of social arrangements, which includes doctor, receptionist, and patients. It's a social-cultural-historic object, and real as can be.

Creating-Discovering

Is mathematics created or discovered? This old chestnut has been argued forever. The argument is a front in the eternal battle between Platonists and anti-Platonists.

Platonists think mathematical entities can't be created. They already exist, whether we know them or not. We can discover them, but we can't create what's already there.

Formalists and intuitionists, on the contrary, know that mathematics is created by people (mainly mathematicians). It can't be discovered, because nothing's there to discover until we create it.

Each position is internally consistent. They seem incompatible.

Let's not replow this well-trodden ground. Instead, let's listen impartially for "create" and "discover" in nonphilosophical mathematical conversation. Why do both words—"create" and "discover"—seem plausible?

Think about some simple school problem.

"Find the area of a polygon, with vertices such and such and such."

Each student chooses her own method to get the area. Yet their different methods must all yield the same number. If someone gets a different answer, he made a mistake. After the mistakes are found and corrected (and any new mistakes made in the correction process also corrected), the whole class ends up with the same number.

This is the canonical, paradigmatic, fundamental experience of mathematical problem solving. That's why we say, "discover." We're not free to answer according to our fancy! The answer is the answer, like it or not. So we're convinced the answer is somewhere *there.*

If a problem is clearly and definitely formulated, we take it for granted that it has a solution.

(Sometimes the solution is a proof that no solution of the sought-for type is possible.)

When we solve the problem, we say the solution is "found" or "discovered." Not *created*—because the solution was already determined by the statement of the problem, and the known properties of the mathematical objects on which the solution depends.

The curvature of a circle of radius 7 is completely determined by the notion of "circle" and the notion of "curvature." We know there's a unique answer before we know it's 1/7.

When a problem is solved, the solution is brought from below the surface. It moves from the implicit to the explicit, from the hidden to the accessible. Before being discovered or *un*covered, it lay hidden, like an image on exposed, undeveloped photographic film. Once it's found—*developed*—we see where it lay hidden in the known mathematical context. And everybody agrees, this is the solution.

But solving well-stated problems isn't the only way mathematics advances. We must also invent concepts and create theories. Indeed, our greatest praise goes to those like Gauss, Riemann, Euler, who *created* new fields of mathematics.

A well-known classification of mathematicians is problem-solvers and theory builders. When speaking of a *theory*—Galois's theory of algebraic number fields,

Cantor's theory of infinite sets, Robinson's theory of nonstandard analysis, Schwartz's theory of generalized functions—we don't say it was "discovered." The theory is *in part* predetermined by existing knowledge, and *in part* a free creation of its inventor. We perceive an intellectual leap, as in a great novel or a great symphony.

In the 1930s and 1940s a need was recognized for a generalized calculus that would include singular functions like Dirac's delta function** (see "Change" in this chapter). Theories of generalized functions were created by Solomon Bochner, Sobolev, and Kurt Friedrichs. Meant to serve similar purposes, they were not identical in the way that different answers to a calculus problem are identical. Sobolev spaces and the "distribution theory" of Laurent Schwartz won general acceptance.

When several mathematicians solve a well-stated problem, their answers are identical. They all *discover* that answer. But when they create theories to fulfill some need, their theories aren't identical. They *create* different theories.

Such was the case with Gibbs's vector analysis versus Hamilton's quaternions (see section on change).

The distinction between inventing and discovering is the distinction between two kinds of mathematical advance. Discovering seems to be completely determined. Inventing seems to come from an idea that just wasn't there before its inventor thought of it. But then, after you *invent* a new theory, you must *discover* its properties, by solving precisely formulated mathematical questions. So inventing leads to discovering.

Invention doesn't happen only in theory-building. In a well-stated problem, the answers must be the same, but the methods of different problem-solvers may be different. It may be necessary to do something new, to bring in some new trick. You may have to *invent* a new trick to *discover* the solution. Again, inventing is part of discovering.

Maybe your new trick will help other people to solve their problems. Then it will receive a name, and be studied for its own sake. To *discover* its properties, *inventing* still other new tricks may be necessary.

For example, Fourier invented Fourier analysis to solve the linear equations of heat and vibration. Then some natural-seeming questions about Fourier series turned out to be very difficult. So Fourier analysis became a field of research on its own. The needs of Fourier analysis (and other parts of mathematics) led to infinite-dimensional linear spaces (Hilbert space and spaces of generalized functions). These new spaces themselves are now big fields of research, which need new ideas in linear operator theory and functional analysis. And so on, and on, and on.

Were the natural numbers 1, 2, 3 . . . discovered or invented? One can't help recalling the diktat of Ludwig Kronecker: "The integers were created by God; all else is the work of man." Since Kronecker was a believer, it's possible he meant this literally. But when mathematicians quote it nowadays, "God" is a figure of speech. We interpret it to mean: "The integers are discovered, all else is invented." Such a statement is an avowal of Platonism, at least as regards the integers or the natural numbers.

How do I as a humanist answer it?

I recall the distinction between Counting Numbers—adjectives applied to collections of physical objects—and pure numbers—objects, ideas in the shared consciousness of a portion of humanity. (See Chapter 1, "A Way Out.")

Counting numbers are discovered. Pure numbers were invented.

Is mathematics created or discovered? Both, in a dialectical interaction and alternation. This is not a compromise; it is a reinterpretation and synthesis.

Finite/Infinite

All the numbers calculated since the formation of the earth are less than $10^{(10^{(10^{10})})}$ (or some higher iterate of iterates.) In other words, they're all finite.

Yet mathematics is full of the infinite. The line R^1 is infinite; the space R^3 is infinite; N, the set of natural numbers, is infinite. There are infinitely many infinite series. There are points "at infinity" on the real line, in the complex plane, in projective space, and, of course, Cantor's hierarchies of infinite sets, infinite ordinal numbers, infinite cardinal numbers.

Where do these infinites come from? Not from observation and not from physical experience. If you don't believe in a separate spiritual or transcendental universe, they must be born in human mind/brains.

Poincaré said arithmetic is based on "the mind's" conviction that what it has done once, it can do again. You needn't search far to find minds holding no such conviction. That "conviction" is trivially false. Any thought or action can be repeated only finitely many times. The person (or animal or machine) that's repeating the thought or action will wear out or die.

The brain is a finite object. It can't contain anything infinite. But we do have ideas of the infinite. It's not the infinite that our mind/brains generate, but *notions* of the infinite.

Logic doesn't force us to bring infinity into mathematics. Euclid had finite line segments, never an infinite line.

In set theory it's the axiom of infinity that provides an infinite set. Without adopting that axiom, Frege and Russell would have had only finite sets.

Sometimes we consciously exclude the infinite. A convergent infinite series is interpreted as a sequence of finite partial sums. We still *call* the series infinite, but we're really interested in the finite partial sums. Yet the "meaningless" intuition of summing infinitely many terms is still the core meaning of "infinite series."

People ask what kind of mathematics would be produced by an alien intelligence (big green critters from Sirius). Maybe they would have no infinity, since it comes out of our heads, not from physical reality.

The Hungarian logician Rózsa Péter wrote a survey of mathematics called *Playing with Infinity*. For her, playing with infinity was the essence of mathematics.

Donald Knuth is an accomplished mathematician and a world champion computer scientist. His *Art of Computer Programming* is called a "bible." Knuth makes vivid a point often scored against constructivists or intuitionists. From Brouwer to Bishop, they can't stomach infinite sets as completed objects, yet they balk at no finite set, no matter how huge. Knuth shows us finite sets so vast that they're just as obscure as the infinite.

The reader may have noticed that nearly all finite sets are vast and huge. To check this obvious fact, just pick a huge number, call it M. How many sets are smaller than M? Very many, but still, finitely many. How many sets are bigger than M? Infinitely many. Infinitely many is *way more* than finitely many, so nearly all finite sets are bigger than M, no matter how big M may be. If you pick a finite set at random, it's almost sure to be bigger than M, no matter how big M might be. Its finitude is small comfort.

In fact, infinite sets are often brought in because they're *simpler* than the given finite sets. Integrating is usually simpler than summing a huge finite number of terms. Differential equations are usually easier than the corresponding finite difference equations.

Moral: If you allow *all* finite sets, you have no excuse to refuse the infinite. The infinite is unintuitive and metaphysical? So are nearly all finite sets. If you want all math to be concrete and intuitive, stick with small finite numbers and small finite sets.

How small? Hard to say, because if n is small, so's $n + 1$. There's no sharp line between small and large, no smallest big number or biggest small number. You could use the M you just picked, and increase Peano's five axioms of arithmetic** by a sixth:

"For all n in N, $n < M$ or $n = M$."

Sad to say, this set of six axioms is inconsistent. To restore consistency, you could modify the first five. For instance,

"If $M/2 < A < M$, then $A + A$ is out of bounds."

Redefine $A + A$ any way you please, or leave it undefined (it doesn't "exist"). Either way, the old arithmetic is out the window. Perhaps we're better off with our infinite number system.

The huge mystery of infinity is an artifact of Platonism. In some transcendental realm, do infinite sets exist? This is the wrong question. Number systems are invented for the convenience of human beings. The appropriate questions about infinity are:

Is it good for anything?

Is it interesting?

Mathematicians have long since answered, yes!

Object/Process

This section strays far from the philosophy of mathematics. In the end it returns to the main subject, with an insight picked up along the way.

Frege said numbers are abstract *objects*. Plato's Ideas, including numbers, were objects of some sort. But intuitionists and formalists deny that numbers are objects.

Kreisel and Putnam say what's needed aren't objects, but objectivity: Numbers are *objective*. We needn't trouble whether they're objects.

Some people say, "Numbers exist, that's plain as the chair I'm sitting on."

Others say, "It's obvious that numbers don't exist in the sense that this chair exists."

The two statements aren't contradictory. Both are true.

If I have no prior explication of "exist," my knowledge of numbers isn't increased if you tell me they exist or don't exist. In his great polemic on the foundations of arithmetic, Frege refrains from explaining the meaning of "exist" or the meaning of "object."

Is a cloud an object? Sort of yes.

A roaring fire on the prairie? No, more like a process.

My grandchild's temper fit? Definitely a process. $10^{(10^{10})}$? An object a la Frege, or just "something objective" a la Kreisel-Putnam?

Does "object" mean "something like a rock"? Something with definite shape and volume? Then only solids would be objects. Melt a piece of ice, and you turn object into nonobject. So stringent an objecthood is too special, too transient, too conditional for philosophy.

Sometimes "object" seems to mean any physical entity, with volume and shape or without. Then atoms are objects. So are electrons, photons, and quarks. But at this level, the distinction between object and process vanishes. Depending on how they're observed, electrons, protons, and photons have particle-like (object) or wavelike (process) behavior.

A wider meaning of "object" is, "independent of my consciousness." Maybe this is what Kreisel-Putnam mean by objective. Does it mean independent of *anyone's* consciousness?

Are you and I objects to each other, but subjects to ourselves? Are we both just objects? Both just subjects? Or objective subjects? Subjective objects?

A related meaning of "object" is, "anything that can affect me." Such a definition was used by Paul Benacerraf in a widely referenced paper. A tree could affect me if I drove into it, and a germ could affect me by giving me the sniffles. An attack of paranoia could make me very sick. Would that attack be an object?

Still another meaning is by opposition to "process." "'Object' is noun, 'process' is verb. Object acts or is acted on. Process is the action."

In all these explications, objects are independent of individual consciousness. They have relatively permanent qualities. They can be observed or experienced by anyone with the appropriate sense organ, the appropriate training of eye and brain, the appropriate scientific instrument.

Niagara Falls is an example of the dialectic interplay of object and process. Niagara Falls is the outlet of Lake Ontario. It's been there for thousands of years. It's popular for honeymoons. To a travel agent, it's an object.

But from the viewpoint of a droplet passing through, it's a process, an adventure:

over the cliff!
fall free!
hit bottom!
flow on!

Seen in the large, an object; felt in the small, a process. (Prof. Robert Thomas informs me that the Falls move a few feet or so, roughly every thousand years.)

Movies show vividly two opposite transformations:

A. Speeding up time turns an object into a process.
B. Slowing down time turns a process into an object.

To accomplish (A), use time-lapse photography. Set your camera in a quiet meadow. Take a picture every half hour. Compose those stills into a movie.

Plant stalks leap out of the ground, blossom, and fall away!
Clouds fly past at hurricane speed!
Seasons come and go in a quarter of an hour!
Speeding up time transforms meadow-object to meadow-process.

To accomplish (B), use high-speed photography to freeze the instant.

A milk drop splashed on a table is transformed to a diadem—a circlet carrying spikes, topped by tiny spheres. By slowing down time, splash-process is transformed to splash-object.

A human body is ordinarily a recognizable, well-defined *object*. But physiologists tell us our tissues are flowing rivers. The molecules pass in and out of flesh and bone. Your friend returns after a year's absence. You recognize her, yet no particle of her now was in her when she left. Large-scale, object—small-scale, process.

In the social-cultural-historic domain, the continuity between object and process is blatant, even though some institutions, beliefs, and practices seem eternal. All institutions change. If they change slowly, over centuries—slavery, piracy, royalty, private property, female subjection—they are thought of as objects. If they change daily—clothing fashions, stock market prices, opinion polls—they are thought of as processes.

A nation defends or alters borders, signs treaties, makes war. It claims to be an object. Yet it's also a horde of individual persons being born, dying, immigrating, and emigrating. A few individuals decide the fate of nations, and nations decide the fate of many individuals. A nation is like a waterfall: an object in the large, a process in the small.

In computing machines, the difference between object (hardware) and process (software) is almost arbitrary. The designer decides which functions to embody in hardware, which in software.

Some features of mind/brain are object-like, life-long. Some change by the minute. A psychotherapist tries to overcome a patient's belief that his neurotic symptom is an object. A step toward cure is convincing him that it is a process.

Before Galileo turned his telescope to the sun, philosophy thought earth was change and process, and the sun was a changeless object. We learned that the "changeless" sun is a bonfire. In time it must fade and die, or explode. The dead moon too had a birth and a history.

For the geologist, using a long time scale, earth is a process. Looking at the huge earth from a short distance, we ordinarily don't see it as a whole, as an object. But when astronauts stood on the moon, they saw earth from a long distance for a short time. They saw an object.

In Einstein's equation

$$E = mc^2$$

where m is mass: *object*, and E is energy: *process*; and c is the speed of light in vacuum. (This tremendous number must be *squared*!)

This equation tells how at Hiroshima and Nagasaki mass-object-bombs transformed into energy-process holocausts. In stars as in nuclear explosions, mass is converted to energy. It's equally sensible or senseless to say a star is an object or a process.

When Edwin Hubble photographed his red shift, and later when Arno Penzias picked up his 15 billion year-old echo, we learned that the cosmos passes through history, from a Big Bang to a Big Crunch or a Cold Death. Cosmology became historical, like biology and geology.

Presently, elementary particles called hadrons and bosons seem to compose the universe. But those particles exist only during one stage of cosmic evolution, when the configuration of the Cosmos as a whole permits that existence. Physics itself takes on a historical, evolutionary aspect.

High-speed and time-lapse photography show that the object-process polarities are ends of a continuum. Any phenomenon is seen as an object or a process, depending on the scale of time, the scale of distance, and human purposes. Consider nine different time-scales—astronomical time, geologic time, evolutionary time, historic time, human lifetime, daily time, firing-squad time, switching time for a microchip, unstable particle lifetime.

A smaller scale in space, or speeding up time, turns an object into a process. A larger scale in space, or slowing down time, turns a process into an object.

In brief, *an object is a slow process. A process is a speedy object.*

What about mathematics? In mathematics don't we have "abstract objects," not located in time or space? Like "the" equilateral triangle? or "the" number 9?

From Pythagoras to Frege, philosophy gave mathematical objects an idyllic existence, free of blemishes such as temporality, impermanence, and indefiniteness. They were thought to be absolutely still, free of any process properties. Plato's Ideas were pure objects. Frege's abstract objects are the nineteenth century version of Plato's Ideas. Instead of timeless, they're "tenseless."

The example of numbers and triangles is supposed to prove that "abstract objects" exist. On the other hand, the puzzle about the mode of existence of numbers and triangles is supposed to be solved by calling them "abstract objects."

What's the meaning of "abstract"? Abstract objects are *not* mental, *not* physical, *not* historical, *not* social or intersubjective.

How do I get acquainted with them? No answer.

We can see that mathematical objects aren't mental or physical. It would be an over-polysyllabic tongue-twister to call them "nonmental-nonphysical objects." So make an abbreviation: "abstract objects." A more honest name would be "transcendental objects."

Nowhere in heaven or on earth is there a pure object, totally free of process aspects—free of change.

Only in mathematics we think we have pure objects. There, it is thought, we find *nothing but* pure objects. Infinitely many of them!

Could this thinking come from seeing mathematics in too short a time scale? Wouldn't a view that encompassed centuries show mathematics evolving—a process?

This book argues that mathematics is a social-historical-cultural phenomenon, without need of anything abstract-Platonic-nonhuman, without need of formalist or intuitionist reduction, but this view implies that mathematical objects, like other objects, are also processes. They change, whether in plain sight or too slowly to be noticed.

The next section takes the number 2 as a case study of a changing mathematical object.

Change

From Pythagoras to us, 2 has changed, as surely as music has changed and religion has changed.

In some ways 2 has eroded; in many ways it has expanded.

$$1 + 2 = 3$$

is true for us. It was true for Pythagoras. Yet his meaning wasn't identical with ours. "1" was not just our 1. It also was God, unity. 2 was the female principle. 3 was the male principle.

$$1 + 2 = 3$$

was more than a rule of making change in the market. It was a religio-philosophical cosmic truth.

The Pythagoreans discovered: "$\sqrt{2}$ does not exist." That is to say, there's no pair of whole numbers, p and q, such that

$$(p/q)^2 = 2.**$$

Today, the real numbers are available to help understand a natural number like 2. Your calculator tells you $\sqrt{2}$ is some close approximation to an irrational number

$$1.414213562\ldots$$

So now $\sqrt{2}$ exists, and consequently the meaning of 2 has changed.

Since Pythagoras mathematics has been regarded as unchanging and eternal.

$$2 + 2$$

always was and always must be 4. Euclid knew it, and so (hopefully) will our descendants millennia hence. How can I claim that

$$2 + 2 = 4$$

isn't timeless?

First a small point. Euclid never saw the formula $2 + 2 = 4$. The symbols "2," "4," "+," and "="—and the practice of putting symbols together to make formulas—were alien to his time.

But never mind the formula, let's stick to the facts. When Euclid went to market, he knew that two oboli plus two oboli was worth four oboli. If we're talking about "2" and "4" as ***adjectives*** modifying "oboli," and + and = as a commercial operation and a commercial relation, we and Euclid agree, just as we would agree that the sun rises in the East. But if we're talking about the ***nouns*** "2" and "4"—meaning some sort of autonomous objects—and operations and relations on ***them***, there are differences between us and Euclid. (See the section on Certainty in Chapter 4.)

2 is no longer only a counting (natural) number. Now it has an additive inverse, -2. (It's an integer.)

2 is no longer isolated. Now it's a rational number. As such, an element in a dense ordered set. 2 is now a point on the continuous number line.

2 is even a point in the complex plane, participating in analytic functions and conformal maps.**

2 has matured into a complex creature, with resonances, possibilities, and connections of unfathomed subtlety.

And most remarkable—it has a *pair* of square roots—nonexistent for Pythagoras and Euclid!

As we add new structures, we embed the old ones into them. It's more efficient to make the natural numbers, the integers, the rational numbers, the real numbers subsets of the complex numbers, rather than separate objects isomorphic to subsets of the complex numbers.

To appease any fussy formalists who join our conversation, we might try more explicit notations: 2_N for the natural number 2, 2_I for the integer 2, 2_Q for the rational number 2, 2_R for the real number 2, 2_C for the complex number 2. But no one working with 2 would bother with this. It would just slow you down. What's the Platonist's alternative? Are there uncountably many undiscovered twos still waiting to be discovered? Or is there and was there, already in Pythagoras's time, one majestic, unique, eternal 2, already an integer, already a rational number, already a real number, already a complex number, ***and*** who knows what else? These are fables you can believe if you want to.

As mathematics grows and changes, the numbers change.

Euclid had line segments and we have line segments, but they're not the same. Euclid's is a simple thing. It has two endpoints and "lies evenly on itself." Ours is a grand mystery, a set with an uncountable infinity of elements (points) and subsets of undecidable sizes. It participates in the operations of algebra and analysis, and is linked to an unfamiliar non-Euclidean cousin. Euclid knew what he was talking about, but it wasn't quite what we talk about.

As mathematics grows and changes, geometry changes.

A more current example. Until the midtwentieth century, the "derivative" or "slope" of a function at a point existed only if at that point the graph of the function was smooth—had a definite direction, and no jumps. Now mathematicians have adopted Laurent Schwartz's generalized functions.** *Every* function, no matter how rough, has a derivative.

The Heaviside function H(t) consists of two pieces.
On the left, when t is less than 0, H(t) = 0 identically.
On the right, when t is greater than 0, H(t) = 1 identically.
When t = 0, H(t) jumps from 0 to 1.

Classically, H'(t), the slope of H(t), exists only for t greater than 0 and for t less than 0. At those points the graph of H(t) is flat, and its slope H'(t) is 0.

At t = 0, where H(t) jumps, H'(t) is classically undefined. The derivative doesn't exist there.

Nowadays there's a derivative for H(t) as a whole, *including at the jump.* It's Dirac's "delta function,"** a gadget introduced by the physicist Paul Dirac. By the classical definition of function, the "delta function" isn't a function. We ought to write, not "delta function," but delta "function."

Dirac's delta function is zero for all t except t = 0. When t = 0, it's infinite. Its "graph" is an infinitely thin, infinitely high spike at t = 0. And the area under this weird graph is 1!

Classically, this is nonsense. "Infinite" isn't a number. Dirac's spike would be a "rectangle" with infinite height and thickness 0, so the "area" would be infinity times 0. But infinity times 0 can be anything at all.

Why bother with such peculiar stuff? For Dirac, it came in handy in quantum mechanical calculation. But it's also a model for some ordinary physics. Slam a ping-pong ball with your paddle. Its speed goes instantly from 0 to 1. As a function of time, its speed is a Heaviside function. Its acceleration is Dirac's delta function!

The meaning of differentiation has changed. Newton and Leibniz's differentiation operator has become something more general. Our generalized differentiation includes the old differentiation, and it's much more powerful.

As mathematics grows and changes, functions and operators change.

A familiar cliché says that while other sciences throw away old theories, mathematics throws away nothing. But the old mathematics isn't preserved intact. Mathematics is intensely interconnected and self-interactive. The new is vitally linked to the old. The old is revitalized, enriched, and complexified by interaction with the new.

An excerpt from Mary Tiles (p. 151):

> The introduction of ideal elements does not leave the pre-existing domain fixed. The conception of number as the measure of a magnitude is what militates against the admission of either 0 or negative numbers as numbers. If they are admitted as ideal elements, to complete the system, then they have no interpretation as proper numbers (magnitudes.) But as they come to form a single representation system with the positive numbers, they come to have applications. The result is that the concept of number itself is no longer tied to that of magnitude. In this sense even the finitary significance of numbers has not, historically, been without change. Peano's axiomatization of the natural number system, making it essentially a system of entities generated from 0 by repeated application of the successor operation, again opens up a pathway to the field of recursive function theory, the theory of algorithms and the whole modern computational life of the number system. Computation as understood by this route is not what it was before those developments. . . . What emerges is not a static picture of a set of entities on the one hand with a determinate and fixed set of reasoning procedures which can, once and for all, be characterized and certified as reliable.

Nonexist/Exist

There are different kinds of existence. There are also different kinds of nonexistence. Speaking of Pegasus, the beautiful winged horse of Greek legend, I can say

There's no such myth (false).
There's no such physical object (true).
There's no such entity in biology (true).
There's no such thought in my mind (false).

In mathematics, nonexistence usually is a matter of impossibility.

"A solution to this problem does not exist" means "It's impossible to solve this problem."

Often nonexistence of one thing is equivalent to existence of another thing. Euclid proves there's no largest prime number—no prime number greater than all other primes. Nonexistence! Today the usual statement of this fact is: "There exist infinitely many primes." Infinite existence!

Back in the early Stone Age when we couldn't count past 20, 20 was an upper bound of our number system. Some easy questions were hard. Like,

$$15 + 15 = ?$$

We overcame this difficulty and enlarged the number system. We reached our system of natural numbers, which has no upper bound. First-graders today know you can go on counting forever. And we can write

$$15 + 15 = 30.$$

In imagination we can add any two numbers, no matter how big. For this enlargement we pay a penalty. Problems arise that couldn't be imagined in a system bounded by 20. We have a number system which is ultimately unknowable.

Two old conjectures we can't prove or disprove are: "Every even number is the sum of two primes" (Goldbach); and "There are infinitely many prime pairs"

(like $\{11,13\}$, $\{29,31\}$, $\{41,43\}$).

We enlarge a mathematical system to make it simpler in some sense, and we thereby create worse complications in some other senses. The new complications generate new problems, which drive us to enlarge the system still further.

Mathematics is the only science in which "exist" is a technical term. Contention about mathematics between philosophical schools is mainly about existence of mathematical objects. An advanced mathematics graduate student can tell whether, in her specialty, existence has been proved. But what is *meant* by existence? *In what sense* does something exist?

Two kinds of mathematical existence are sometimes distinguished—constructive and indirect. Constructive existence means that the object is gotten from known objects by a finite number of steps, or within arbitrarily small error by a finite number of steps.

To "construct" the rationals, I say, "Consider the set of ordered pairs (a b), etc." Voilà! The rationals have been constructed!**

Step back from the math for a moment. Isn't this amazing? We've actually constructed infinitely many distinct entities, in a few minutes of pencil-pushing, at little effort, and no expense! Isn't there something strange about that?

On the other hand, most people think all these numbers existed way back in time, long before we started this conversation. If that's so, have we constructed anything?

Yet this kind of "construction" is considered straightforward. There is controversy about indirect proof, in which you prove "A must be true" by proving "not-A is impossible." Indirect proofs accepted by classical mathematicians are rejected by intuitionists and constructivists. (See Brouwer and Bishop in Chapter 9, and "Finite/Infinite" above.)

How Mathematics Grows

Predicting the growth of mathematics would require knowing what mathematicians are trying to do, not just what they have done so far. A judgment on whether their goals are attainable would help prediction. In 1900 David Hilbert laid 23 problems before the mathematical world. One could have tried to guess which problems would be attractive and approachable, and thereby made a short-term prediction about future mathematics.

A deeper attack on the prediction problem would look, not only at what mathematicians have done recently, and what they're now attempting, but at what the present state of mathematical knowledge naturally calls for—what we *should* be trying to do. That's what Hilbert did to make his list of problems. Because he presented it in 1900 at an International Congress, and because he was David Hilbert, it was a self-fulfilling prophecy.

Some mathematicians say that if history started all over, mathematics would evolve into much the same thing, in much the same order. The opportunities and questions arising from what we know decide what advances we make, what we'll know next year. Igor Shafarevich, a Russian mystic and anti-Semitic algebraic geometer, says mathematicians don't make mathematics, they're instruments for mathematics to make itself. This strange-sounding theory is supported by many instances of repeated or simultaneous discovery.

Desargues discovered projective geometry in the seventeenth century. In the shadow of the analytic geometry of Fermat and Descartes it was overlooked and forgotten. In the nineteenth century Monge and Poncelet rediscovered it.

Gauss discovered Abelian integrals before Abel and the method of least squares before Legendre.

The "Argand diagram" of complex numbers was found by Caspar Wessel before Argand, by Gauss after Argand, and by Euler before any of them.

The Ascoli theorem of the 1920s became the Arzela-Ascoli theorem when it was found in a paper Arzela had published 39 years before Ascoli.

Polya's counting formula was published by Redfield 30 years before Polya.

Jesse Douglas and Tibor Radó had a priority quarrel about the Plateau problem.

And so on.

These examples suggest that mathematical discoveries force themselves on mathematicians. Mathematics unfolds, like a flower opening, or a tree plunging roots deeper and crown higher.

Michael Dummett said that mathematics grows as if it moved on rails projecting a little way into the future. This means short-run prediction can work, long-run prediction can't. I distinguish two kinds of "rails" into the future of mathematics. One kind is solving problems already recognized and stated. (One kind of "solution" would be a proof that the problem is unsolvable.) Since these problems are under attack today, their solutions, though yet unknown, are already determined. Once a problem is solved, we see how the shape of the solution was predetermined by the statement of the problem. This kind of rail is rigid but not long—few problems stay unsolved more than a few dozen years.

Dummett's idea of rails that go only a little way forward is an appealing metaphor. How seriously can we take it? What are the rails? How far do they go? Some people talk as if math is already determined forever by the axioms of set theory—because all theorems are in principle already determined by the axioms and rules of inference. But the mathematics that will ever be known and used is a tiny fraction of all the "in principle" consequences. Real-life mathematics can only include consequences that somebody thought useful or interesting. To what extent do notions of useful or interesting endure? (v. Polanyi). Some theories lose interest. New ideas appear unexpectedly.

A different kind of "rail" is created by unrecognized potentialities of what we know. The whole theory of analytic functions was predetermined when sixteenth-century algebraists introduced the square root of -1. Some of today's familiar concepts are pregnant with beautiful developments, which will seem inevitable once we notice them.

New theories allow more individual leeway than solutions of explicitly stated problems. They're less rigid but stick further into the future.

Mathematics is made of theories in the first place, of objects only secondarily. We study the theory of numbers, not numbers per se. The theory of distributions was created by Laurent Schwartz in the 1930s. A Platonist would say Schwartz's distributions always existed in the set theoretic universe. It's more interesting to see how aspects of his theory were foreshadowed, for example, by Hadamard's "finite part" of a divergent integral. Saying distributions already existed in an abstract set universe "somewhere" is of little interest to mathematicians.

We have to distinguish two senses of "exist"—potential and actual. The statue of David already existed potentially. Michelangelo gave it actual existence. The passage from potential to actual is the interesting event.

The distinction between potential and actual existence isn't limited to mathematics. It arises whenever you consider the future. Possible futures have potential existence. Some will be realized, many will not. Whatever is ultimately actualized must already have been possible (potential). Related notions are common in physics, biology, and political science.

In *Mathematical Notes and Comments*, in the section on "Imaginary Becomes Reality," we give examples of one way mathematics grows—by creating new entities or theories, to make unsolvable problems solvable.

Is the philosophy of mathematics concerned with potential existence or actual existence? Frege and Gödel, and set-theoretic Platonists in general, seem to think of the potential as if it were already actual.

Before Socrates, Parmenides taught that whatever isn't contradictory already exists. Brouwer's intuitionism, on the other hand, defines mathematics as the thinking of the Creative Mathematician. It focuses on the actual. One could class many formalists with intuitionists in this respect. The present work emphasizes both the distinction and the connection between potential and actual. Mathematical discovery or creation is transformation of potential to actual.

Part Two

six

Mainstream Before the Crisis

The next four chapters tell the mainstream philosophies of mathematics. They try to be entertaining and relevant, not definitive or exhaustive. Most of the facts are well known. The arrangement and some interpretation are novel.

The name "foundationism" was invented by a prolific name-giver, Imre Lakatos. It refers to Gottlob Frege in his prime, Bertrand Russell in his full logicist phase, Luitjens Brouwer, guru of intuitionism, and David Hilbert, prime advocate of formalism. Lakatos saw that despite their disagreements, they all were hooked on the same delusion: *Mathematics must have a firm foundation.* They differ on what the foundation should be.

Foundationism has ancient roots. Behind Frege, Hilbert, and Brouwer stands Immanuel Kant. Behind Kant, Gottfried Leibniz. Behind Leibniz, Baruch Spinoza, and René Descartes. Behind all of them, Thomas Aquinas, Augustine of Hippo, Plato, and the great grandfather of foundationism—Pythagoras.

We will find that the roots of foundationism are tangled with religion and theology. In Pythagoras and Plato, this intimacy is public. In Kant, it's half covered. In Frege, it's out of sight. Then in Georg Cantor, Bertrand Russell, David Hilbert, and Luitjens Brouwer, it pops up like a jack-in-the-box.

In the twentieth century, we look at Russell, Brouwer, Hilbert, Edmund Husserl, Ludwig Wittgenstein, Kurt Gödel, Rudolph Carnap, Willard V. O. Quine, and a small sample of today's authors. Philip Kitcher said the philosophy of mathematics is generally supposed to begin with Frege—before Frege there was only "prehistory." Frege transformed the issues constituting philosophy of mathematics. In that sense earlier philosophy can be called prehistoric. But to understand Frege you must see him as a Kantian. To understand Kant you must see his response to Newton, Leibniz, and Hume. Those three go back to Descartes, and through him to Plato. Plato was a Pythagorean. The thread from Pythagoras to Hilbert and Gödel is unbroken. I aim to tell a connected story from Pythagoras to the present—where foundationism came from, where it left us.

Instead of going straight through from Pythagoras, I've split the story into two parallel streams—the first section is about the "Mainstream." The second is about the "humanists and mavericks."

For the Mainstream, mathematics is superhuman—abstract, ideal, infallible, eternal. So many great names: Pythagoras, Plato, Descartes, Spinoza, Leibniz, Kant, Frege, Russell, Carnap. (For Kant, membership in this group is partial.)

Humanists see mathematics as a human activity, a human creation. Aristotle was a humanist in that sense, as were Locke, Hume, and Mill. Modern philosophers outside the Russell tradition—mavericks—include Peirce, Dewey, Roy Sellars, Wittgenstein, Popper, Lakatos, Wang, Tymoczko, and Kitcher (a self-styled maverick). There are some interesting authors who aren't labeled philosophers: psychologist Jean Piaget; anthropologist Leslie White; sociologist David Bloor; chemist Michael Polányi; physicist Mario Bunge; educationists Paul Ernest, Gila Hanna, Anna Sfard; mathematicians Henri Poincaré, Alfréd Rényi, George Pólya, Raymond Wilder, Phil Davis, and Brian Rotman.

First we honor the great grandfather of the mainstream.

Pythagoras (fl. 540–510 B.C.)

Religion Is Mathematics—Mathematics Is Religion

No one knows when mathematics started. Koehler reports on studies by comparative psychologists of counting by jackdaws and pigeons. Human arithmetic goes back at least as far as possession of property—clam shells, fish heads, whatever. Geometry in the most elementary sense goes back even further, to the animal that must run straight toward a prey or straight away from a predator.

The philosophy of mathematics, on the other hand, seems to start with Pythagoras (about 572–479 B.C. or a little later; Heath, 1981, p. 67).

Schoolchildren know his namesake, the indispensable relation between the sides of a right triangle. Whatever Pythagoras had to do with that formula, it was known centuries before him, in China and Babylon. More than about Pythagoras, we know about the Pythagoreans, the secret society he supposedly founded. They combined mysticism and superstition with geometry and arithmetic in a way incomprehensible today. According to Heath (1981, p. 66):

> It is difficult to disentangle the portions of the Pythagorean philosophy which can safely be attributed to the founder of the school. Aristotle evidently felt this difficulty; . . . when he speaks of the Pythagorean system, he always refers it to "the Pythagoreans," sometimes even to "the so-called Pythagoreans."

But

> It is certain that the Theory of Numbers originated in the school of Pythagoras; and with regard to Pythagoras himself, we are told by Aristoxenus that "he seems to have attached supreme importance to the study of arithmetic, which he advanced and took out of the region of commercial utility."

Boyer and Merzbach (1991, p. 53) tell us that "Many early civilizations shared various aspects of numerology, but the Pythagoreans carried number worship to its extreme, basing their philosophy and their way of life upon it. The number one, they argued, is the generator of numbers and the number of reason; the number two is the first even or female number, the number of opinion; three is the first true male number, the number of harmony, being composed of unity and diversity; four is the number of justice or retribution, indicating the squaring of accounts; five is the number of marriage, the union of the first true male and female numbers; and six is the number of creation. Each number had its peculiar attributes. The holiest of all was the number ten, or the *tetractys*, for it represented the number of the universe, including the sum of all possible dimensions. [See also Heath, 1981, p. 75.] A single point is the generator of dimensions, two points determine a line of dimension one, three points S (not on a line) determine a triangle with area of dimension two, and four points (not in a plane) determine a tetrahedron with volume of dimension three; the sum of the numbers representing all dimensions, therefore, is . . . ten. It is a tribute to the abstraction of Pythagorean mathematics that the veneration of the number ten evidently was not dictated by anatomy of the human hand or foot."

R. Tarnas writes, in *The Passion of the Western Mind* p. 46, "For Pythagoreans, as later for Platonists, the mathematical patterns discoverable in the natural world secreted, as it were, a deeper meaning that led the philosopher beyond the material level of reality. To uncover the regulative mathematical forms in nature was to reveal the divine intelligence itself, governing its creation with transcendent perfection and order. The Pythagorean discovery that the harmonics of music were mathematical, that harmonious tones were produced by strings whose measurements were determined by simple numerical ratios, was regarded as a religious revelation. Those mathematical harmonies maintained a timeless existence as spiritual exemplars, from which all audible musical tones derived. The Pythagoreans believed that the universe in its entirety, especially the heavens, was ordered according to esoteric principles of harmony, mathematical configurations that expressed a celestial music. To understand mathematics was to have found the key to the divine creative wisdom. . . . Through intellectual and moral discipline, the human mind can arrive at the existence and properties of the mathematical Forms, and then begin to unravel the mysteries of nature and the human soul."

Heath (1981, pp. 68–69) thinks that in Pythagorean doctrine "the number in the heavens" means the number of visible stars. He asks, "may this not be the origin of the theory that all things are numbers, a theory which of course would be confirmed when the further capital discovery was made that musical harmonies depend on numerical ratios, the octave representing the ratio 2:1 in length of string, the fifth 3:2, and the fourth 4:3?"

Plutarch (40 A.D.–120 A.D.) connects the Pythagoreans to the Isis cult of Egypt, blending sacred history and mathematical theorems: "The Egyptians relate that the death of Osiris occurred on the seventeenth (of the month), when the full moon is most obviously waning. Therefore the Pythagoreans call this day the 'barricading' and they entirely abominate this number. For the number seventeen, intervening between the square number sixteen and the rectangular number eighteen, two numbers which alone of plane numbers have their perimeters equal to the areas enclosed by them (*proving this makes a nice problem for the reader with pencil and paper handy*), bars, discretes, and separates them from one another, being divided into unequal parts in the ratio of nine to eight. The number of twenty-eight years is said by some to have been the extent of the life of Osiris, by others of his reign; for such is the number of the moon's illuminations and in so many days does it revolve through its own cycle. When they cut the wood in the so-called burials of Osiris, they prepare a crescent-shaped chest because the moon, whenever it approaches the sun, becomes crescent-shaped and suffers eclipse. The dismemberment of Osiris into fourteen parts is interpreted in relation to the days in which the planet wanes after the full moon until a new moon occurs."

Nicomachus of Gerasa (c. 100 A.D.) was a later Pythagorean. (Gerasa is in Judaea east of the Jordan; Heath, 1981, p. 97). His *Introductio Arithmetica* came out in several versions. After a religio-philosophical introduction, it's a straightforward account of number theory as known to the Pythagoreans and to Euclid. He writes (p.187) "arithmetic . . . existed before all the others [of the quadrivium: music, geometry, and astronomy] in the mind of the creating God like some universal and exemplary plan, relying upon which as a design and archetypal example the creator of the universe sets in order his material creations and makes them attain to their proper ends" (p. 189). "All that has by nature with systematic method been arranged in the universe seems both in part and as a whole to have been determined and ordered in accordance with number, by the forethought and the mind of him that created all things; for the pattern was fixed, like a preliminary sketch, by the domination of number preexistent in the mind of the world-creating God, number conceptual only and immaterial in every way, but at the same time the true and the eternal essence, so that with reference to it, as to an artistic plan, should be created all these things time, motion, the heavens, the stars. . . ."

Heath thinks (1981, pp. 97–99) that the success of Nicomachus's book is "difficult to explain except on the hypothesis that it was at first read by philosophers rather than mathematicians, and afterward became generally popular at a time when there were no mathematicians left, but only philosophers who incidentally took an interest in mathematics." Van der Waerden, on the other hand, (p. 97) finds Nicomachus entertaining.

The Pythagoreans are not talked about in philosophy courses. Their thinking seems unworthy to be called philosophy. But Cornford says (p. 194): "Parmenides, the discoverer of logic, was an offshoot of Pythagoreanism, and Plato himself [found] in Pythagoreanism the chief source of his inspiration."

Bertrand Russell (1945) says, "Pythagoras was intellectually one of the most important men that ever lived. . . . The influence of mathematics on philosophy, partly owing to him, has, ever since his time, been both profound and unfortunate (p. 29). . . . He may be described, briefly, as a combination of Einstein and Mary Baker Eddy. He founded a religion, of which the main tenets were the transmigration of souls and the sinfulness of eating beans" (p. 31).

"The combination of mathematics and theology, which began with Pythagoras, characterized religious philosophy in Greece, in the Middle Ages, and in modern times down to Kant. [My addition: and also down to Frege and Russell] . . . in Plato, Saint Augustine, Thomas Aquinas, Descartes, Spinoza, and Kant there is an intimate blending of religion and reasoning, of moral aspiration with logical admiration of what is timeless, which comes from Pythagoras, and distinguishes the intellectualized theology of Europe from the more straightforward mysticism of Asia. . . . I do not know of any other man who has been as influential as he was in the sphere of thought . . . what appears as Platonism is . . . in essence Pythagoreanism. The whole conception of an eternal world, revealed to the intellect but not to the senses, is derived from him" (p. 37).

Plato (428-7–348-7 B.C.)

Pythagoras Is Refined—Numbers Live in Heaven

R. Tarnas in *Passion of the Western Mind*, p. 10 ff., writes, "The paradigmatic example of Ideas for Plato was mathematics. Following the Pythagoreans, with whose philosophy he seems to have been especially intimate, Plato understood the physical universe to be organized in accordance with the mathematical Ideas of number and geometry. These Ideas are invisible, apprehensible by intelligence only, and yet can be discovered to be the formative causes and regulators of all empirical visible objects and processes. But again, the Platonic and Pythagorean conception of mathematical ordering principles in nature was essentially different from the conventional modern view. In Plato's understanding, circles, triangles, and numbers are not merely formal or quantitative structures imposed by the human mind on natural phenomena, nor are they only mechanically present in phenomena as a brute fact of their concrete being. Rather, they are numinous and transcendent entities, existing independently of both the phenomena they order and the human mind that perceives them. While the concrete phenomena are transient and imperfect, the mathematical Ideas ordering those phenomena are perfect, eternal, and changeless. Hence the basic Platonic belief—that there exists a deeper, timeless order of absolutes behind the surface confusion and randomness of the temporal world—found in mathematics, it was thought, a particularly graphic demonstration. The training of the mind in mathematics was therefore deemed by Plato essential to the philosophical enterprise, and according to tradition, above the door to his Academy were placed the words, 'Let no one unacquainted with geometry enter here.'

"It had become evident that several celestial bodies did not move with the same eternal regularity as did the rest, but instead they "wandered" (the Greek root for the word "planet," *planetes*, means "wanderer" and signified the Sun and Moon as well as the other five visible planets—Mercury, Venus, Mars, Jupiter, and Saturn.) Not only did the Sun (in the course of a year) and the Moon (in a month) move gradually eastward across the starry sphere in an opposite direction from the westward diurnal movement of the entire heavens. More puzzling, the other five planets had glaringly inconsistent cycles in which they complete those eastward orbits, periodically appearing to speed up or slow down relative to the fixed stars, and sometimes to stop altogether and reverse direction while emitting varying degrees of brightness. The planets were inexplicably defying the perfect symmetry and circular uniformity of the heavenly motions.

"Because of his equation of divinity with order, or intelligence and soul with perfect mathematical regularity, the paradox of the planetary movements seems to have been felt most acutely by Plato, who first articulated the problem and gave directions for its solution.

"To Plato the proof of divinity in the universe was of the utmost importance for only with such certainty could human ethical and political activity have a firm foundation. In the *Laws*, he cited two reasons for belief in divinity—his theory of the soul (that all being and motion is caused by soul, which is immortal and superior to the physical things it animates) and his conception of the heavens as divine bodies governed by a supreme intelligence and world soul. The planetary irregularities and multiple wanderings seemingly contradicted that perfect divine order, thereby endangering human faith in the divinity of the universe. Therein lay the significance of the problem. Part of the religious bulwark of Platonic philosophy was at stake. Indeed, Plato considered it blasphemous to call any celestial bodies 'wanderers.'

"But Plato not only isolated the problem and defined its significance. He also advanced, with remarkable confidence, a specific—and in the long run extremely fruitful—hypothesis: namely that the planets, in apparent contradiction to the empirical evidence, actually move in single uniform orbits of perfect regularity. Although there would seem to have been little but Plato's faith in mathematics and the heavens' divinity that could have supported such a belief, he enjoined future philosophers to grapple with the planetary data and find 'what are the uniform and ordered movements by the assumption of which the apparent movements of the planets can be accounted for'—to discover the ideal mathematical forms that would resolve the empirical discrepancies and reveal the true motions" (Heath, 1913; and Plato's *Laws*, 1961, pp. 821–22).

Unlike Pythagoras, Plato left us plenty of written records. As a youth, he studied with Socrates, who was convicted in King Archon's court for impiety and corrupting young men. He was executed by drinking hemlock in 399 B.C., when Plato was about 31 (Ryle). We know little of Socrates's life. Plato made him the central figure in most of his *Dialogues*. We can only guess which of Socrates's speeches are Socrates's thoughts, which the thoughts of Plato.

The *Dialogues* aren't systematic philosophical expositions. Nowhere is there a coherent statement of Plato's philosophy of mathematics, which we could compare with "Platonism" today. They achieve trenchant philosophical analysis by lively conversation among living, breathing Athenians. Plato's views changed over the years when the *Dialogues* were written. In midlife, on a trip to Syracuse, he became a Pythagorean. It's said that unknown to the young Plato Socrates too had belonged to a Pythagorean cell. The Pythagoreans were a political as well as a religio-philosophical group, and there was reason for secrecy.

Plato's *Republic* is a manifesto for dictatorship by philosopher-kings, trained under Plato's principles.

It contains the famous metaphor of the cave. Here is the end of Socrates's concluding speech about the cave:

"What appears to me is, that in the world of the known last of all is the idea of the good, and with what toil to be seen! And seen, this must be inferred to be the cause of all right and beautiful things for all, which gives birth to light and the king of light in the world of sight, and, in the world of mind, herself the queen produces truth and reason; and she must be seen by one who is to act with reason publicly or privately."

Here are rules for educating "guardians" or philosopher-kings, from Book 7 of *The Republic* ("Great Dialogues of Plato," pp. 315–16, 323–31). Plato gave Socrates the first person, "I."

> My dear Glaucon, what study could draw the soul from the world of becoming to the world of being? . . . this, which they all have in common, which is used in addition by all arts and all sciences and ways of thinking, which is one of the first things every man must learn of necessity."
>
> "What's that?" he asked again.
>
> "Just this trifle, I said—to distinguish between one and two and three: I mean, in short, number and calculation . . ."

"Number, then, appears to lead towards the truth?"

"That is abundantly clear."

"Then, as it seems, this would be one of the studies we seek; for this is necessary for the soldier to learn because of arranging his troops, and for the philosopher, because he must rise up out of the world of becoming and lay hold of real being or he will never become a reckoner."

"That is true," said he.

"Again, our guardian is really both soldier and philosopher."

"Certainly."

"Then, my dear Glaucon, it is proper to lay down that study by law, and to persuade those who are to share in the highest things in the city to go for and tackle the art of calculation, and not as amateurs; they must keep hold of it until they are led to contemplate the very nature of numbers by thought alone, practicing it not for the purpose of buying and selling like merchants or hucksters, but for war, and for the soul itself, to make easier the change from the world of becoming to real being and truth."

"Excellently said," he answered.

"And besides," I said, "it comes into my mind, now the study of calculations has been mentioned, how refined that is and useful to us in many ways for what we want, if it is followed for the sake of knowledge and not for chaffering."

"How so?" he asked.

"In this way, as we said just now; how it leads the soul forcibly into some upper region and compels it to debate about numbers in themselves; it nowhere accepts any account of numbers as having tacked onto them bodies which can be seen or touched. . . . I think they are speaking of what can only be conceived in the mind, which it is impossible to deal with in any other way."

"You see then, my friend, said I, that really this seems to be the study we need, since it clearly compels the soul to use pure reason in order to find out the truth."

"So it most certainly does. . . .

"For all these reasons, the best natures must be trained in it."

After arithmetic, Glaucon and Socrates consider geometry.

Says Socrates, "The knowledge the geometricians seek is not knowledge of something which comes into being and passes, but knowledge of what always is."

"Agreed with all my heart, said he, for geometrical knowledge is of that which always is."

> "A generous admission! Then it would attract the soul toward truth, and work out the philosopher's mind so as to direct upwards what we now improperly keep downwards."
>
> After arithmetic and plane geometry, Glaucon proposes astronomy as the third subject in the curriculum of the Guardians. Socrates objects; solid geometry is more appropriate, he says.
>
> "Quite so," says Glaucon, "but it seems that those problems have not yet been solved."
>
> "For two reasons," I said, "because no city holds them in honour, they are weakly pursued, being difficult. Again, the seekers lack a guide, without whom they could not discover; it is hard to find one in the first place, and if they could, as things now are, the seekers in these matters would be too conceited to obey him. But if any whole city should hold these things honourable and take a united lead and supervise, they would obey, and solutions sought constantly and earnestly would become clear. Indeed even now, although dishonoured by the multitude, and held back by the seekers themselves having no conception of the objects for which they are useful, these things do nevertheless force on and grow against all this by their own charm, and I should not be too surprised if they should really come to light. . . .
>
> "Let us put astronomy as the fourth study, assuming that solid geometry, which we leave aside now, is there for us if only the city would support it."

Plato didn't have a "philosophy of mathematics" as we understand that phrase today. Mathematics is central in his philosophy. His believes the physical world of visible, changeable entities is illusion. What's real is invisible, immaterial, eternal. Mathematics is real *because* it's immaterial and eternal. It's tied to religion, as a stepping stone in one's ascent toward "the good," the loftiest aspect of invisible reality. A challenge to his notion of mathematics would be a challenge to his religion.

In a famous mathematical episode in Plato's *Meno*, Socrates is working, as usual, on the problem of virtue.

What is virtue?

How can we know it?

Examples show it's not learned by observation or from teachers. Socrates makes an analogy between virtue and mathematics. To make his point about the nature of mathematics, a "slave boy" is brought in. (His name is never mentioned.) Socrates makes the slave boy draw a small square inscribed at the midpoints of the sides of a larger square. What's the area of the small square? Socrates wants to make the slave boy *remember* the answer.

Socrates (to Meno): I shall do nothing more than ask questions and not teach him. Watch whether you find me teaching and explaining things to him instead of asking for his opinion.

(to the slave boy): You tell me, is this not a four-foot figure? You understand?

Slave boy: I do.

Socrates: We add to it this figure which is equal to it?

Slave boy: Yes.

Socrates: And we add this third figure equal to each of them?

Slave boy: Yes.

Socrates: Could we then fill in the space in the corner?

Slave boy: Certainly.

Socrates: So we have these four equal figures?

Slave boy: Yes.

Socrates: Well then, how many times is the whole figure larger than this one?

Slave boy: Four times.

By more questioning Socrates gets the slave boy to correct his mistake, and to say that the small square has area, not one fourth, but one half that of the big square.

Socrates: What do you think, Meno? Has he, in his answers, expressed any opinion that was not his own?

Meno: No, they were all his own . . .

Socrates: And he will know it without having been taught but only questioned, and find the knowledge within himself.

Meno: Yes

Socrates: And is not finding knowledge within oneself recollection?

Meno: Certainly.

Socrates: Must he not either have at some time acquired the knowledge he now possesses, or else have always possessed it?

Meno: Yes.

Socrates: If he always had it, he would always have known. If he acquired it, he cannot have done so in his present life. Or has someone taught him geometry? . . . You should know, especially as he has been born and brought up in your house.

Meno: But I know that no one has taught him . . .

Socrates: If he has not acquired them in his present life, is it not clear that he had them and had learned them at some other time?

Meno: It seems so.

Socrates: Then that was the time he was not a human being?

Meno: Yes.

You expect Socrates to argue next that knowledge of virtue, like knowledge of geometry, is remembered from Heaven. But he says something a little different:

"Virtue appears to be present in those of us who may possess it as a gift from the gods."

Bertrand Russell (1945) wrote, "When . . . in the *Meno*, [Socrates] applies his method to geometrical problems, he has to ask leading questions which any judge would disallow" (*A History of Western Philosophy*, p. 92).

The *Meno* shows that Plato understood geometrical reasoning. As proof that we remember geometry from before birth, it's a hoax. It is one of the most popular of Plato's *Dialogs*. But its popularity is from literary merits, not philosophical ones.

"Tradition is the forgetting of the origins"—Edmund Husserl, quoted by Philip J. Davis.

Once, long ago, men and women first captured fire, from a grass fire started by lightning. Lightning was the thunderbolt of the gods, so it seemed they had stolen fire from the gods.

Millennia passed. The true origin of fire was forgotten. In its place, an origin myth was created.

"The demigod Prometheus loved mankind. He stole fire from the gods, and gave it to us. In revenge, Jupiter chained him to a rock, where an eagle eternally plucks at his liver."

Number, like fire, can seem a divine gift to mankind. We forgot how we captured fire, and we forgot how we invented number. The *Meno* is an origin myth for numbers, a variant on the Prometheus myth for the origin of fire.

Modern Platonism is a descendant of Plato's Platonism. It too is a kind of origin myth.

An even more influential dialogue of Plato was the *Timaeus.* It's a creation story, like Genesis in the *Old Testament*—a myth of how a Demiurge or Divine Architect created the universe. In Cornford's book, *Plato's Cosmology*, he explains how in this dialogue Plato makes the "series" 1, 2, 4, 8 and 1, 3, 9, 27 "the basis for the harmony of the world-soul" ("harmony" in both musical and spiritual senses). This is Pythagorean thinking. In the Middle Ages *Timaeus* was the most popular Platonic dialogue, judging by copies found in old abbeys and monasteries. It transmitted Pythagorean numerology to the Middle Ages, and thence to modern times.

There's a well-known puzzle about Plato's teaching. The reports by Aristotle and other followers of Plato's teaching don't match what we read in Plato's

Dialogues. Yet "Aristotle, who was 37 when Plato died, had belonged to Plato's Academy for the last twenty years of Plato's life" (Ryle). It is believed that there was an Unwritten Doctrine, presented by Plato only in speech, perhaps because it never attained a satisfactory final form, perhaps because of cultic secrecy. In the Unwritten Doctrine, numbers are not merely examples or samples of the Ideas, the unmaterial eternal realities. *All* the Ideas are numbers! Perhaps just the first four numbers (1, 2, 3, 4), or perhaps even just the first two numbers (1, 2). If this seems inconceivable, remember that in twentieth-century set-theoretic Platonism, the set-theoretic superstructure, which contains all present, past and future mathematics, is generated *from the empty set.***

The Unwritten Doctrine has two kinds of numbers, Ideals and Mathematicals. The Ideals are what we understand from *The Republic*—there's an Ideal 5, an Ideal 6, an Ideal 7, 8, and 9. The Mathematicals are "lower" than the Ideals. In the equation,

$$5 + 5 = 10,$$

we see on the left of the equal sign *two* 5s. But there is only *one* Ideal 5, so what can these *two* 5s be?

When we notice that there's only one Ideal Horse but many material horses, we say the material horses are physical manifestations or exemplifications of the Ideal Horse. But the 5s in our equation are not material quintuples, they are ideal or abstract 5s, yet there are two of them! To escape this dilemma, Plato invented "Mathematicals." They are immaterial entities below the Ideal. This is an early example of the Philosophers' Rule: "When stopped by a contradiction, invent a distinction."

These matters are discussed by Wedberg, Findlay and Dillon.

Neoplatonists—Still in Heaven

The Neoplatonists, mystics in the Hellenistic world of the late Roman Empire, are interesting mathematically as a link between Plato-Pythagoras and Saint Augustine. I quote a few secondary sources, to give the flavor. From Saunders, 218 ff., Philo Judaeus of Alexandria (20? B.C.–54? A.D.):

"I doubt whether anyone could adequately celebrate the properties of the number 7, for they are beyond all words. . . . So August is the dignity inherent by nature in the number 7, that it has a unique relation distinguishing it from all the other numbers, for of these some beget without being begotten, some are begotten but do not beget, some do both these, both beget and are begotten: 7 alone is found in no such category. We must establish this assertion by giving proof of it."

(He means, if n is between 1 and 10, and neither n/2 nor 2n is between 1 and 10, then n = 7. *What about n = 9??* He must mean, if no multiple of n is between 1 and 10, and no factor of n is between 1 and 10, then n = 7. Which is correct.)

From Pistorius, p. 160, Plotinus:

"The only practical direction to the philosopher [given by Plotinus (204–270)] is a course in mathematics to train him in abstract thought, and in a faith in the unembodied."

From Cornford, p. 204 ff.: "The whole nature of things, all the essential properties of physis, were believed by the Pythagoreans to be contained in the tetractys of the decad . . . the Pythagorean One, or Monad, splits into two principles, male and female, the Even and the Odd, which are the elements of all numbers and so of the universe. The analogy reminds us that the One is not simply a numerical unit, which gives rise to other numbers by a process of addition. . . . We must think of the One (which is not itself a number at all) as the primary, undifferentiated group-soul, or physis, of the universe, and numbers must arise from it by a process of differentiation or separating out. . . . Similarly, each of these numbers is not a collection of units, built up by addition, but itself a sort of minor group-soul—a distinct 'nature', with various mystical properties."

Aurelius Augustinus, Bishop of Hippo Regius (354–430)

Pythagoras Is Christianized

In St. Augustine, neo-Platonist philosophy is wedded to Catholic theology. His life is told in his eloquent *Confessions.* From the age of 17 until 32 he lived with a dearly beloved concubine. Then he was converted to Christianity by his saintly mother. He wrote: "My concubine was torn from my side as a hindrance to my marriage; my heart which clave to her was torn and bleeding. And she returned to Africa, vowing unto Thee never to know any other man, leaving with me my son by her."

Thereafter he devoted himself to God. His obsession with human depravity, and his insistence that salvation comes only by Divine Grace, make him the forerunner of Calvinists and Jansenists a thousand years later.

One of his sentences is alarming: "Those impostors then, whom they style Mathematicians, I consulted without scruple; because they seemed to use no sacrifice, nor to pray to any spirit for their divinations; which art, however, Christian and true piety consistently rejects and condemns" (*Confessions,* p. 50). This quote is sometimes misused by failing to explain that by "mathematicians" Augustine meant "astrologers." In fact, he was deeply interested in arithmetic. Chapter VIII of *On Free Choice of the Will* (Mourant, p. 89) is headed: "The reason of numbers is perceived by no bodily sense; by anyone who understands it, it is perceived as one and immutable." In following paragraphs we find these arguments:

"Seven and three are ten, not only now but always; nor was there ever a time when seven and three were not ten, nor will ever be a time when seven and three will not be ten. I say, therefore, that this incorruptible truth of number is common to me and to any reasoning person whatsoever. . . .

"We cannot perceive the number one by our bodily senses. For whatever reaches us through such a sense is clearly seen to be not one but many, for it is body, and therefore has innumerable parts. . . . Moreover, if we do not perceive one by a sense of the body, we do not perceive any number by that sense . . . for there is no one of them that does not get its name from the number of times it contains one; and one is not perceived by the bodily sense. . . . The *number* that we call two because it contains twice that which is simply one, its half part, is something which is itself simply one, and cannot have a half or a third or any part whatsoever, because it is simply and truly one. . . .

"After one we get two, which comparing with one we find to be its double; the double of two does not follow in immediate succession, but four, which is the double of two, follows after the interposition of three. And this rule extends to all the other numbers by a most certain and immutable law . . . that the double of any number is just as many numbers after it as that number itself is from the very beginning. [*Today we say, for all* x, $2x - x = x - 0$.]

"But now, when we perceive this thing to be for all numbers fixed and inviolate, whence comes this perception? For no one has touched all numbers by any sense of the body, for they are innumerable. Whence then do we know this for all . . . this truth of number throughout things innumerable, if we do not perceive it by that inner light which the sense of the body knows not? By this and many like proofs, those to whom God has given an inquiring mind and who are not blinded by obstinacy are compelled to acknowledge that the reason and truth of numbers is not related to the bodily senses. . . . For not for nothing is number joined to wisdom in the Sacred Books, where it is said: "I and my heart have gone round, that I might know, and consider, and inquire the wisdom and the number" (Ecclesiastes, 7:25).

From Chapter 11, pp. 97–99: "I would much like to know whether these two, wisdom and number, are contained in some one genus. . . . For I should not have ventured to say that wisdom arises from number or consists in number, for wisdom strikes me as far more venerable than number, because I have known many calculators or computers, or whatever else you may call them, who reckon superbly and marvelously—but of wise men only a very few or possibly none. . . .

"When I reflect upon the immutable truth of numbers and upon its lair, as it were, or shrine or region or whatever else we may appropriately call the dwelling place and seat as it were of numbers, I am far removed from the body; and finding maybe something which I can think, but not finding anything which I can put into words, I return as if wearied to this world of ours. . . .

"Even if it cannot be clear to us whether number is in wisdom or from wisdom, or wisdom itself is in number or from number, or whether it can be shown that they are names for one thing; it is certainly manifest that both are true and true immutably."

Instead of this elevated Platonism, his *On the Trinity* contains naked Pythagorean number worship. Chapter IV is headed, "The ratio of the single to the double comes from the perfection of the senary number. The perfection of the senary number is commended in the scriptures. The year abounds in senary numbers."

The "senary number" is 6. In Pythagorean arithmetic, "perfect" means equal to the sum of the factors. Six is perfect, because $1 + 2 + 3 = 6$. Perfect numbers are scarce, and they are still a topic in recreational number theory. To Augustine, "perfect" has religious as well as arithmetical meaning.

Chapter V continues his praise of 6. It is headed, "The number six is also commended in the building up of the body of Christ and of the temple in Jerusalem."

It seems Solomon's temple was forty-six years in building. "And six times forty-six makes two hundred and seventy-six. And this number of days completes nine months and six days . . . the perfection of the body of the Lord is found to have been brought in so many days to the birth. . . . For He is believed to have been conceived on the 25th of March. . . . But he was born, according to tradition, upon December the 25th. If then, you reckon from that day to this you find two hundred and seventy-six days, which is forty-six times six. And in this number of years the temple was built, because in that number of sixes the body of the Lord was perfect, which being destroyed by the suffering of death, He raised again on the third day . . . 'the Son of man be three days and three nights in the heart of the earth.' From the evening of the burial to the dawn of the resurrection are thirty-six hours which is six squared . . . there is no one surely so foolish or so absurd as to contend that [these numbers] are so put in the Scriptures for no purpose at all, and that there are no mystical reasons why those numbers are there mentioned."

Augustine and Plato share the vision of mathematics as an aspect of the Divine. Said Augustine, "Among the disciples of Socrates, Plato was the one who shone with a glory which far excelled that of the others. . . . Why discuss with the other philosophers? It is evident that none come nearer to us than the Platonists" (1950, p. 23).

An impressive chapter in *On the Trinity* (Chapter XII, Book XV, pp. 849–50) uses a recursion argument that is unmistakably mathematical. It could have inspired Descartes's immortal one-liner, "I think, therefore I am."

"The knowledge . . . that we live is the most inward of all knowledge, of which even the Academic cannot insinuate: perhaps you are asleep, and do not know it, and you see things in your sleep . . . [Where Augustine says 'Academic' we say 'skeptic.'] he who is certain of the knowledge of his own life, does not say, I know I am awake, but, I know I am alive; therefore, whether he be asleep or awake, he is alive. . . . Nor can the Academic again say, in confutation of knowledge: perhaps you are mad, and do not know it: for what madmen see is precisely like what they also see who are sane; but he who is mad is alive. Nor does he answer the Academic by saying, I know I am not mad, but, I know I am alive. Therefore he who says he knows he is alive, can neither be deceived nor lie. . . . For he who says, I know I am alive, says that he knows one single thing. Further, if he says, I know that I know I am alive, now there are two; but that he knows these two is a third thing to know. And so he can add a fourth, and a fifth, and innumerable others, if he holds out. But since he cannot either comprehend an innumerable number by additions of units, or say a thing innumerable times, he comprehends this at least, and with perfect certainty, viz. that this is both true and so innumerable that he cannot truly comprehend and say its infinite number."

Emperor, Pope, and Magic Number

Arithmetic Is Sanctified

To the medieval mind, a sacred number was part of the divine spiritual order.

In 1240, western Europe was disturbed by rumors of a great king in the Far East who was making his way relentlessly westward. One Islamic kingdom after another had fallen to him. Some thought the news meant the coming of the legendary Prester John, who would join the Christian kings and destroy Islam. European Jews thought the Eastern monarch was David's son King Messiah, and prepared to meet him joyously. Actually, the eastern king was Batu, son of Genghis Khan.

Why was it thought that the Messiah was arriving? The year 1240 A.D. corresponded to the Jewish year 5000. Some thought the Messiah should come at year 5000.

One type of number mysticism is "gematria." It associates a number with a word by adding up the numbers of its letters. The name "Innocentius Papa" (Pope Innocent IV) has the number 666. This is the Number of the Beast in Revelations 13:18. Hence Pope Innocent is Antichrist!

Thomas Aquinas (1225?—1274)

Aristotle Is Christianized

St. Thomas is the principal author of Christian theology and philosophy. He led in incorporating Aristotle into Church doctrine. Mathematics is one of the few major topics on which he didn't write at length. It's hard to find any but passing reference to mathematics in his voluminous works or those of his commentators. In *The Pocket Aquinas* I found some tantalizing crumbs.

The Exposition of Boethius on the Trinity, Book 5, Chapter 1, has a passage classifying the sciences. To distinguish mathematics from physics and natural science, which deal with material objects, and theology or metaphysics, which deal with immaterial objects, Aquinas explains, "There are some things which, though they depend on matter according to their act of being, do not do so when they are understood, because sensible matter is not included within their definitions: for instance, lines and numbers. Now, *mathematics* deals with these objects."

The same work, Book 6, Chapter 1: "In mathematical science, the process [of demonstrative reasoning] works through those items that pertain merely to the essence of a thing, since these sciences demonstrate only by means of the formal cause. So, in them something is not demonstrated about one thing by means of another thing, but by the proper definition of that thing. For, although some demonstrations are given concerning the circle from the triangle, or conversely, this is done only because the triangle is potentially in the circle, or the converse . . . to proceed according to a learning method is characteristic of mathematical science; not that it alone progresses by learning but because this is especially appropriate to it. Indeed, since to learn simply means to get scientific knowledge from another, we are said to use the method of learning when our procedure leads to such knowledge which is called science. Now this happens chiefly in the mathematical sciences.

"Since mathematics is intermediate between natural and divine science, it is more certain than either of them; more so than the natural because its thinking is cut off from motion and matter; while the thinking of the natural scientist is directed to matter and motion. . . .

"Moreover, the procedure of mathematics is more certain than that of divine science because the things that divine science studies are more removed from the objects of sensation from which our knowledge takes its start . . . the objects of mathematics do fall with sense experience and are subjects for the imagination; for instance, figures, lines, numbers, and the like. Thus, human understanding grasps the knowledge of these items from the phantasms more easily and more certainly than the knowledge about an intelligence, or even than about a substantial essence, act, potency, and other similar items. So, it is clear that mathematical thinking is easier and more certain than physical or theological, and much more so than that of the other operative sciences."

From the *Exposition of the Book of Causes*, Lecture 1: "[The philosophers] put the science of the first causes last in the order of learning, and assigned its consideration to the last period in a person's life. They began first with logic, for it teaches the method of the sciences. Then they went to mathematics, which even boys can grasp." Here Aquinas is misrepresenting Plato, who explicitly included women in his school for guardians.

From the *Exposition of Aristotle's Physics*, Book 1: "A curve, though it cannot exist except in sensible matter, does not, however, include sensible matter in its definition. This is the way that all mathematicals are, for instance, number, size, and figure. . . . Mathematics is concerned with these things which depend on sensible matter for their actual being but not for their rational meaning. The natural science that is called physics is concerned with those things which depend on matter not only for actual being but also for their rational meaning."

Thinking of Aquinas as a Catholic philosopher, and seeing him place mathematics between physics and theology, it may seem he is a Platonist. But his analysis of mathematics is Aristotelian, not Platonic. Mathematics studies real objects, defined to exclude their materiality. This is a restatement of Aristotle's analysis of mathematical objects as abstractions from real objects. Aristotle says: to get a mathematical object, ignore all qualities of the real object except mathematical qualities. Aquinas says: define the mathematical object to be free of the material properties of the real object.

I'm tempted to move Aquinas over to the Humanists and Mavericks, but if anybody is Mainstream, he is. The classification between mainstream philosophers and humanist or maverick philosophers (like any classification) is not always à propos.

Cardinal Nicholas Krebs of Cusa (1401–1464)

Infinity in Theology

In the *Encyclopedia of Philosophy* Nicholas Krebs is called "theologian, philosopher, mathematician." In addition to many religious-political and theological-philosophical works, Maurer lists four mathematical-scientific books: *De Staticis Experimentis* (1450), *De Transmutationibus Geometricis* (1450), *De Mathematicis Complementis* (1453), and *De Mathematica Perfectione* (1458). Jasper says Cusa produced eleven mathematical writings from 1445 to 1459. Among his subjects: "the quadrature of the circle, the reform of the calendar, the improvement of the Alfonsine Tables (planetary tables which improved those left by Ptolemy), the heliocentric theory of the universe (he looked on it as a paradox, not a scientific probability) and the theory of numbers. Wallis said Nicholas was the first writer known to have worked on the cycloid, but this is not supported by the evidence" (Davis).

The son of a fisherman, Nicholas rose to become a diplomat and counselor for the Church. "He was a member of the commission sent to Constantinople to negotiate with the Eastern church for reunion with Rome, which was temporarily effected at the Council of Florence (1439)." In 1448 he became cardinal and governor of Rome.

Cusa was not a philosopher of mathematics. He was a philosopher whose thinking was imbued with mathematical images, so that he used mathematics to teach theology. He knew that there are different degrees of infinity. He said, amazingly, that the physical universe is finite but unbounded. He showed that a geometric figure can be both a maximum and a minimum, depending on how it's parametrized.

Again from the *Encyclopedia*, "According to Cusa, a man is wise only if he is aware of the limits of the mind in knowing the truth. . . . Knowledge is learned ignorance (*docta ignorantia*). Endowed with a natural desire for truth, man seeks it through rational inquiry, which is a movement of the reason from something presupposed as certain to a conclusion that is still in doubt. . . . As a polygon inscribed in a circle increases in number of sides but never becomes a circle, so the mind approximates to truth but never coincides with it. . . . Thus knowledge at best is conjecture (*coniectura*)."

Cusa was a Platonist at a time when Aristotelians were dominant. "He constantly criticized the Aristotelians for insisting on the principle of noncontradiction and stubbornly refusing to admit the compatibility of contradictories in reality. It takes almost a miracle, he complained, to get them to admit this; and yet without this admission the ascent of mystical theology is impossible. . . . He constantly strove to see unity and simplicity where the Aristotelians could see only plurality and contradiction.

"Cusa was most concerned with showing the coincidence of opposites in God. God is the absolute maximum or infinite being, in the sense that he has the fullness of perfection. There is nothing outside him to oppose him or to limit him. He is the all. He is also the maximum, but not in the sense of the supreme degree in a series. As infinite being he does not enter into relation or proportion with finite beings. As the absolute, he excludes all degrees. If we say he is the maximum, we can also say he is the minimum. He is at once all extremes. . . . The coincidence of the maximum and minimum in infinity is illustrated by mathematical figures. For example, imagine a circle with a finite diameter. As the size of the circle is increased, the curvature of the circumference decreases. When the diameter is infinite, the circumference is an absolutely straight line. Thus, in infinity the maximum of straightness is identical with the minimum of curvature. . . .

"Cusa denied that the universe is positively infinite; only God, in his view, could be described in these terms. But he asserted that the universe has no circumference, and consequently that it is boundless or undetermined—a revolutionary notion in cosmology. . . . Just as the universe has no circumference, said Cusa, so it has no fixed center. The earth is not at the center of the universe, nor is it absolutely at rest. Like everything else, it moves in space with a motion that is not absolute but is relative to the observer. . . .

"Beneath the oppositions and contradictions of Christianity and other religions, he believed there is a fundamental unity and harmony, which, when it is recognized by all men, will be the basis of universal peace."

To these quotes from the *Encyclopedia of Philosophy*, I adjoin an excerpt from Cusa's *De Mente*. This is a dialogue between Layman and Philosopher. Layman is speaking, p. 59: "Just as from God's viewpoint the plurality of things comes from the divine mind, so from our perspective the plurality of things proceeds from our mind. Mind alone counts; if mind is removed, distinct numbers do not exist. Because mind grasps one and the same thing individually and in signs and we consider this itself (for we say something is one from the fact that mind understands a single thing once and individually), mind is truly the equality of unity. But when mind grasps a single thing individually and by multiplying, we judge that there are many things by speaking of a two because the mind grasps singly one and the same thing twice or by doubling it; and so on for the rest. . . . The plurality of things arises from this, that the mind of God understands one thing in a certain way and a second in another way. If you attend sharply you will discover that the plurality of things is no more than the way the divine mind understands. I conjecture that one can say without blame that the first exemplar of things in the mind of their maker is number. The delight and beauty inherent in all things demonstrates this, for it consists in proportion, and proportion in number. So number is the principal clue which leads to wisdom."

Philosopher. First the Pythagoreans, then the Platonists said that, and Severinus Boethius followed them.

Layman. In like manner I say number is the exemplar of our mental concepts. Mind can do nothing without number.

In the *Monadology* Leibniz mentions Cusa as one of his sources.

René Descartes (1596–1650)

Skepticism Is Refuted—Geometry Is Born Again

Philosophy students are supposed to read Descartes's *Discourse on Method* (1637). They don't realize that the complete *Discourse* includes Descartes' mathematical masterpiece, the *Geometry*. The full title is *Discourse on Method, Optics, Geometry, and Meteorology.*

On the other hand, mathematics students also are miseducated. They're supposed to know that Descartes was a founder of analytic geometry, but not that his *Geometry* was part of a great work on philosophy.

Descartes's Method consists of four simple rules:

> The first of these was to accept nothing as true which I did not clearly recognize to be so: that is to say, carefully to avoid precipitation and prejudice in judgments, and to accept in them nothing more than what was presented to my mind so clearly and distinctly that I could have no occasion to doubt it.
>
> The second was to divide up each of the difficulties which I examined into as many parts as possible, and as seemed requisite in order that it might be resolved in the best manner possible.
>
> The third was to carry on my reflections in due order, commencing with objects that were the most simple and easy to understand, in order to rise little by

> little, or by degrees, to knowledge of the most complex, assuming an order, even if a fictitious one, among those which do not follow a natural sequence relatively to one another.
>
> The last was in all cases to make enumerations so complete and reviews so general that I should be certain of having omitted nothing" (Vol. I, p. 92, of Haldane and Ross).

Leibniz sneered "famously" that these "rules" amount to little more than "Take what you need, do what you should, and you will get what you want." But what's left out of the rules is more important than what is stated. By omission, Descartes advises the investigator to respect no authorities, neither Aristotle nor Church Fathers nor Holy Scripture! His Method proclaims individual autonomy in the search for truth.

D'Alembert wrote that it was Descartes who first "dared . . . to show intelligent minds how to throw off the yoke of scholasticism, of opinion, of authority—in a word, of prejudices and barbarism. . . . He can be thought of as a leader of conspirators who, before anyone else, had the courage to arise against a despotic and arbitrary power, and who, in preparing a resounding revolution, laid the foundations of a more just and happier government which he himself was not able to see established."

Philosophers of the scholastic persuasion pointed to the dangerous parallel between Descartes's scientific individualism and the outlawed Protestant heresy. Descartes said individual thinkers could find scientific truth; Protestants said individual souls could find direct communion with the Almighty. But the Holy Roman Catholic Church knew that individual souls and thinkers could be deceived. It took the experience and wisdom of the Church to prevent the seeker from wandering astray. Despite such scholastic criticism, Descartes quickly came to dominate West European intellectual life.

For Descartes, mathematics was the central subject. "I was delighted with Mathematics because of the certainty of its demonstrations and the evidence of its reasoning. . . . I was astonished that, seeing how firm and solid was its basis, no loftier edifice had been reared thereupon" (Vol. I, p. 85).

He wrote (p. 7), "Arithmetic and geometry alone deal with an object so pure and uncomplicated, that they need make no assumptions at all which experience renders uncertain, but wholly consist in the rational deduction of consequences. They are on that account much the easiest and clearest of all . . . in them it is scarce humanly possible for anyone to err except by inadvertence . . . in our search for the direct road towards truth we should busy ourselves with no object about which we cannot attain a certitude equal to that of the demonstrations of Arithmetic and Geometry" (Vol. I, p. 3). Isaac Beeckman visited Descartes in 1628, and wrote: "He told me that insofar as arithmetic and geometry were concerned, he had nothing more to discover, for in these branches during the past nine years he had made as much progress as was possible for the human mind. He gave me decisive proofs of this affirmation and promised to send me shortly his Algebra, which he said was finished and by which not only had he arrived at a perfect knowledge of geometry but also he claimed to embrace the whole of human thought" (Vrooman, p. 78).

This lofty goal is hardly to be attained by one mind! Descartes's mathematical tools were certainly insufficient for this grandiose purpose. Like Galileo, Descartes recognized mathematics as the principal tool for revealing truths of nature. He was more explicit than Galileo about how to do it. In every scientific problem, said Descartes, find an algebraic equation relating an unknown variable to a known one. Then solve the algebraic equation! With the development of calculus, Descartes's doctrine was essentially justified. Today we don't say "find an algebraic equation." We say "construct a mathematical model." This is only a technical generalization of Descartes's idea. Our scientific technology is an inheritance from Descartes.

In *Rules for the Direction of the Mind*, Descartes wrote: "The first principles themselves are given by intuition alone, while, on the contrary, the remote conclusions are furnished only by deduction. . . . These two methods are the most certain routes to knowledge, and the mind should admit no others. All the rest should be rejected as suspect of error and dangerous."

Descartes was embracing the Euclidean ideal: Start from self-evident axioms, proceed by infallible deductions. But in his own research, Descartes forgot the Euclidean ideal. Nowhere in the *Geometry* do we find the label Axiom, Theorem, or Proof.

In classical Greece, and again in the Renaissance and after, mathematicians distinguished two ways of proceeding—the "synthetic" and the "analytic." The synthetic way was Euclid's: from axioms through deductions to theorems. In the analytic mode, you start with a problem and "analyze" it to find a solution. Today we might call this a "heuristic" or "problem-solving" approach.

In formal presentation of academic mathematics, the synthetic was and still is the norm. Foundationist schools of the nineteenth and twentieth centuries identify mathematics with its synthetic mode—true axioms followed by correct deductions to yield guaranteed true conclusions.

In his *Rules for the Direction of the Mind*, Descartes insists on the synthetic method. But his own research, in the *Geometry*, uses *only* the analytic mode. He solves problems. He finds efficient methods for solving problems. Never does he bother with axioms.

Descartes's conviction of the certainty of mathematics might lead readers to expect that at least Descartes's own mathematics is error-free. But of course, as we will see, the *Geometry*, like every other math book, has mistakes. Certitude is only a goal.

In the *Geometry* Descartes makes no pretense of unquestionable axioms or irrefutable reasoning. He follows the heuristic, pragmatic style normal in mathematical research. This starkly contradicts his *Method*.

Replying to criticism Descartes wrote, "Analysis shows the true way by which a thing was methodically discovered and derived, as it were effect from cause. . . . Synthesis contrariwise employs an opposite procedure. . . . It was this synthesis alone that the ancient Geometers employed in their writings, not because they were wholly ignorant of the analytic method, but, in my opinion, because they set so high a value on it that they wished to keep it to themselves as an important secret" (Vol. II, pp. 48–49).

He admits that some of his arguments are deficient: "I have not yet said on the basis of what reasons I venture to assure you that a thing is or is not possible. But if you take note that, with the method I use, everything falling under the geometer's consideration can be reduced to a single class of problem—namely, that of looking for the value of the roots of a certain equation—you will clearly see that it is not difficult to enumerate all the ways through which they can be found, which is sufficient to prove that the simplest and the most general one has been chosen" (Vol. I, Book 3, p. 251).

It is strange that in the vast body of writing about Descartes accumulated in three centuries, almost no one seems to have called attention to this bizarre misfit—Euclidean certainty boldly advertised in the *Method* and shamelessly ditched in the *Geometry*. The Dutch mathematician Willem Kuyk is the only author I know who has noticed this remarkable fact. Maybe philosophers don't read the *Geometry*, and mathematicians don't read the *Method*.

The *Method* is an essential link in the chain from Pythagoras to modern foundationists. Descartes's ignoring it in his own research casts doubt on it as a viable, realistic methodology.

You won't find in the *Geometry* the method we teach nowadays as Cartesian or "analytic" geometry. Our analytic geometry is based on rectangular coordinates (which we call "Cartesian"). To every point in the plane we associate a pair of real numbers, the "x" and "y" coordinates of the point. To an equation relating x and y corresponds a "graph"—the set of points whose x and y coordinates satisfy the equation. For an equation of first degree, the graph is a straight line. For an equation of second degree, it's a circle or other conic section. Our idea is to solve geometric problems by reducing them to algebra. Nowhere in Descartes's book do we see these familiar horizontal and vertical axes! Boyer says it was Newton who first used orthogonal coordinate axes in analytic geometry.

The conceptual essence of analytic geometry, the "isomorphism" or exact translation between algebra and geometry, was understood more clearly by Fermat than by Descartes. Fermat's analytic geometry predated Descartes's, but it wasn't published until 1679. The modern formulation comes from a long development. Fermat and Descartes were the first steps. Instead of systematically developing the technique of orthogonal coordinate axes, the *Geometry* studies a group of problems centering around a problem of Pappus of Alexandria (third century A.D.). To solve Pappus's problem Descartes develops an algebraic-geometric procedure. First he derives an algebraic equation relating known and unknown lengths in the problem. But he doesn't then look for an algebraic or numerical solution, as we would do. He is faithful to the Greek conception, that by a solution to a geometric problem is meant a construction with specified instruments. When possible Descartes uses the Euclidean straight edge and compass. When necessary, he brings in his own instrument, an apparatus of hinged rulers. Algebra is an intermediate device, in going from geometric problem to geometric solution. Its role is to reduce a complicated curve to a simpler one whose construction is known. He solves third and fourth-degree equations by reducing them to second degree—to conic sections. He solves certain fifth- and sixth-degree equations by reducing them to third degree. A modern reader knows that the general equation of fifth degree can't be solved by extraction of roots. So he's skeptical about Descartes's claim that his hinged rulers can solve equations of degree six and higher. Descartes was mistaken on several points. In themselves, these are of little interest today. But they discredit his claim of absolute certainty. *Descartes's mathematics refutes his epistemology.*

Emily Grosholz and Carl Boyer point out errors in the *Geometry*. "When he turns his attention to the locus of five lines, he considers only a few cases, not bothering to complete the task, because, as he says, his method furnishes a way to describe them. But Descartes could not have completed the task, which amounted to giving a catalogue of the cubics. . . . Newton, because he was able to move with confidence between graph and equation, first attempted a catalogue of the cubics; he distinguished seventy-two species of cubics, and even then omitted six" (Grosholz, referring to Whiteside).

"Descartes stratified his hierarchy into levels of pairs of degrees, since (so he thought) from curves of degree n and n + 1 his apparatus of hinged rulers produced curves of degree n + 2 and n + 3. Fermat gave a counter-example to this generalization.

"Descartes classification into orders of two degrees each was based on the fact that the algebraic solution of the quartic leads to a resolvent cubic from which Descartes rashly concluded—incorrectly, as Hudde later showed—that an equation of degree 2n would in all cases lead to a resolvent of degree 2n − 1. . . . The method Descartes proposed for the study of the properties of a space curve is to project it upon two mutually perpendicular planes and to consider the two curves of projection. Unfortunately, the only illustrative property given here is erroneous for one reads that the normal to a curve in three-space at a point P on the curve is the line of intersection of the two planes through P, determined by the normal lines to the curves of projection at the points corresponding to P. This would be true of the tangent line, but does not in general hold for a normal. . . . Descartes, in these casual remarks, seems not to have been aware of the fact that for space of more than two dimensions a normal is not uniquely determined for a point on a curve."

From these descriptions by Boyer and Grosholz, it's clear that Descartes's mistakes weren't merely inadvertent. We can hardly imagine Descartes nodding asleep as he composes the *Geometry*! These are *conceptual* errors. Descartes failed to understand certain aspects of his subject, and consequently made false statements while in full possession of his senses. If mathematics were indubitable axioms followed by infallible reasoning, this would be impossible. Even when mathematics is presented in the synthetic mode, mistakes happen. But the analytic mode used by Descartes is the usual mode of mathematical thinking. It isn't infallible, in Descartes's hands, yours, or mine. Therefore any philosophical claim based on the infallibility of mathematics is discredited.

Back to Boyer: "One gets the impression that Descartes wrote *La géometrie* not to explain, but to boast about the power of his method. He built it about a difficult problem and the most important part of his method is presented, all too concisely, in the middle of the treatise for the reason that it was necesary for the solution of this problem."

Do you think Boyer is unfair? In a letter to Florimond De Beaune, Descartes wrote, on February 20, 1639 (Adam and Tannery, Vol. 2, p. 511): "In the case of the tangents, I have only given a simple example of analysis, taken indeed from a rather difficult aspect and I have left out many of the things which could have been added so as to make the practice of the analysis more easy. I can assure you, nevertheless, that I have omitted all that quite deliberately, except in the case of the asymptote, which I forgot. But I felt sure that certain people, who boast that they know everything, would not miss the chance of saying that they knew already what I had written, if I had made myself easily intelligible to them. I should not then have had the pleasure, which I have since enjoyed, of noting the irrelevance of their objections."

Boyer goes on, "In concluding the work, Descartes justifies the inadequacy of exposition by the incongruous remark that he had left much unsaid in order not to rob the reader of the joy of discovery" (Boyer, p. 104). Here is the quote to which Boyer refers: "Finally, I have not demonstrated here most of what I have said, because the demonstrations seem to me so simple that, provided you take the pains to see methodically whether I have been mistaken, they will present themselves to you; and it will be of much more value to you to learn them this way than by reading them" (Vol. I, p. 244).

Says Boyer: "Either this was sarcasm or else the author grossly misjudged the abilities of his readers to profit by what he had written. It is no wonder that the number of editions of his *Geometrie* was relatively small during the seventeenth century and has been still smaller since then" (Boyer, p. 104).

Descartes claimed his Method was infallible in science and mathematics. He was more cautious with religion. He didn't derive Holy Scripture or divine revelation by self-evident axioms and infallible deductions. When he heard that Galileo's *Dialogue on the Two Chief Systems* was condemned by the Holy Church, he suppressed his first book, *Le Monde,* even though he was living in Holland, safe from the Church. (Galileo was kept under house arrest at first. For three years he had to recite the seven penitential psalms every week.) Descartes wrote to Father Mersenne, "I would not want for anything in the world to be the author of a work where there was the slightest word of which the Church might disapprove."

Following the four rules of the *Method* is an addendum: "But this does not prevent us from believing matters that have been divinely revealed as being more certain than our surest knowledge, since belief in these things, as all faith in obscure matters, is an action not of our intelligence but of our will. They should be heeded also since, if they have any basis in our understanding, they can and ought to be, more than all things else, discovered by one or other of the ways above-mentioned, as we hope perhaps to show at greater length on some future opportunity" (Vol I., p. 11).

Like Pascal, Newton, and Leibniz, Descartes may have valued his contributions to theology above his mathematics. His struggle against skeptics and heretics is the major half of his philosophy, more explicit than his battles with scholastics. "In Descartes' reply to the objections of Father Bourdin, he announced that he was the first of all men to overthrow the doubts of the Sceptics . . . he discovered how the best minds of the day either spent their time advocating scepticism, or accepted only probable and possibly uncertain views, instead of seeking absolute truth. . . . It was in the light of this awakening to the sceptical menace, that when he was in Paris Descartes set in motion his philosophical revolution by discovering something so certain and so assured that all the most extravagant suppositions brought forward by the sceptics were incapable of shaking . . . in the tradition of the greatest medieval minds, (he) sought to secure man's natural knowledge to the strongest possible foundation, the all-powerful eternal God" (Popkin, p. 72).

The essence of the *Meditations* is a proof that the world exists by first proving Descartes exists, and then, by contemplating Descartes's thoughts, proving that God exists *and is not a deceiver.* Once a nondeceiving God exists, everything else is easy.

In the fifth meditation Descartes gives a "mathematical" proof that God exists, based on the geometry of the triangle. In essence it's exactly the argument that Plato originated, and passed on to Locke, Berkeley, and Leibniz: The certainty of mathematics implies the certainty of religion.

Here it is: "When I imagine a triangle, although there may nowhere in the world be such a figure outside my thought, or ever have been, there is nevertheless in this figure a certain determinate nature, form, or essence, which is immutable and eternal, which I have not invented, and which in no wise depends on my mind, as appears from the fact that diverse properties of that triangle can be demonstrated, viz. that its three angles are equal to two right angles, that the greatest side is subtended by the greatest angle, and the like, which now, whether I wish it or do not wish it, I recognize very clearly as pertaining to it, although I never thought of the matter at all when I imagined a triangle for the first time, and which therefore cannot be said to have been invented by me . . . but now, if just because I can draw the idea of something from my thought, it follows that all which I know clearly and distinctly as pertaining to this object does really belong to it, may I not derive from this an argument demonstrating the existence of God? It is certain that I no less find the idea of God, that is to say, the idea of a supremely perfect Being, in me, than that of any figure or number whatever it is; and I do not know any less clearly and distinctly that an actual and eternal existence pertains to this nature than I know that all which I am able to demonstrate of some figure or number truly pertains to the nature of this figure or number, and therefore although all that I concluded in the preceding Meditations were found to be false, the existence of God would pass with me as at least as certain as I have ever held the truths of mathematics to be. . . . I clearly see that existence can no more be separated from the essence of God than can its having its three angles equal to two right angles be separated from the essence of a rectilinear triangle, or the idea of a mountain from the idea of a valley; and so there is not any less repugnance to our conceiving a God (that is, a Being supremely perfect) to whom existence is lacking (that is to say, to whom a certain perfection is lacking) than to conceive of a mountain without a valley. . . . But after I have recognized that there is a God—because at the same time I have also recognized that all things depend upon Him, and that He is not a deceiver, and from that have inferred that what I perceive clearly and distinctly cannot fail to be true. . . . And this same knowledge extends likewise to all other things which I recollect having formerly demonstrated, such as the truths of geometry and the like. . . . And now that I know him I have the means of acquiring a perfect knowledge of an infinitude of things, not only of those which relate to God Himself, and other intellectual matters, but also of those which pertain to corporeal nature in so far as it is the object of pure mathematics" ("Meditation V, Vol. 1, pp. 180 ff.).

There is an obvious difficulty with this and the other antiskeptical arguments Descartes invented. Skeptics like Pierre Gassendi immediately pointed out that Descartes argues from what he sees in *his own* mind. Others find things otherwise in their minds. To Descartes it may have seemed that his use of the Method in geometry and in theology was all of a piece. Today, we see two different men: Descartes the mathematician, inventive and ingenious; and Descartes the theologian, self-duped by trivially fallacious arguments.

The two aspects of Descartes split apart in his intellectual heritage. His *Geometry* instructed Newton and Leibniz, and is forever integrated into the living body of mathematics. Mathematicians remember him with the name "Cartesian product" for ordered pairs in set theory and geometry.

In theology, the story is more complex. It would be unfair to suppose Descartes' bows to the Church were insincere. He really was a devout Catholic. His commitment to philosophy and science originated in a dream-vision of the Blessed Virgin.

Descartes was very considerate of the Church's worries. Still some denounced him as a skeptic or crypto-skeptic. His First Meditation raises profound doubt. Does the Third Meditation really dispel it? Will his "clear and distinct idea" really revive faith, once his doubt has shaken it? In 1663 he was put on the Index (Vrooman, p. 252). In 1679 Leibniz wrote of Descartes's philosophy, "I do not hesitate to say absolutely that it leads to atheism" (Leibniz, p. 1).

In the following century, nevertheless, Cartesianism became popular among Church apologists. But Descartes's follower, the "God-intoxicated" Spinoza, was denounced as an atheist, and "Spinozism" became a synonym for atheism. Cartesians were prominent denouncers of Spinoza (Balz, pp. 218–41).

seven

Mainstream Philosophy at Its Peak

Baruch Spinoza (1632–1677)

The Higher Criticism. Ethics à l'Euclide.

Benedictus de Spinoza, born Baruch Spinoza, was excommunicated by the Jews of Amsterdam, but never baptized Christian. His native language was Portuguese; he spoke Dutch with difficulty.

The Amsterdam Jews had fled to Holland from the Holy Inquisition. Like similar communities today, they were grateful for refuge, and eager not to disturb their Protestant hosts. The bill of excommunication, proclaimed in Portuguese on July 27, 1656, reads: "The chiefs of the council, having long known the evil opinions and works of Baruch de Espinoza . . . have had every day more knowledge of the abominable heresies practiced and taught by him. . . . With the judgment of the angels and of the saints we excommunicate, cut off, curse, and anathematize Baruch de Espinoza. . . . Cursed be he by day and cursed be he by night. Cursed be he in sleeping and cursed be he in waking, cursed in going out and cursed in coming in . . . none may speak with him by word of mouth nor by writing, nor show any favour to him, nor be under one roof with him, nor come within four cubits of him, nor read any paper composed or written by him." Before the excommunication he had been attacked with a dagger (Pollock, pp. 17–18).

Spinoza learned of the anathema while visiting a friend in "the small dissenting community of Remonstrants or Collegiants, heretics anathematized by the Synod of Dort . . . men who without priests or set forms of worship carried out the precepts of simple piety" (Pollock, p. 17). It seems Spinoza was something like a Quaker! (See Fell, 1987.)

His first book, *Tractatus Theologico-Politicus*, scandalized Amsterdam's Orthodox. Like Descartes, Spinoza believed that reason should guide him in understanding the world. Unlike Descartes, who tried not to tread on the Church's toes, Spinoza followed reason without regard to consequences. He made no exception of Scripture and revealed religion. His *Tractatus* is the first example of what later became known as "higher criticism" of the Bible. Spinoza saw the *Old Testament* as an incomplete, corrupted collection of documents about the history of the Jews.

He wrote: "It is thus clearer than the sun at noonday that the Pentateuch was not written by Moses, but by someone who lived long after Moses. . . . The history relates not only the manner of Moses death and burial and the thirty days mourning of the Hebrews, but further compares him with all the prophets who came after him, and states that he surpassed them all. . . . Such testimony cannot have been given of Moses by himself" (VII, p. 124).

"Moses . . . conceived God as a ruler, a legislator, a king, as merciful, just, &c., whereas such qualities are simply attributes of human nature and utterly alien from the nature of the Deity" (IV, p. 63).

"What pretension will not people in their folly advance! They have no single sound idea concerning either God or human nature, they confound God's decrees with human decrees, they conceive nature as so limited that they believe man to be its chief part!" (VI, p. 81).

(Richard Popkin has called attention to a book by the Quaker Samuel Fisher. He was part of the Quaker missionary effort in Holland, and acquainted with Spinoza. Fisher's book, published ten years before Spinoza's, has remarkably similar arguments and examples.)

"The States of Holland and West Friesland, being satisfied that the book . . . entitled 'B.D.S. Opera Posthuma' 'labefactated' various essential articles of the faith and 'vilipended the authority of miracles', expressed 'the highest indignation' at the disseminating thereof, declared it profane, atheistic, and blasphemous, and forbad printing, selling, and dealing in it, on pain of their high displeasure."

"The stupid Cartesians," Spinoza wrote to Oldenburg, "being suspected of favoring me, endeavored to remove the aspersion by abusing everywhere my opinions and writings, a course which they still pursue" (Ratner, 1927).

Spinoza took up lens-grinding. The rest of his life he maintained himself in poverty by that trade. He was offered a philosophy chair at Heidelberg, a pension from Louis XIV of France, and annuities by friends in the Netherlands. He always declined, valuing intellectual independence above physical comfort.

After meeting with a Frenchman during the war between the Netherlands and France, he was accused of spying. His host was in fear of an attack on his home. He told his host: "So soon as the crowd makes the least noise at your door I will go out and make straight for them though they should serve me as they have done the unhappy De Witts. I am a good republican, and never had any aim but the honour and welfare of the State." The De Witts were two brothers, one a friend of Spinoza, who were lynched by a mob at the Hague (Pollock, pp. 36–37). The feared attack did not take place.

Spinoza died in 1677 of consumption, aggravated by inhaling glass particles while grinding lenses. "The funeral took place on the 25th February, being attended by many illustrious persons, and followed by six coaches" (Spinoza, 1895, p. xx).

Spinoza didn't write explicitly about philosophy of mathematics, but he was a major influence in general philosophy, and it's interesting to surmise what he thought about mathematics.

When Spinoza wanted to say something was indubitable, he said it was as certain as that the angle sum of a triangle equals two right angles. This familiar theorem of Euclid was his paradigm of the indubitable. (*Ethics*, p. 55, for example.) But this "indubitable fact" is negated in non-Euclidean geometry. (See the section on Certainty in Chapter 4.) What to Spinoza was indubitable, today is dubitable for every mathematics student.

The Ethics is the only work of Spinoza read in philosophy class today. It has five parts: "Concerning God," "Of the Nature and Origin of the Mind," "On the Origin and Nature of the Emotions," "Of Human Bondage or the Strength of the Emotions," and "On the Power of the Understanding, or of Human Freedom." Spinoza treats these matters "geometrically," that is, in the style of Euclid's *Elements*. Each part starts with Axioms and Definitions and produces Propositions.

In Part I, Definition I is: "By that which is self-caused, I mean that of which the essence involves existence, or that of which the nature is only conceivable as existent." Axiom I of Part I is "Everything which exists, exists either in itself or in something else."

From such ingredients, Spinoza obtains Proposition 33: "Things could not have been brought into being by God in any manner or in any order different from that which has in fact obtained."

Here's the proof: "All things necessarily follow from the nature of God (Proposition xvi), and by the nature of God are conditioned to exist and act in a particular way (Proposition 29). If things, therefore, could have been of a different nature, or have been conditioned to act in a different way, so that the order of nature would have been different, God's nature would also have been able to be different from what it now is; and therefore (by Proposition 11) that different nature also would have perforce existed, and consequently there would have been able to be two or more Gods. This (by Proposition 14, Corollary 1) is absurd. Therefore things could not have been brought into being by God in any other manner, etc. Q.E.D."

To a modern reader this argumentation is embarrassing. Spinoza's proof is not what we call a proof. Roth (41) thinks that "It is . . . not a method of proof, but an order of presentation, as may be proved . . . by the fact that Spinoza proposed to deal in precisely the same way with the intricacies of Hebrew Grammar. In the *Ethics* itself the geometric form, even as an order, is dropped at convenience. . . . He adopted it for a definite reason, and that was its impersonality. Mathematics recognizes and has no place for personal prejudice. It neither laughs nor weeps at the objects of its study, because its aim is to understand them. . . . The mathematical method, therefore, meant to Spinoza the free unprejudiced inquiry of the human mind, uncramped by the veto of theology and theological philosophy" (Roth, p. 43).

Spinoza wrote, in the preface to Part III of the *Ethics*: "Persons who would rather abuse or deride human emotions than understand them . . . will doubtless think it strange that I should attempt to treat of human vice and folly geometrically . . . the passions of hatred, anger, envy, and so on, considered in themselves, follow from [the] necessity and efficacy of nature. . . . I shall therefore treat of the nature and strength of the emotions . . . in exactly the same manner, as though I were concerned with lines, planes, and solids."

Another glimpse at Spinoza's idea of mathematics, as cited by Roth, is in the appendix to the *Ethics*, pp. 72–73. "[Superstitious people] laid down as an axiom, that God's judgments far transcend human understanding. Such a doctrine might well have sufficed to conceal the truth from the human race for all eternity, if mathematics had not furnished another standard of verity in considering solely the essences and properties of figures without regard to their final causes."

In the century after Spinoza's death, he suffered universal condemnation and near oblivion. Protestants and Catholics attacked him for rejecting their dogmas. Free thinkers and liberals repudiated him for fear of guilt by association. Since Spinoza's only openly published work was an exposition of Descartes, the Cartesian sect was particularly active in denouncing Spinozism.

Toward the end of the eighteenth century the German playwright Lessing privately confessed to being a Spinozist. Then Goethe openly praised and studied Spinoza. By the time of Hegel, you didn't know philosophy if you didn't know Spinoza. Samuel Coleridge, the poet, critic, and importer of advanced German culture, awoke Spinozism in England.

Although Spinoza didn't directly enter into the philosophy of mathematics, his indirect influence is important. Until the late eighteenth century, religion was rarely questioned in scholarly writing. The philosophy of mathematics and theology were twin trees holding each other up. Plato's *Meno* and Descartes's fifth meditation show mathematics shoring up religion. Writings of Berkeley and Leibniz show religion supporting the philosophy of mathematics. The notion that nonphysical reality or even all reality, physical and nonphysical, is located in the mind of God was a complete answer to the question of the nature of mathematics. Recent troubles in philosophy of mathematics are ultimately a consequence of the banishing of religion from science.

Religion in general is not decaying. Fundamentalist religion is spreading like a noxious weed. In intellectual circles you meet disciples of Buber, Kierkegaard,

Maharishi, and Nagarjuna. But religion is no longer granted a role in science. Newton said his discoveries redounded to the glory of God, but Laplace told Napoleon he had found God to be an unnecessary hypothesis. Today's scientist may attend church or *shul*, but he wouldn't dream of using the will of God to explain a puzzle in microbiology or particle physics.

Kant is the dividing line. As we shall see, he still has God in his philosophy. But his principal work, *The Critique of Pure Reason* is secular. God appears in the *Critique of Practical Reason*, in second place, so to speak.

Following Kant, mainstream philosophy became secular. Perhaps theology had reached the stage of self-sufficient independence, and, like psychology, broke off from philosophy. Perhaps philosophers, yearning for respect from scientists, followed the example of scientists in ghettoizing religion.

The important role of Spinoza in the philosophy of mathematics is his contribution to discrediting Scripture and Revelation, and ultimately secularizing science. By so doing he helped create the modern dilemma of the philosophy of mathematics.

Gottfried Wilhelm Freiherr von Leibniz (1646–1716)

Monads;Infinitesimals.The Differential Calculus.

Leibniz was a giant in both mathematics and philosophy. He's the founder of German idealism, the father of Hegel and the Hegelians. Christian von Wolff produced a popularization of Leibnizism, which became virtually official religion in Germany in the late eighteenth century.

(In his *Preliminary Discourse on Philosophy in General*, Wolff writes: "mathematical knowledge must be joined to philosophy if you desire the highest possible certitude. For this reason, we also grant a place in philosophy to mathematical knowledge, even though we have distinguished it from philosophy. For we hold that nothing is more important than certitude.")

Leibniz was the great polymath, expert and active in every field of learning and human affairs. He was rich, lived where and how he pleased. Yet by his own ambitions he was a failure.

At age 22 he produced a Latin tractate, *ordine geometrico demonstrata* (a l'Euclide), supporting the aspiration of the Count Palatine, Philip William of Neuburg, for the vacant throne of Poland. Philip William lost the election.

In 1672 Louis XIV's wars of conquest were turning Europe upside-down. Leibniz "urged the Christian powers of Europe instead of fighting one another to combine against the infidel, and suggested that in such a united war against the Turks, Egypt, one of the best situated lands in the world, would fall to France . . . the same idea commended itself more than a century later to Napoleon . . . [who discovered] Leibniz's memorial when he took possession of Hanover in 1803" (Carr, p. 12). Louis did not follow Leibniz's urging. But in

Paris Leibniz met Christian Huygens (1629–1695) who became his mathematics tutor, and he also met leading thinkers Antoine Arnauld, Jacques Bossuet (1627–1704), and Nicolas Malebranche (1638–1715). He visited London and was elected to the Royal Society.

At age 27 he accepted the post of librarian to John Frederick, of the House of Brunswick-Luneburg, Duke of Hanover. "In his correspondence he always refers to its head as 'my prince' " (Carr, p. 13). He continued to circulate in Paris, Vienna, and Rome, with ideas and projects, notably the reunion of Protestant and Catholic churches (he was Protestant).

Duke George (successor to Ernest Augustus, who was successor to John Frederick) grew vexed with Leibniz's slowness on the *History of the Guelph Family.* In 1714 when the duke was called to London to become George I, he commanded 68-year-old Leibniz to stay in Hannover until he finished the *History.* Leibniz died two years later, having got to the year 1005. The *History of the Guelph Family* is still incomplete.

Leibniz was a compulsive writer. He published only a few short works, but left a large secret cabinet in Hannover with "a great mass of hardly legible drafts" (Meyer, pp. 1–4). Johann Edward Erdmann spent years in the Leibniz archive. He wrote, "I leave to those who come after me not honeycombs, but pure wax." The Berlin Academy, which Leibniz founded, plans a 40-volume *Collected Works.* A dozen or so have appeared.

Leibniz was co-inventor, with Isaac Newton, of the infinitesimal calculus. Their priority fight is the most famous of its kind. As a consequence, English mathematicians ignored continental mathematics for a century, to their own great detriment.

Unlike Newton, Leibniz worked with actual infinitesimals,** though he couldn't explain coherently what they were. Cavalieri and others had calculated areas by dividing regions into infinitely many strips, each having infinitesimal positive area. unfortunately, as Huygens showed, this method could give wrong answers.

Leibniz had intimations of a formal language, but not in connection with infinitesimals. He dreamed of a language in which all correct reasoning would be straightforward calculation. He wrote (Nidditch, pp. 19–23) that "the invention of an ABC of man's thoughts was needed, and by putting together the letters of this ABC and by taking to bits the words made up from them, we would have an instrument for the discovery and testing of everything. . . . If we had a body of signs that were right for the purpose of our talking about all our ideas as clearly and in as true and as detailed a way as numbers are talked about in Arithmetic or lines are talked about in the Geometry of Analysis, we would be able to do for every question, in so far as it is under the control of reasoning, all that one is able to do in Arithmetic and Geometry. . . . It would not be necessary for our heads to be broken in hard work as much as they are now and we would certainly be able to get all the knowledge possible from the material given. In addition, we would have everyone in agreement about whatever would have been worked out, because it would be simple to have the working gone into by doing it again or by attempting tests like that of 'putting out nines' in Arithmetic. And if anyone had doubts of one of my statements I would say to him: let us do the question by using numbers, and in this way, taking pen and ink we would quickly come to an answer."

What a naive dream! But calculating machines were known to Leibniz. Pascal invented an adder while he was a teenager, and Leibniz taught it to multiply. Leibniz's unpublished thoughts were a forerunner of modern logic.

Leibniz's metaphysics is a most fascinating fantasy. At the request of Prince Eugene of Savoy (Carr, p. 3), Leibniz wrote his *Monadology.* Leibniz's idealistic atomism makes him think there must be "simple" parts or "monads" out of which the world is made. Therefore, apart from these monads there can't be anything real. It follows that their relations can't be real. They can't see each other. They're "windowless"!

Leibniz was aware that things do seem to interact. He explained this in a most beautiful and incredible way. Imagine two clocks that always show the same time. It might seem that they are communicating with each other. But they aren't. They're just separately keeping correct time. There's no connection, only a "preestablished harmony." So, said Leibniz, body and soul are two separate windowless monads, keeping the same time. When you and I converse, we each speak independently, as predetermined by God. He sees our seeming conversation. His knowledge of it is its only reality.

Leibniz's world is beautifully described by Gottfried Martin (p. 1 ff.). "All monads are first defined as living. By a bold leap the whole knowable universe is then filled with a dense sea of living things, that is to say with a dense sea of monads. Everything is alive; everything unfruitful, everything sterile, everything dead in the universe is only outward illusion. . . . In this great ocean of living things there are no empty places. Wherever one looks there surges an infinite world of creatures, of living beings, of animals, of entelechies, of souls. Every single particle of matter, however small, is a garden full of plants, a pond full of fishes, and every twig of these plants in this garden and every drop in the blood of these fishes is another garden full of plants, another pond full of fishes, and so forth to infinity. Everywhere, in the infinitely great and in the infinitely small, there is life, everywhere there are monads. Every monad perceives and every monad wills . . . [My comment: Infinitely small monads are reminiscent of Leibniz's mathematical infinitesimal. There's a reverberation between Leibniz's mathematics and his metaphysics.]"

Martin continues, "Outside and between the individual existence of monads there is nonreality. Since only monads and modifications of monads have real existence in this sense, relations cannot have real existence . . . to use an alternative expression which Leibniz often uses, they only have mental existence . . . [relations] include number, time, duration, space, the extension of bodies. . . . But the understanding that thinks relations is the understanding of God. Relations have their substantial reality taken away from them by being referred to an understanding, but because the understanding that carries them is the divine understanding, they . . . receive back again a new reality. . . ."

"Matter," Leibniz keeps repeating, "is a mere phenomenon, and it is even clearer for him that space is a mere phenomenon . . . but a *phaenomenon Dei.*" [Reminiscent of Berkeley, the "British empiricist" who conventionally is classed in opposition to Leibniz the rationalist. For Berkeley all seeming existence, including physical existence, is illusory except as a thought in the mind of God.]

Martin goes on, "The divine thinking also provides the ground of the possibility of mathematics, because in his continuous thinking of possible worlds God also thinks continuously the eternal truths that hold for these worlds, in particular all numerical and spatial relations. This holds especially of Euclidean space which, as the only possible space, is the same for all possible worlds. The being of Euclidean space thus consists primarily in being thought without contradiction by God. . . .

"The propositions of three-dimensional Euclidean geometry are primarily and continuously thought by God, and when the mathematician discovers, understands, and proves an adequately formulated proposition in geometry, this knowing is a repetition of the primary divine knowing. . . ."

[I can't help repeating a remark in the last section: What a beautifully simple solution to the ontology of mathematics! We needn't "break our heads" figuring out how numbers have existence independent of human thought. The thought is God's! The present trouble with the ontology of mathematics is an after-effect of the spread of atheism. But mysticism still hasn't vanished from mathematics. See Paul Erdös and "The Book" in Chapter 1.]

"The profound consequences of Leibniz's connections with Plato . . . begin to unfold at this point. Plato pronounced repeatedly that thinking the truth means becoming like God. Leibniz may have met the ancient Platonic thesis . . . in Malebranche, in the proposition that we see all things in God . . . Leibniz was carrying on a great tradition, following Plato, Plotinus, Augustine, and Malebranche. . ." (Martin).

In his *Discourse on Metaphysics*, section 26, Leibniz summarizes Plato's dialogue, the *Meno*. (See the section above on Plato.) In this dialogue Socrates coaches a "slave boy" to discover that a small square inscribed at the midpoints of the sides of a large square has half the area of the large square. Socrates claims that in so doing he proved that the "slave boy" remembers this bit of geometry from before his birth, in Heaven. Leibniz calls the *Meno* "a beautiful experiment." He thinks Plato actually makes a strong case for his doctrine of "reminiscence." A fine example of intertwining between religion and philosophy of mathematics!

Some of Leibniz's pieties are worthy of his parody, Voltaire's Dr. Pangloss. "There will be no good action unrewarded and no evil action unpunished; everything must turn out for the well-being of the good; that is to say of those who are not disaffected . . . if we were able to understand sufficiently well the order of the universe, we should find that it surpasses all the desires of the wisest of us, and that it is impossible to render it better than it is, not only for all in general, but also for each one of us in particular, provided that we have the proper attachment for the author of all . . . who alone can make us happy" (1992, Vol. IV; 1965, p. 90).

As Kant stands behind Frege, Hilbert, and Brouwer, Leibniz and Hume stand behind Kant. Kant's metaphysics is sometimes described as an effort to wed Leibniz's rationalism to Hume's empiricism.

The "pre-critical" Kant (before the *Critique of Pure Reason*) had been soaked in Christian von Wolff's popular Leibnizism. Reading Hume "woke him from dogmatic slumber." He responded by trying to rehabilitate Leibniz's rationalism from Hume's attack.

"Kant can therefore justly say, looking back on his philosophical life-work, 'The *Critique of Pure Reason* is, after all, the real apologia for Leibniz, even against his followers who exalt him with praises which do him no honor' " (Martin).

Bishop George Berkeley (1685—1753)

Ghosts of Departed Quantities

George Berkeley said something interesting about calculus. His book *The Analyst* (1734) is the only philosophical critique that challenges mathematicians as mathematicians. *The Analyst* is never taught in Philosophy 101, perhaps because philosophy professors don't think it's philosophy, perhaps because calculus isn't a prerequisite for the students or the professor.

Students learn that Berkeley was the empiricist between Locke and Hume. This is misleading. Berkeley's empiricism proved that nothing exists, except in the mind of God—a conclusion far from those of Locke or Hume.

The Analyst is an attack on the differential calculus of Newton and Leibniz.** Its subtitle is, "A Discourse Addressed to an Infidel Mathematician, Wherein it is examined whether the object, principles, and inferences of the modern Analysis are more distinctly conceived, or more evidently deduced, than religious Mysteries and points of Faith." The Infidel Mathematician was the Astronomer Royal, Edmund Halley, who gave his name to a comet, and paid for publishing Newton's *Principia*.

According to Berkeley's brother, Berkeley wrote *The Analyst* in response to the circumstances of the death of the King's physician, Dr. Garth. Joseph Addison, the essayist, visited Garth to remind him to "prepare for his approaching dissolution." Said the dying man, "Surely, Addison, I have good reason not to believe those trifles, since my friend Dr. Halley who has dealt so much in demonstration has assured me that the doctrines of Christianity are incomprehensible, and the religion itself an imposture" (Berkeley, *Works*, Vol. 4, Introduction).

Halley deprived Garth of eternal bliss!

Berkeley counter-attacked, not by defending religion, but by showing that the mathematics of Newton and Leibniz is more obscure than the Church's deepest mystery.

"I shall claim the privilege of a Free-thinker," he wrote, "and take the liberty to inquire into the object, principles, and method of demonstration admitted by the mathematicians of the present date, with the same freedom that you presume to treat the principles and mysteries of Religion."

Bishop Gibson of London wrote gratefully to Bishop Berkeley: "the men of science (a conceited generation) are the greatest sticklers against revealed religion . . . we are much obliged to your Lordship for retorting their arguments upon them."

Berkeley's attack on mathematics went beyond calculus. He also denounced the mathematicians' claim that a line segment can be divided into arbitrarily short subsegments. For then the line segment would have infinitely many parts. But that is inconceivable to any "man of sense." What any man of sense cannot conceive is impossible.

Berkeley didn't say what was the shortest possible length, or even claim it could be known. His was a purely existential proof by contradiction. Infinite divisibility is inconceivable, so, by the law of the excluded middle, divisibility must be finite, so there must be "atoms" of length. That would mean that finding instantaneous velocities is impossible in principle. Instead of instantaneous velocity, we would have minimal-distance velocity—the minimal distance divided by the time elapsed in traversing it. Berkeley didn't go into these ramifications. But he did reject Euclid's construction for bisecting a line segment. For if the segment happens to be made of an odd number of indivisible length-atoms, it obviously can't be bisected.

Neither did he accept the diagonal of the unit square. All line segments are made of length-atoms, so they're all commensurable. Since no rational number equals $\sqrt{2}$, there's no line segment of length $\sqrt{2}$. So there's no line segment connecting opposite corners of a unit square. The supposed diagonal misses one corner or the other by a little bit.

I don't think Berkeley considered whether time is also discrete. Hume later proved that both time and space are discrete. This idea has been revived in our time, in an effort to make sense of quantum mechanics.

Berkeley also had an original idea about arithmetic: Numbers don't exist. Numerals are meaningless symbols. This was part of his doctrine that abstractions and general concepts don't exist. In today's philosophical lingo, he was a strict nominalist. More than that, he said that the sun, the moon, the teeth in your mouth, all are mere appearances. They, we, and all else are only thoughts in the mind of God.

He was a remarkable anomaly, a "British empiricist" who attacked a whole tribe of scientists—the mathematicians. Plato, Descartes, and Spinoza wanted to use the supposed certainty of mathematics to advance religion. Berkeley used the *deficiency* of mathematics to advance religion. His attack on mathematicians is unique since St. Augustine.

Immanuel Kant (1724–1804)

Synthetic a priori. Non-Euclidean Geometry.

Classical philosophy reached its peak at the end of the eighteenth century in Kant. Kant's metaphysics is a continuation of the Platonic search for certainty and timelessness in human knowledge. He wanted to rebut Hume's denial of certainty. To do so, he made a sharp distinction between noumena, things in themselves, which we can never know, and phenomena, appearances, which our senses tell us. His goal was knowledge a priori—knowledge timeless and independent of experience.

He distinguished two kinds of a priori knowledge. The "analytic a priori" is the kind we know by logical analysis, by the meanings of the terms being used. Like the rationalists, Kant believed we also possess a priori knowledge that is not logical truism. This is his "synthetic a priori." Our intuitions of time and space are such knowledge, he believed. He explained their a priori nature by saying they're intuitions—inherent properties of the human mind. Our intuition of time is systematized in arithmetic, based on the intuition of *succession*. Our intuition of space is systematized in geometry. For Kant, as for all earlier thinkers, there's only one geometry—the one we call Euclidean. The truths of geometry and arithmetic are forced on us by the way our minds work; this explains why they are (supposedly) true for everyone, independent of experience. The intuitions of time and space, on which arithmetic and geometry are based, are objective in the sense that they're valid for every human mind. No claim is made for existence outside the human mind. Yet the Euclid myth (see below) remains central in Kantian philosophy.

Indeed, mathematics is central for Kant. His *Prolegomena to any Future Metaphysics Which Will Be Able to Come Forth as a Science*, has three parts. Part One is, "How Is Pure Mathematics Possible?" (a question I discussed in Chapter 1).

Kant's fundamental presupposition is that "contentful knowledge independent of experience (the 'synthetic a priori') can be established on the basis of universal human intuition." In *The Critique of Pure Reason*, he gives the two examples already mentioned: (1) space intuition, the foundation of geometry, and (2) time intuition, the foundation of arithmetic. In *The Critique of Practical Reason*, without using the term "synthetic a priori," he gives a third intuition: (3) moral intuition, the foundation of religion.

I will discuss intuitions (1) and (3), although (2) is still interesting today. It deals with the question, "Are primitive counting notions universal and invariable?" Most writers think so. Wittgenstein disagreed. Frege also disagreed; he thought arithmetic was analytic a priori—based on logic—rather than synthetic a priori—based on time intuition.

I discuss three topics on Kant:

1. Synthetic a priori. Intuition of space and time.
2. Effect of non-Euclidean geometry on Kant's theory of space intuition.
3. Intuition of duty, God, and the parallel with intuitions of space and time.

1. The *Prolegomena*, p. 21: "Weary therefore of dogmatism [Leibniz], which teaches us nothing, and of skepticism [Hume], which does not even promise us anything—even the quiet state of a contented ignorance—disquieted by the importance of knowledge so much needed, and rendered suspicious by long experience of all knowledge which we believe we possess or which offers itself in the name of pure reason, there remains but one critical question, on the answer to which our future procedure depends, namely, "Is metaphysics at all possible?". . . The *Prolegomena* must therefore rest upon something already known as trustworthy, from which we can set out with confidence and ascend to sources as yet unknown, the discovery of which will not only explain to us what we knew but exhibit a sphere of many cognitions which all spring from the same sources. The method of prolegomena, especially of those designed as a preparation for future metaphysics, is consequently analytical.

"But it happens, fortunately, that though we cannot assume metaphysics to be an actual science, we can say with confidence that there is actually given certain pure *a priori* synthetical cognitions, pure mathematics and pure physics; for both contain propositions which are unanimously recognized, partly apodictically certain by mere reason, partly by general consent arising from experience and yet as independent of experience. We have therefore at least some uncontested synthetical knowledge *a priori*, and need not ask *whether* it be possible, for, it is actual, but *how* it is possible, in order that we deduce from the principle which makes the given knowledge possible the possibility of the rest."

Against synthetic (contentful) knowledge he contrasted analytic knowledge, which is derived from logic and the meaning of words.

Again from the *Prolegomena*:

"Analytical judgments express nothing in the predicate but what has been already actually thought in the concept of the subject, though not so distinctly or with the same (full) consciousness. When I say, 'All bodies are extended,' I have not amplified in the least my concept of body, but have only analyzed it, as extension was really thought to belong to that concept before the judgment was made, though it was not expressed. This judgment is therefore analytical. On the contrary, this judgment, "All bodies have weight," contains in its predicate something not actually thought in the universal concept of body. It amplifies my knowledge by adding something to my concept, and must therefore be called synthetical.

"First of all, we must observe that all strictly mathematical judgments are a priori, and not empirical, because they carry with them necessity, which cannot be obtained from experience. . . . It must at first be thought that the proposition $7 + 5 = 12$ is a mere analytical judgment, following from the concept of the sum, of seven and five, according to the law of contradiction. But on close examination it appears that the concept of the sum of $7 + 5$ contains merely their union in a single number . . . Just as little is any principle of geometry analytical." (This is the point at which Frege turned away from Kant.)

Richard Tarnas writes (p. 342): "The clarity and strict necessity of mathematical truth had long provided the rationalists—above all Descartes, Spinoza and Leibniz—with the assurance that, in the world of modern doubt the human mind had at least one solid basis for attaining certain knowledge. Kant himself had long been convinced that natural science was scientific to the precise extent that it approximated to the ideal of mathematics. . . . By Hume's reasoning, with which Kant had to agree, the certain laws of Euclidean geometry could not have been derived from empirical observation. Yet Newtonian science was explicitly based upon Euclidean geometry. . . . Kant began by noting that if all content that could be derived from experience was withdrawn from mathematical judgments, the ideas of space and time will remain. From this he inferred that any event experienced by the senses is located automatically in a framework of spatial and temporal relations. Space and time are 'a priori forms of human sensibility': They condition whatever is apprehended through the senses. Mathematics could accurately describe the empirical world because mathematical principles necessarily involve a context of space and time, and space and time lay at the basis of all sensory experience: they condition and structure any empirical observation. . . . Because [geometrical] propositions are based on direct intuitions of spatial relations, they are '*a priori*'—constructed by the mind and not derived from experience—and yet they are also valid for experience, which will by necessity conform to the *a priori* form of space."

Kant's intuitions are supposed to explain, not how we might or could, but how we *actually do* conceive of time and space. There's no claim that they correspond to an objective reality. They're properties of Mind.

For Kant and his predecessors, mathematics and Mind are unchanging, eternal, and universal. Kant's intuitions are supposed to be eternal, universal features of Mind. But the Mind Kant knows is the mind of eighteenth-century Europe, plus the books in his library. He assumes this constitutes all human thinking.

2. Kant's views came to dominate West European philosophy, in spite of a development in geometry that made Kant's account of space untenable. That development was non-Euclidean geometry.**

The fifth axiom of Euclid's *Elements*, the parallel postulate, for centuries was considered a blot on the fair cheek of geometry. This postulate says: "If a line A crossing two lines B and C makes the sum of the interior angles on one side of A less than two right angles, then B and C meet on that side."

An equivalent axiom, the usual one in geometry books, is Playfair's: "Through a point not on a given line passes one parallel to the line."

This parallel axiom, everybody agreed, is intuitively true. Yet it isn't as "self-evident" as the other axioms. It says something happens at a point that possibly is very remote, where our intuition isn't as firm as nearby. Mathematicians wanted it proved, not assumed as Euclid did. Many tried, no one succeeded.

Then, as I mentioned in Chapter 4, Gauss, Bolyai, and Lobachevsky had the same brilliant idea: Suppose the fifth postulate *is* false, and then see what happens! Each of them got a new geometry! A possibility never before conceived.

Later Beltrami, Klein, and Poincaré showed that Euclidean and non-Euclidean geometry are "equiconsistent." If either is consistent, so is the other. Since no one doubts that Euclidean geometry is consistent, non-Euclidean also is believed to be consistent.

Kant's theory of spatial intuition meant Euclidean geometry was inescapable. But the establishment of non-Euclidean geometry gives us choices. Which geometry works best in physics? The question becomes empirical, to be settled by observation.

In 1915, Einstein published his theory of general relativity. The cosmos is a non-Euclidean curved space-time, more general than the hyperbolic space of Gauss, Lobatchevsky, and Bolyai. So non-Euclidean geometry is not just consistent, it governs the universe! Non-Euclidean geometry is used to represent relativistic velocity vectors. Physics doesn't prefer Euclid to non-Euclid.

Our intuitive notion of space is learned on a small scale, compared to the universe as a whole. Locally, "in the small," the difference between Euclidean and non-Euclidean geometries is too tiny to notice. The belief that the Euclidean angle sum theorem is "indubitable" or "absolute" is based on belief in an infallible spatial intuition. That belief is discredited by non-Euclidean geometry and general relativity.

Decades before non-Euclidean geometry was discovered by Kant's countryman Karl Friedrich Gauss, it was "almost known" to Johann Heinrich Lambert (1728–1777), a German mathematician who was actually an acquaintance or friend of Kant! Lambert came to a crucial recognition—that *if* the "postulate of the acute angle" were true it would lead to a strange new geometry. This already would have refuted Kant's theory that Euclidean geometry is an unavoidable innate intuition of the human mind.

Did Kant know Lambert's work? Martin thinks he did, but disregarded it as a "mere abstraction."

Körner says Kant didn't deny the abstract conceivability of non-Euclidean geometries; he thought they could never be realized in real time and space. This idea was wiped out by the advance of science.

Even though Kant's philosophy of space had already been exploded by non-Euclidean geometry, Philip Kitcher shows that all three foundationist gurus—Frege, Hilbert, and Brouwer—were Kantians. That was a consequence of the dominance of Kantianism in their early milieus, and the usual tendency of research mathematicians toward an idealist viewpoint. When they became disturbed by the "crisis in foundations" they couldn't help thinking in Kant's categories, in particular, his analytic and synthetic a priori. But instead of talking about the synthetic a priori, they talked about restoring the indubitability of mathematics—building or finding a solid foundation.

Non-Euclidean geometry makes Kant's philosophy of space untenable. But mathematicians avoid philosophical disputation by not mentioning the issue. To this day, texts on non-Euclidean geometry ignore its revolutionary philosophical implications. The first direct statement of the contradiction seems to be by Hermann Helmholtz, in *Mind* in 1877 (the birth year of that august journal.) In the next volume of *Mind* a Dutch philosopher, H. K. Land, replied that, by the nature of things, nothing in mathematics could be relevant to Kant's theory. Modern philosophy texts and lecturers on Kant seem to follow Land's principle. They don't mention non-Euclidean geometry.

3. Kant may have been the last philosopher or mathematician in the chain from Pythagoras to the present who explicitly made theology part of his philosophy. There's a half-hidden connection between Kant's a prioristic philosophy of mathematics and his moral-intuition version of Christianity.

In the *Critique of Practical Reason* he demolishes the three standard proofs of the existence of God. "Ontological": By definition, God is Perfect. Nonexistence would be an imperfection. "Cosmological": Every event has a cause. To avoid infinite regress, there had to have been a First Cause (God). "Teleological": A watch has a watch-maker. The World is more intricate than a watch, so it has a World-Maker (God).

Kant tears these proofs to shreds. He says they're the only proofs "speculative reason" (Leibnizian rationalism) could ever give. Kant isn't doubting God's existence. He's showing the superiority of his own proof, based on intuition. Not so different from his intuitions of time and space. Everyone has an intuition of duty, Kant thinks, of right and wrong. He doesn't say this *proves* God exists. He says it *justifies the postulate* "God exists."

"The moral law leads us to postulate not only the immortality of the soul, but the existence of God. . . . This second postulate of the existence of God rests upon the necessity of presupposing the existence of a cause adequate to the effect which has to be explained . . . a being who is a part of the world and is dependent upon it . . . ought to seek to promote the highest good, and therefore the highest good must be possible. . . . There is therefore implied, in the idea of the highest good, a being who is the supreme cause of nature, and who is the cause or author of nature through his intelligence and will, that is, God . . . or, in other words, it is morally necessary to hold the existence of God."

And in the *Prolegomena*, paras. 354–55, p. 103: "We must therefore think an immaterial being, a world of understanding, and a Supreme Being (all mere noumena) because in them only, as things in themselves, reason finds that completion and satisfaction which it can never hope for in the derivation of appearances from their homogeneous grounds, and because these actually have reference to something distinct from them (and totally heterogeneous), as appearances always presuppose an object in itself, and therefore suggest its existence whether we can know more of it or not."

Tarnas again (p. 350): "It is clear that at heart Kant believed that the laws moving the planets and stars ultimately stood in some fundamental harmonious relation to the moral imperatives he experienced within himself. 'Two things fill the heart with ever new and always increasing awe and admiration: the starry heavens above me and the moral law within me.' But Kant also knew he could not prove that relation, and in his delimitation of human knowledge to appearances, the Cartesian schism between the human mind and the material cosmos continued in a new and deepened form.

"In the subsequent course of Western thought, it was to be Kant's fate that, as regards both religion and science, the power of his epistemological critique tended to outweigh his positive affirmations. On the one hand, the room he made for religious belief began to resemble a vacuum, since religious faith had now lost any external support from either the empirical world or pure reason, and increasingly seemed to lack internal plausibility and appropriateness for secular modern man's psychological character. On the other hand, the certainty of scientific knowledge, already unsupported by any external mind-independent necessity after Hume and Kant, became unsupported as well by any internal cognitive necessity with the dramatic controversion by twentieth century physics of the Newtonian and Euclidean categories which Kant had assumed were absolute" (Tarnas, p. 350).

As the universal intuition of space is refuted by non-Euclidean geometry, the universal intuition of duty is refuted by history. For Winston Churchill and Harry Truman, fire-bombing German and Japanese civilians was duty. In the police stations of the world, torturing prisoners is duty. In Nazi Germany, genocide was duty.

What's the connection between Kant's philosophy of mathematics and his moral-intuition version of religion? Unlike Descartes and Leibniz, Kant does not use the certainty of mathematics (time and space) to support the certainty of God's existence. He considers the intuition of duty independently of the intuitions of time or space. He keeps his theory of God separate from his theory of mathematics. But they both have the same logic. Both rely on intuition: knowledge coming, not from the senses, study, or learning, but from the nature of Mind. Right and wrong, like time and space, are universal intuitions. Our space intuition leads to geometry, our time intuition leads to arithmetic, our duty intuition leads to Divinity.

In God's mind, the difficulties and puzzles in philosophy of mathematics disappear. How do numbers exist? Why do mathematical facts seem certain and timeless? Why does mathematics work in the "real world"?

In the mind of God, it's no problem.

The trouble with today's Platonism is that it gives up God, but wants to keep mathematics a thought in the mind of God.

Euclid as a Myth. Nobody's Perfect.

The myth of Euclid is the belief that Euclid's *Elements* contain indubitable truths about the universe. Even today, most educated people still believe the Euclid myth. Up to the middle or late nineteenth century, the myth was unquestioned. It has been the major support for metaphysical philosophy—philosophy that sought a priori certainty about the nature of reality.

The roots of our philosophy of mathematics are in classical Greece. For the Greeks, mathematics was geometry. In Plato and Aristotle, philosophy of mathematics is philosophy of geometry.

Rationalism served science by denying the intellectual supremacy of religious authority, while defending the truth of religion. This equivocation gave science room to grow without being strangled as a rebel. It claimed for science the right to independence from the Church. Yet this independence didn't threaten the Church, since science was the study of God's handiwork. "The heavens proclaim the glory of God and the firmament showeth His handiwork."

The existence of mathematical objects as ideas independent of human minds was no problem for Newton or Leibniz; they took for granted the existence of a Divine Mind. In that belief, the problem is rather to account for the existence of nonideal, material objects.

After rationalism displaced medieval scholasticism, it was challenged by materialism and empiricism; by Locke and Hobbes in Britain, by the encyclopedists in France. The advance of science on the basis of the experimental method gave the victory to empiricism. The conventional wisdom became: "The material universe is the fundamental reality. Experiment and observation are the only legitimate means of studying it."

The empiricists held that all knowledge *except mathematical* comes from observation. They usually didn't try to explain how mathematical knowledge originates. In the controversies, first between rationalism and scholasticism, later between rationalism and empiricism, the sanctity of geometry was unchallenged.

Philosophers disputed whether we proceed from Reason (a gift from the Divine) to discover the properties of the world, or whether only our bodily senses can do so. Both sides took it for granted that geometrical knowledge is not problematical, even if all other knowledge is. Hume exempted books of mathematics and of natural science from his outcry, "Commit it to the flames."

For rationalists, mathematics was the main example to confirm their view of the world. For empiricists, it was an embarrassing counter-example, which had to be ignored or explained away. If, as seemed obvious, mathematics contains knowledge independent of sense perception, then empiricism is inadequate as an explanation of all human knowledge. This embarrassment is still with us; it's a reason for the difficulties of philosophy of mathematics.

Mathematics always had a special place in the battle between rationalism and empiricism. The mathematician-in-the-street, with his common-sense belief in mathematics as knowledge, is the last vestige of rationalism.

The modern scientific outlook took ascendancy in the nineteenth century. By the time of Russell and Whitehead, only logic and mathematics could still claim to be nonempirical knowledge, obtained directly by Reason.

From the customary viewpoint among scientists now, the Platonism of most mathematicians is an anomaly. For many years the accepted assumptions in science have been materialism in ontology, empiricism in epistemology. The world is all one stuff, "matter," which physics studies. If matter gets into certain complicated configurations, it falls under a special science with its own methodology—chemistry, geology, and biology. We learn about the world by looking at it and thinking about what we see. Until we look, we have nothing to think about.

Yet in mathematics we have knowledge of things we can never observe. At least, this is the natural point of view when we aren't trying to be philosophical.

Until well into the nineteenth century, the Euclid myth was universal among mathematicians as well as philosophers. Geometry was the firmest, most reliable branch of knowledge. Mathematical analysis—calculus and its extensions and ramifications—derived legitimacy from its link with geometry. We needn't say "Euclidean geometry." The qualifier became necessary only after non-Euclidean geometry had been recognized. Before that, geometry was simply geometry—the study of the properties of space. These were exact, eternal, and knowable with certainty by the human mind.

eight

Mainstream Since the Crisis

How Did We Get Here? Can We Get Out?

Vacillation between two unacceptable philosophies wasn't always the prevalent mode. Where did it come from?

Until the nineteenth century, geometry was regarded by everybody, *including mathematicians*, as the most reliable branch of knowledge. Analysis got its meaning and its legitimacy from its link with geometry.

In the nineteenth century, two disasters befell. One was the recognition that there's more than one thinkable geometry. This was a consequence of the discovery of non-Euclidean geometries.

A second disaster was the overtaking of geometrical intuition by analysis. Space-filling curves** and continuous nowhere-differentiable curves** were shocking surprises. They exposed the fallibility of the geometric intuition on which mathematics rested.

The situation was intolerable. Geometry served from the time of Plato as proof that certainty is possible in human knowledge—including religious certainty. Descartes and Spinoza followed the geometrical style in establishing the existence of God. Loss of certainty in geometry threatened loss of all certainty.

Mathematicians of the nineteenth century rose to the challenge. Led by Dedekind and Weierstrass, they replaced geometry with arithmetic as a foundation for mathematics. This required constructing the continuum—the unbroken line segment—from the natural numbers. Dedekind,** Cantor, and Weierstrass found ways to do this. It turned out that no matter how it was done, building the continuum out of the natural numbers required new mathematical entities—infinite sets.

Foundationism—Our Inheritance

The textbook picture of the philosophy of mathematics is strangely fragmentary. You get the impression that the subject popped up in the late nineteenth century because of difficulties in Cantor's set theory. There was talk of a "crisis in the foundations." To repair the foundations, three schools appeared. They spent thirty or forty years quarreling. But none of the three could fix the foundations. The story ends some sixty years ago. Whitehead and Russell abandoned logicism. Gödel's incompleteness theorem checkmated Hilbert's formalism. Brouwer remained in Amsterdam, preaching constructivism, ignored by most of the mathematical world.

This episode was a critical period in the philosophy of mathematics. By a striking shift in meaning of words, the domination of philosophy of mathematics by foundationism became the *identification* of philosophy of mathematics with foundations. We're left with a peculiar impression: The philosophy of mathematics was awakened by contradictions in set theory. It was active for forty or fifty years. Then it went back to sleep.

Of course there has always been a philosophical background to mathematical thinking. In the foundationist period, leading mathematicians engaged in public controversy about philosophical issues. To make sense of that period, look at what went before and after. Two strands of history have to be followed, philosophy of mathematics and mathematics itself. The "crisis" manifested a long-standing discrepancy between the Euclid myth, and the reality, the actual practice of mathematicians.

In discussions of foundations three dogmas are presented: Platonism, formalism, and intuitionism. Platonism was described in Chapter 1. I remind the reader what it says: "Mathematical objects are real. Their existence is an objective fact, independent of our knowledge of them. Infinite sets, uncountably infinite sets, infinite-dimensional manifolds, space-filling curves—all the denizens of the mathematical zoo—are definite objects, with definite properties. Some of their properties are known, some are unknown. These objects aren't physical or material. They're outside space and time. They're immutable. They're uncreated. A meaningful statement about one of these objects is true or false, whether we know it or not. Mathematicians are empirical scientists, like botanists. We can't invent anything; it's there already. We try to discover."

In recent times Platonism has sometimes been identified with logicism. *If* a Platonist makes an effort to explain the nature of his nonhuman mathematical objects, it's usually in terms of logic and/or set theory.

According to formalism, on the other hand, there are *no* mathematical objects. Mathematics is *axioms, definitions, and theorems*—in brief, formulas. A strong version of formalism says that there are rules to derive one formula from another, but the formulas aren't *about* anything. They're strings of meaningless symbols. Of course the formalist knows that mathematical formulas are being applied to physics. When a formula gets a physical interpretation, *then* it acquires meaning. *Then* it can be true or false. But the truth or falsity refers only to the physical interpretation. As a mathematical formula apart from any interpretation, it has no meaning and can be neither true nor false.

The difference between formalist and Platonist is clear in their attitudes to Cantor's continuum hypothesis. Cantor conjectured that there's no infinite cardinal number greater than $\aleph_0$ (the cardinality of the integers) and smaller than c (the cardinality of the real numbers). Kurt Gödel and Paul J. Cohen showed that on the basis of the Zermelo—Fraenkel axioms of set theory, the continuum hypothesis can neither be disproved (Gödel, 1937) nor proved (Cohen, 1964). To the Platonist, this means our axioms of sets are incomplete. The continuum hypothesis *is* either true or false. We just don't understand the real numbers well enough to tell which is the case.

To the formalist, the Platonist interpretation makes no sense, because there *is* no real number system, except as we "create" it by laying down axioms to describe it. We're free to change these axioms, for convenience, usefulness, or any criterion that appeals to us. But the criterion can't be better correspondence with reality, because there's no reality to correspond with.

Formalists and Platonists take opposite sides on existence and reality. On the principles of mathematical proof, they have no quarrel. Opposed to both of them are the constructivists. Constructivists accept only mathematics that's obtained from the natural numbers by a finite construction. The set of real numbers, and any other infinite set, cannot be so obtained. Consequently, the constructivist accepts neither the Platonist not the formalist view of Cantor's hypothesis. Cantor's hypothesis is meaningless. Any answer is a waste of breath.

Today some mathematicians still call themselves formalists or constructivists, but in philosophical circles one speaks more often of Platonists versus fictionalists. Fictionalists reject Platonism. They can be formalists, constructivists, or something else (see Chapter 10).

Philosophers like to call their arguments and counter-arguments "moves." It's a standard move to finesse a dispute by declaring it meaningless. This was favored by the logical positivists in days of yore. They decreed that the meaning of any statement is no more or less than its truth conditions. It followed, where logical positivists were in charge, that metaphysics, ethics, and much else was thrown out of philosophy. In time, by the same rule, logical positivism was thrown out too.

Two large facts about mathematics are hardly doubted.

Fact one: Mathematics is a human product.

It may seem unfair to expect a Platonist to admit this; it may seem like asking him to give the game away. Nevertheless, contributions to mathematics are made every day, by specific, particular human beings. Many contributions are signed by an author or authors. No one questions the claim of these authors for their results.

The quibble between discovery and invention or creation was discussed in Chapter 5. But no one doubts that the mathematics we know comes from the work of human beings. In fact, it is sometimes possible to account for features of a mathematical discovery by the interests, tastes, and attitudes of the discoverer and sometimes also by the needs or traditions of his country. This fact is the bulwark of "fictionalism." In its way of coming into being, in the way in which it's thought of by its creators, mathematics is like an art such as fiction or sculpture.

Fact two: We can choose a problem to work on, but we can't choose what the answer should be.

When you resolve a mathematical difficulty, you sense that the answer was already "there," waiting to be found. Even if the answer is, "There's no answer," as in Cantor's continuum problem, *that* is the answer, like it or not.

The number 6,785,123,080,772,901,001 is either prime or composite. I don't know which. But I know that any method I use will give the same answer—prime or composite. 6,785,123,080,772,901,001 is what it is, regardless of what I think or know. In this respect numbers are independent of their creators. Did I just bring 6,785,123,080,772,901 into existence? Or was it waiting and ready, some*how* if not some*where*, along with billions, quadrillions, and quintillions of cousins?

The philosophers more impressed by the *objectivity* of mathematics are Platonists. They say numbers exist apart from human consciousness. Those more impressed with the *human role* in creating mathematics are anti-Platonists. Depending on what they offer to replace Platonism, they may be fictionalists, formalists, constructivists, intuitionists, conventionalists.

What Is logic? What Should It Be?

Is it the rules of correct thinking?

Everyday experience, and ample study by psychologists, show that most of our thinking doesn't follow logic.

This might mean most human thinking is wrong. Or it might mean the scope of logic is too narrow.

Computing machines do almost always obey logic.

That's the answer! Logic is the rules of computing machinery! Logic also applies to people when they try to be computing machines.

Once upon a time logic and mathematics were separate. Then George Boole figured out how to make logic part of mathematics.

Russell claimed the opposite—that mathematics is nothing but logic. But the paradoxes made that idea unpalatable. Far from a solid foundation for mathematics, set theory/logic is now a branch of mathematics, and the least trustworthy branch at that.

Like other branches, logic has expanded greatly in scope and power. It offers problems and challenges, techniques and tools to other parts of mathematics. And it renounces any desire or duty to check up on other branches of mathematics, or to tell people how to think. For today's mathematical logician, logic is just another branch of mathematics like geometry or number theory. He disowns philosophical responsibilities.

"This book does not propose to teach the reader how to think. The word 'logic' is sometimes used to refer to remedial thinking, but not by us" (Enderton).

In U.S. philosophy departments, on the other hand, "analytic philosophy," a kind of left-over from logicism and logical positivism, lingers on. Kitcher gave it the fitting sobriquet "neo-Fregeanism" (see Carnap and Quine in Chapter 9).

Analytic philosophers mustn't be confused with mathematical logicians. A few outstanding logicians do encompass both mathematical and philosophical logic; that is, they are competent by both mathematical and philosophical standards.

Of course logical blunders aren't acceptable in mathematical reasoning. In that sense, mathematicians (and other scientists) are subject to logic. This isn't the business of logicians. It's the business of the mathematician and her referees.

Gottlob Frege (1848–1925)

Grandpa of the Mainstream

Frege is the first *full-time* philosopher of mathematics. According to Baum, "Although Frege is sometimes spoken of as being the first philosopher of mathematics, he was at most the initiator of the recent period of intensive concentration on this area by specialists using the tools of mathematical logic. Frege considered himself to be working entirely within the tradition of Plato, Descartes, Leibniz, etc. with regard to his work on the philosophy of mathematics" (Baum, p. 263). Frege's greatest contribution to learning is the *Begriffsschrift* (*Idea Script*), where he introduced quantifiers—symbols for "there exists" (now written backward E) and for "for all" (now written upside-down A). Quantifiers were independently invented by O. H. Mitchell, a student of Charles Sanders Peirce (Lewis, 1918; Putnam, 1990). Frege's introduction of quantifiers is considered the birth of modern logic. His technical logic is a means to a philosophical end. He wants to establish arithmetic as a part of logic.

It's believed that logic with quantifiers (usually called "predicate calculus") can express any reasoning mathematicians use in strict, formal proof. (We also do heuristic, intuitive, informal reasoning.) In principle, using Frege's notation or others developed later, it seems possible to write any complete mathematical proof in a form that a computer can check.

Kant thought geometry is based on space intuition, and arithmetic on time intuition. That made both geometry and arithmetic "synthetic a priori." About geometry, Frege agreed with Kant that it is a synthetic intuition. About arithmetic, he agreed with Leibniz: It is not *synthetic* but *analytic*. That is, it doesn't depend on an intuition of time. It comes from logic. For Frege and Leibniz, "logic" means the intuitively obvious rules of correct reasoning. These are supposed to be certain and indubitable, independent of anybody's thought or experience. Deriving arithmetic from logic would make arithmetic equally certain and indubitable.

One really cannot speak of Frege's philosophy of mathematics. He had a philosophy of arithmetic, and a different philosophy of geometry. Arithmetic is logic; geometry is space intuition. Fitting them together is as awkward as yoking an ape and an alligator. If arithmetic is part of logic, why not geometry as well (since we construct spaces from numbers by using coordinates)? On the other hand, if geometry is space intuition, why may not arithmetic be time intuition, as Kant had it?

Frege's Grundlagen. Logicism's Koran

In his *Grundlagen der Arithmetik* (Foundations of Arithmetic) Frege constructed the natural numbers out of logic. This achievement was ignored for 16 years, until Bertrand Russell took up the same project and made Frege known to the world. "Today, Frege's *Grundlagen* is widely appreciated as a philosophical masterpiece. In retrospect the mathematicians who ignored it appear as men who failed to recognize a pioneering work" (Kitcher, "Frege, Dedekind . . ."). Since all classical mathematics can be built from the natural numbers, Russell claimed that *all mathematics* is logic. This is called logicism.

Before giving his definition of number in the *Grundlagen*, Frege tries to demolish all previous definitions. He carries out a merciless, hilarious campaign against psychologism (numbers are ideas in someone's head—Berkeley, Schloemilch); historicism (numbers evolve); and empiricism (numbers are things in the physical world—Mill). To this day, philosophers of mathematics hardly dare contemplate psychologism or historicism.

First Frege trounces Mill, who based arithmetic on empirical experience (Frege, 1980, pp. 9–10): "The number 3 . . . consists, according to him, in this, that collections of objects exist, which while they impress the senses thus,

0 0 ,
 0

they may be separated into two parts, thus,

00   0.

What a mercy, then, that not everything in the world is nailed down; for if it were, we should not be able to bring off this separation, and 2 + 1 would not be 3!

What a pity that Mill did not also illustrate the physical facts underlying the number 0 and 1! . . . From this we can see that it is really incorrect to speak of three strokes when the clock strikes three, or to call sweet, sour and bitter three sensations of taste, and equally unwarrantable is the expression 'three methods of solving an equation'. For none of these is a parcel which ever impresses the senses thus,

0 0."
 0

(A bit unfair! Mill actually mentions strokes of the clock as an example of counting. See the article on Mill in the next chapter.)

Mill's lack of precision makes him an easy mark for Frege. Nevertheless, Mill is right to say number has something to do with physical reality. Every child learns arithmetic from the pebbles and ginger snaps Frege laughs at. If our ancestors didn't need to keep track of coconuts or fish heads, they wouldn't have invented arithmetic. Much deeper is the discovery, many times repeated, that mathematics is the language of nature. Mill grapples with the relation between numbers and physical reality. Frege brushes it aside. In that respect, Mill did a service to human understanding, Frege a disservice.

Next is Hermann Hankel. Frege writes, "the first question to be faced is whether number is definable." Hankel thought not, and expressed himself in this unfortunate manner: "What we mean by thinking or putting a thing once, twice, three times, and so on, cannot be defined, because of the simplicity in principle of the concept of putting." Replies Frege, "But the point is surely not so much the putting as the once, twice, and three times. If this could be defined, the indefinability of putting would scarcely worry us" (p. 26).

Next Frege turns on George Berkeley (p. 33), whom he quotes: "Number . . . is nothing fixed and settled, really existing in things themselves. It is entirely the creature of the mind. . . . We call a window one, a chimney one, and yet a house in which there are many windows, and many chimneys, hath an equal right to be called one, and many houses go to the making of one city." In fact, as mentioned in an earlier chapter, Berkeley thought numbers don't exist. To him, numerals were meaningless symbols.

Frege's answer: "This line of thought may easily lead us to regard number as something subjective . . . number is no whit more an object of psychology or a product of mental processes than, let us say, the North Sea is. The objectivity of the North Sea is not affected by the fact that it is a matter of our arbitrary choice which part of all the water on the earth's surface we mark off and elect to call the North Sea. This is no reason for deciding to investigate the North Sea by psychological methods. In the same way, number too, is something objective. If we say 'The North Sea is 10,000 square miles in extent' then neither by the 'North Sea' nor by '10,000' do we refer to any state of or process in our minds: on the contrary, we assert something quite objective, which is independent of our ideas and everything of the sort."

His attack on psychologism means to prove that numbers aren't ideas. To do so, he assumes surreptitiously that ideas are property only of individuals, uncorrelated with other people's ideas or with physical reality. An indefensible assumption!

Here is his next assault, against Schloemilch: "I cannot agree with Schloemilch either (p. 36), when he calls number the idea of the position of an item in a series. If number were an idea, then arithmetic would be psychology. But arithmetic is no more psychology than, say astronomy is. Astronomy is concerned, not with ideas of the planets, but with the planets themselves, and by the same token the objects of arithmetic are not ideas either. If the number two were an idea, then it would have straight away to be private to me only. [No! No!] Another man's idea is, *ex vi termini*, another idea. We should then have it might be many millions of twos on our hands. We should have to speak of my two and your two, of one two and all twos. If we accept latent or unconscious ideas, we would have unconscious twos among them, which would then return subsequently to consciousness. As new generations of children grew up new generations of twos would continually be being born and in the course of millennia these might evolve, for all we could tell, to such a pitch that two of them would make five. Yet, in spite of all this, it would still be doubtful whether there existed infinitely many numbers, as we ordinarily suppose. 10^{10}, perhaps, might be only an empty symbol, and there might exist no idea at all, in any being whatever, to answer to the name.

"Weird and wonderful, as we see, are the results of taking seriously the suggestion that number is an idea. And we are driven to the conclusion that number is neither spatial and physical, like Mill's piles of pebbles and gingersnaps, nor yet subjective, like ideas, but non-sensible and objective. Now objectivity cannot, of course, be based on any sense-impression, which as an affection of our mind is entirely subjective, but only, so far as I can see, on the reason. It would be strange if the most exact of all the sciences had to seek support from psychology, which is still feeling its way none too surely."

It's too late to defend Mill or Hankel or Schloemilch. But we must reject Frege's argument against "psychologism"—the belief that mathematical objects are ideas. He says 2 cannot be an idea, because different people have different ideas, and there is only one 2.

Frege is confounding private and public senses of "idea." It's not unusual to say "We have the same idea." Frege assumes an idea resides only in one person's head (private ideas). But ideas can be shared by several people, even millions of people (public ideas). Cheap and dear, legal and illegal, sacred and profane, patriotic and treasonous—all ideas, but not ideas of a particular person. Public ideas, part of society, history, and culture. (Philosophers say "intersubjective" to avoid "society" and "culture.") The existence of language, society, and all social institutions prove that people sometimes do have the same idea. Not in the sense of subjective inner consciousnesses; in the sense of verbal and practical understanding and agreement. A piece of green paper called a dollar is worth a quart of milk because many people agree that it is. All these people have the same idea—the equal value of a quart of milk and a dollar bill.

Whoever told Frege, "2 is an idea" intended the public meaning of "idea." Frege replaced that with the private meaning of "idea," and then had fun throwing stones at Schloemilch. Could Frege prove that number is *not* an intersubjective, social-cultural object? No. His sarcasm about psychologism has no bearing on my proposal that mathematical objects are ideas on the social level.

After Frege makes mincemeat of empiricism, historicism, psychologism, and (in another book, the *Grundgesetze*) formalism, you are ready for his solution: *NUMBERS ARE ABSTRACT OBJECTS.* Objects which are real, but not physically, not psychologically, real in an *abstract* sense.

What is "abstract"?

Evidently, *not* mental or physical. What qualities are possessed by abstract objects? They're timeless or tenseless. Aren't born, do not die.

What an astonishing kinship to Plato's Ideas! They were neither mental nor physical, but eternal and changeless. Frege is a Platonist as well as a Kantian.

Frege's abstract objects include numbers. Everything else has been proved wrong, so this must be right. But the elimination argument isn't valid, because Frege hasn't considered the alternative we offer: mathematics as part of the social-cultural-historical side of human knowledge. (He did attack the notion of numbers as historical entities.)

Frege's argument against formalism, psychologism, and empiricism comes down to a declaration: "Anyone can see that $7 + 3 = 10$. There's no possible doubt of it. Clearly it's true a priori, now and forever, certainly and indubitably." The same argument was given 1,300 years earlier by Augustine, Bishop of Hippo: "Seven and three are ten, not only now but always; nor was there ever a time when seven and three were not ten, nor will ever be a time when seven and three will not be ten. I say, therefore, that this incorruptible truth of number is common to me and to any reasoning person whatsoever."

Augustine's arithmetical Platonism went with his theology. The certainty of mathematics supported the certainty of religion. By Frege's time, the association between Christian theology and mathematical Platonism had gone underground. The success of secular science made it bad form to bring religion into logic or mathematics. Even so, David Hilbert and Bertrand Russell, unlike Frege, were frank about their religious motives.

The important thing is Frege's analysis of number. *Numbers are classes.* More precisely, *they're equivalence classes of classes*, under the equivalence relation of one-to-one mappings. For example, *two is the class of all pairs.* This class of all pairs exists objectively, timelessly, independently of us. It's an abstract object.

Some readers haven't seen arithmetic built up from logic. It's easy. Frege's logic uses the concept of "class." This is almost what we mean today by set. A set is defined by who are its *members* ("extension"). A class is defined by the *property* that decides whether or not you're a member ("intension").

Frege says "Two is the class of all pairs. Three is the class of all triplets. And so on." The point of this construction is to prove the "a priority" of the numbers, their independence of experience.

One slight problem: These definitions are circular. Knowing what's a "pair" is already knowing what's 2. A better statement is: 2 is by definition the class of all classes equivalent to {A,B}. 2 is called the cardinality of any such class (commonly known as a pair). 3 is the class of all classes equivalent to {A,B,C}. 3 is called the "cardinality" of any such class (commonly known as a triplet).

What's "equivalent"? Classes are equivalent if their members can be matched with nothing left over. Any two pairs are equivalent. Any two triples are equivalent. Any natural number, including 0 and 1, is a class of equivalent classes. The reader is encouraged to think through these two special cases.

A little explanation will help you sympathize with Frege. When he defines 2 as the equivalence class of all pairs, he's assuming that the notion, "equivalence class of all pairs" is free of ambiguity. To *justify* his definition of "2," we have to see if there *is* a "class of all pairs." If it doesn't exist, we needn't bother about it! That such a thing exists may have been crystal clear to Frege. It's not so clear today. Today, with caution learned from Frege's burnt fingers, "the class of all pairs" or "the set of all sets equivalent to {0,1}" would not go down so easily. "Pairs of what?" the students would rightfully demand.

Not pairs of numbers; we're trying to *define* numbers. Not shoes or socks; they're too earthy for a transcendental theory. Probably pairs of *abstract* objects. But where are they and what are they? To create the mathematical universe from scratch, I have no ingredients available. If I want to create numbers as collections, I need something to collect! In today's mathematics, one doesn't take for granted that any specification written in English or German is meaningful to define a set. Nowadays we want a set to be *located*, a subset of a given universal set. You may not simply "define" some infinite set. You must show that the definition isn't self-contradictory. Our caution is due partly to the disaster that befell Frege: the Russell paradox.**

This proposal opened a new direction of thinking in foundations. It was the basis of Frege's plan to make arithmetic part of logic/set theory.

Frege's influence is not so much his semi-Kantian philosophy as his statement of the issue—*establish mathematics on a solid, indubitable foundation.*

To an unprepared mind, Frege's definition of number is bizarre. It explains the clear and simple, number, by the complicated and obscure: infinite equivalence classes. The bizarrerie is mitigated if you remember the point—to reduce arithmetic to logic. Mathematicians know how to build analysis and geometry on arithmetic. Frege and Russell believe logic is rock-solid. If they could have built arithmetic on logic, that would have made *all* mathematics as solid as logic itself. It didn't work out that way.

0 is particularly nice. It's the class of sets equivalent to the set of all objects unequal to themselves! *No* object is unequal to itself, so 0 is the class of all empty sets. But all empty sets have the same members—none! So they're not merely *equivalent* to each other—they're all *the same* set. There's only one empty set! (A set is characterized by its membership list. There's no way to tell one empty membership list from another. Therefore all empty sets are the same thing!)

Once I have *the* empty set, I can use a trick of von Neumann as an alternative way to construct the number 1. Consider the class of *all* empty sets. This class has exactly one member: the unique empty set. It's a singleton. "Out of nothing" I have made a *singleton* set—a "canonical representative" for the cardinal number 1. 1 is the class of all singletons—all sets with but a single element. To avoid circularity: "1 is the class of all sets equivalent to the set [{ }]." In words, 1 is the class of all sets equivalent to the set whose only element is the empty set. Continuing, you get pairs, triplets, and so on. Von Neumann recursively constructs the whole set of natural numbers out of sets of sets of sets of nothing.

Set theory was introduced by Georg Cantor as a fundamental new branch of mathematics. The idea of set—any collection of distinct objects—was so simple and fundamental, it looked like a brick out of which all mathematics could be constructed. Even arithmetic could be downgraded (or upgraded) from primary to secondary rank, for the natural numbers could be constructed, as we have just seen, from nothing—i.e., the empty set—by operations of set theory.

At first set theory seemed to be the same as logic. The set-theoretic relation of inclusion, "A is a subset of B," is the same as the logical relation of implication, "If A, then B." "Logic" here means the fundamental laws of reason, of contradiction and implication—the objective, indubitable bedrock of the universe. To show mathematics is part of logic would show it's objective and indubitable. It would justify Platonism, passing to the rest of mathematics the indubitability of logic.

This was the "logicist program" of Russell and Whitehead's *Principia Mathematica*. The logicist school was philosophically (not technically) similar to Hilbert's formalist school. For the logicists it was logic that was indubitable a priori; for the formalists, it was finite combinatorics. The difference between the logicists and Kant is that they give up his claim that mathematics is synthetic a priori. They settle for analytic a priori. According to Bertrand Russell, mathematics is a vast tautology.

The logicists proposed to redeem all mathematics by injecting it with the soundness of logic. First of all, to reduce arithmetic to rock-solid logic. Is the notion of class rock-solid? Even an infinite class? No. If we include infinite sets, logic isn't rock-solid any more. But the class of singletons is already infinite. Without infinite sets, there's no mathematics.

Frege regarded "set" or "class" as equivalent to "property." To any property corresponds the set of things having that property. To any set corresponds the property of membership in it.

Frege's Fifth Basic Law says that to any properly specified property corresponds a set (Furth, 1964). Definition by properties gives a "concept." To Frege, defining numbers as sets is automatically defining them as concepts—not notions in someone's head, but abstract objects.

Frege was about to publish a monumental work in which arithmetic was reconstructed on the foundation of set theory. His hope was shattered in one of the most poignant episodes in the history of philosophy. Russell found a contradiction in the notion of set as he and Frege used it! After struggling for weeks to escape, he sent Frege a letter (van Heijenoort).

Frege added this postscript to his treatise: "A scientist can hardly meet with anything more undesirable than to have the foundations give way just as the work is finished. In this position I was put by a letter from Mr. Bertrand Russell, as the work was nearly through the press."

The axioms from which Russell and Frege attempted to construct mathematics are contradictory!

"My Basic Law concerning courses-of-values (V) . . . the (unrestricted) Axiom of Set Abstraction states that there exists, for any property we describe via an open formula, a set of things which possess the property. From this Axiom we can easily derive Russell's Paradox" (Musgrave, 1964, p. 101). Russell's paradox is catastrophic because it exhibits a legitimate property that is self-contradictory—a property to which no set can correspond.

The Russell paradox and the other "antinomies" showed that intuitive logic is riskier than classical mathematics, for it led to contradictions in a way that never happens in arithmetic or geometry. This was the "crisis in foundations," the central issue in the famous controversies of the first quarter of this century. Three remedies were proposed—logicism, intuitionism, and formalism. As we have already mentioned, all failed.

The response of "logicism," the school of Frege and Russell, was to reformulate set theory to avoid the Russell paradox, and thereby save the Frege-Russell-Whitehead project of establishing mathematics on logic as a foundation.

Work on this program played a role in the development of logic. But in terms of foundationism, it was a failure. To exclude the paradoxes, set theory had to be patched up into a complicated structure. It acquired new axioms such as the axiom of replacement (a complex recreation of Frege's Axiom 5) and the axiom of infinity (there exists an infinite set).

"There is something profoundly unsatisfactory about the axiom of infinity. It cannot be described as a truth of logic in any reasonable use of that term and so the introduction of it as a primitive proposition amounts in effect to the abandonment of Frege's project of exhibiting arithmetic as a development of logic" (Kneale and Kneale, p. 699).

This patched up set theory could not be identified with logic in the philosophical sense of "rules for correct reasoning." You can build mathematics out of this reformed set theory, but it no longer passes as a foundation, in the sense of justifying the indubitability of mathematics. Mathematics was not shown to be part of logic in the classical sense, as Russell and Whitehead dreamed. It became untenable to claim, as Russell had done, that mathematics is one vast tautology.

"Among all mathematical theories it is just the theory of sets that requires clarification more than any other" (Mostowski).

After Frege's first shock, he continued his foundationalist labors. Russell searched for a way out for a long time. He came up with a modified form of set theory, the theory of types. Zermelo introduced the axiom of foundation, which says any chain of set membership terminates in finitely many steps. This outlaws Russell's paradox by outlawing "Russell sets"—sets that belong to themselves—since the membership relation for a Russell set cycles round ad infinitum. (Recently a British computer scientist, Peter Aczel, published a version of set theory in which Zermelo's axiom of foundation is negated. This theory permits self-membership, and has applications in computer science. It has been proved to be relatively consistent!)

But set theory doctored up with the axiom of infinity, and Zermelo's axiom of foundation was no longer the perspicuous elementary set theory that had aroused foundationist hopes. Russell's paradox was unexpected. Are other paradoxes lurking?

The Russell paradox doomed that hope. Despite this philosophical failure, logico-set theoreticism dominates the philosophy of mathematics today. Philip Kitcher writes that "mathematical philosophy in the last 30 years is a series of footnotes to Frege." This suggests that mathematical philosophy is ready for new ideas and problems. Perhaps ideas and problems rising from today's mathematical practice.

Logicism never recovered from the Russell paradox. Eventually both Frege and Russell gave it up. Set theory had become, not clear and indubitable like elementary logic, but unclear and dubitable. To define a set by a property, I must show the property isn't self-contradictory. Such a demonstration can be harder than the problem it was supposed to clarify. "Reducing" arithmetic to logic was a disappointment. Instead of being anchored to the rock of logic, it was suspended from the balloon of set theory.

Frege's construction of number is defensible. But it's not sufficient to convince doubters that arithmetic is a priori. Its long-range importance wasn't Frege's philosophical goal but the stimulation it gave to logic and foundations. The Frege-Russell definition or the equivalent von Neumann definition let us derive arithmetic from facts about sets. Mainstream mathematicians ignored it for a long time. Frege at Jena was an unknown outsider, but even when the respected Richard Dedekind wrote on the foundation of the natural numbers, he too aroused little interest among mathematicians.

Mathematicians don't regard the natural numbers as a problem. With millennia of experience behind us, and deep, complex problems before us, we're not worrying about elementary arithmetic. Dedekind and Frege may object that we have only vague notions of what's meant by 0 or 1 or 2. Nevertheless, we have no qualms about 0, 1, 2.

Frege and Russell weren't mainly concerned with the opinion of the ordinary, unphilosophical mathematician. They were concerned with establishing mathematics on a solid foundation.

Frege always allowed geometry to rest on space intuition. In his old age he decided arithmetic too was based on geometry and space intuition (1979, pp. 267–81).

Hilbert published his *Grundlagen der Geometrie*, an epoch-making book that led to universal acceptance of the axiomatic method as the right way to present mathematics—in principle. In this book Hilbert (following Pasch and Peano) filled in the gaps in Euclid, making Euclidean geometry for the first time the rigorously logical subject it had always claimed to be. He did more. He showed that the axioms are *independent* (can't be deduced from each other) by giving examples in which all the axioms were satisfied except one.

Frege's Kantian views on geometry led him to attack Hilbert. He told Hilbert that Hilbert didn't know the difference between a definition and an axiom. Hilbert answered Frege's first letter or two (1979, pp. 167–73). Thereafter he ignored him. But Frege continued to crow. He even insinuated that Hilbert's failure to keep up the controversy was because Hilbert was afraid his results might be false!

Musgrave: "By 1924 Frege had come to the conclusion that 'the paradoxes of set theory have destroyed set theory.' He continued: 'The more I thought about it the more convinced I became that arithmetic and geometry grew from the same foundation, indeed from the geometrical one; so that the whole of mathematics is actually geometry.' " (These two remarks are quoted by Bynum in his Introduction to Frege [1972], cf. pp 53–54.)

Bertrand Russell (1872–1970)

A Loss of Faith

It wouldn't be too wrong to say philosophy of science in the twentieth century is mostly Bertrand Russell. Two other leading thinkers—Frege and Wittgenstein—are both Russell proteges. He didn't create them as philosophers, of course. But his enthusiasm for them is in part their compatibility with his logical atomism. In helping them become influential, he indirectly advances his own point of view.

Russell is frank about his motives, so far as he understands them. In philosophy of science, his leading motive is to establish certainty. In this, he confesses, he's seeking to replace the Christian faith he has rejected. He is also continuing an old tradition: Plato, Descartes, Leibniz, Kant. From "Reflections on My Eightieth Birthday" in *Portraits from Memory*:

"I wanted certainty in the kind of way in which people want religious faith. I thought that certainty is more likely to be found in mathematics than elsewhere. But I discovered that many mathematical demonstrations, which my teachers expected me to accept, were full of fallacies, and that, if certainty were indeed discoverable in mathematics, it would be in a new field of mathematics, with more solid foundations than those that had hitherto been thought secure. But as the work proceeded, I was continually reminded of the fable about the elephant and the tortoise. Having constructed an elephant upon which the mathematical world could rest, I found the elephant tottering, and proceeded to construct a tortoise to keep the elephant from falling. But the tortoise was no more secure then the elephant, and after some twenty years of very arduous toil, I came to the conclusion that there was nothing more that I could do in the way of making mathematical knowledge indubitable."

"Mathematics is, I believe," says Russell, "the chief source of the belief in eternal and exact truth, as well as in a super-sensible intelligible world. Geometry deals with exact circles, but no sensible object is *exactly* circular; however carefully we may use our compasses, there will be some imperfections and irregularities. This suggests the view that all exact reasoning applies to ideal as opposed to sensible objects; it is natural to go further, and to argue that thought is nobler than sense, and the objects of thought more real than those of sense-perception. Mystical doctrines as to the relation of time to eternity are also reinforced by pure mathematics, for mathematical objects, such as number, if real at all, are eternal and not in time. Such eternal objects can be conceived as God's thoughts. Hence Plato's doctrine that God is a geometer, and Sir James Jeans' belief that He is addicted to arithmetic. Rationalistic as opposed to apocalyptic religion has been, ever since Pythagoras, and notably ever since Plato, very completely dominated by mathematics and mathematical method.

"So it compels the soul to contemplate being, it is proper; if to contemplate becoming, it is not proper" (*Republic*, p. 326). For Plato, the "becoming" or the "unreal" is anything visible, ponderable, changeable. The "being," the "real," is invisible, immaterial, unchangeable. That means mathematics.

Russell calls himself a "logical atomist," in opposition to both the classical and evolutionist trends in early twentieth-century philosophy.

"Philosophy is to be rendered scientific" (p. 28).

"The philosophy which is to be genuinely inspired by the scientific spirit . . . brings with it—as a new and powerful method of investigation always does—a sense of power and a hope of progress more reliable and better grounded than any that rests on hasty and fallacious generalization as to the nature of the universe at large. . . . Many hopes which inspired philosophers in the past it cannot claim to fulfil; but other hopes, more purely intellectual, it can satisfy more fully than former ages could have deemed possible for human minds" (p. 20).

He's good at understated sarcasm. "The classical tradition in philosophy is the last surviving child of two very diverse parents: the Greek belief in reason, and the medieval belief in the tidiness of the universe. To the schoolmen, who lived amid wars, massacres, and pestilences, nothing appeared so delightful as safety and order . . . the universe of Thomas Aquinas or Dante is as small and neat as a Dutch interior. . . . To us, to whom safety has become monotony . . . the world of dreams is very different . . . the barbaric substratum of human nature, unsatisfied in action, finds an outlet in imagination (Written before August, 1914)."

Alan Musgrave (1977) quotes Russell, *An Essay on the Foundations of Geometry*, 1897, p. 1: "Geometry, throughout the 17th and 18th centuries, remained, in the war against empiricism, an impregnable fortress of the idealists. Those who held—as was generally held on the Continent—that certain knowledge, independent of experience, was possible about the real world, had only to point to Geometry: none but a madman, they said, would throw doubt on its validity, and none but a fool would deny its objective reference. The English Empiricists, in this matter, had, therefore, a somewhat difficult task; either they had to ignore the problem, or, if, like Hume and Mill, they ventured on the assault, they were driven into the apparently paradoxical assertion that Geometry at bottom, had no certainty of a different *kind* from that of Mechanics."

P. H. Nidditch

Frank Talk on Logicism

The logician and historian P. H. Nidditch gave a fair summing up of the logicist struggle to save the foundations of mathematics. "The effect of these discoveries (Russell & Burali-Forti antinomies) on the development of Mathematical Logic has been very great. The fear that the current systems of mathematics might not have consistency has been chiefly responsible for the change in the direction of Mathematical Logic towards metamathematics, for the purpose of becoming free from the disease of doubting if mathematics is resting on a solid base. A special reason for being troubled is that the theory of classes is used in all parts of mathematics; so if it is wrong in some way, they are possibly in error. Further, quite separately from the theory of classes, might not discoveries of opposite theorems in algebra, geometry or Mathematical Analysis suddenly come into view, as the discoveries of Burali-Forti and Russell had done? It has been seen that common sense is not good enough as a lighthouse for keeping one safe from being broken against the overhanging slope of sharp logic. To become certain with good reason that the systems of mathematics are all right it is necessary for the details of these structures to be looked at with care and for demonstrations to be given that, with those structures, consistency is present.

"This last point and the fears and troubled mind that we have been talking about in these lines have been and are common among workers in what is named 'the foundations of mathematics,' that is, axiom systems of logic-classes-and-arithmetic. However, some persons, with whom the present writer is in agreement, have a different opinion. They would say the well being of mathematics is not dependent on its 'foundations.' The value of mathematics is in the fruits of its branches more than in its 'roots'; in the great number of surprising and interesting theorems of algebra, analysis, geometry, topology, theory of numbers and theory of chances more than in attempts to get a bit of arithmetic or topology as simply a development of logic itself. They would say that the name 'foundations of mathematics' is a bad one in so far as it sends a wrong picture into one's mind of the relations between logic and higher mathematics. Higher mathematics is not resting on logic or formed from logic. They would say that the troubles in the theory of classes came from most special examples of classes, and such classes are not used in higher mathematics. They would say further that though to be certain of consistency is to be desired if such certain knowledge is possible to us, a knowledge of the consistency of the theories of mathematics that is probable is generally enough and the only sort of knowledge of consistency that one does in fact generally have. And they would say that such probable knowledge is well supported if the theories of mathematics have been worked out much and opposite theorems in them have not come to light. There is no suggestion in all this that Mathematical Logic is not an important part of mathematics; the view put forward is that there is much more to mathematics than Mathematical Logic is and might ever become. And there is no suggestion that the questions of consistency and like questions, and discovery of ways of answering them are not important; to no small degree Mathematical Logic now is as interesting and important as it is because of its interest in such questions and answers."

Luitjens E. J. Brouwer (1882–1966)

An Angry Topologist

After logicism came intuitionism, the doctrine of the great Dutch topologist L. E. J. Brouwer. The name intuitionism displays its descent from Kant's intuitionist theory of mathematical knowledge. Brouwer followed Kant in saying that mathematics is founded on intuitive truths. Brouwer's impact came not only from the force of his philosophy, but also from his wonderful discoveries in topology, and from his dominating presence, which led some people to see him as a leader.

Here is the manifesto Brouwer called the FIRST ACT OF INTUITIONISM: "Completely separating mathematics from mathematical language and hence from the phenomena of language described by theoretical logic, recognizing that intuitionistic mathematics is an essentially languageless activity of the mind having its origin in the perception of a move of time. This perception of a move of time may be described as the falling apart of a life moment into two distinct things, one of which gives way to the other, but is retained by memory. If the twoity thus born is divested of all quality, [echo of Aritotles's 'abstraction'] it passes into the empty form of the common substratum of all twoities. And it is this common substratum, this empty form, which is the basic intuition of mathematics" (Brouwer, 1981).

According to Brouwer, the natural numbers are given by the fundamental intuition of "a move in time." This intuition is the starting point for all mathematics, and all mathematics must be based *constructively* on the natural numbers. But the notion "constructive" cannot and need not be explained. Supposed mathematical objects not constructively based on the natural numbers are not meaningful. Their existence would not be established, even if it were shown that assuming their nonexistence leads to a contradiction.

From the First Act flows the main dogma that separates intuitionistic mathematics from ordinary "classical" mathematics: "The belief in the universal validity of the principle of the excluded third [excluded middle] in mathematics is considered by the intuitionists as a phenomenon of the history of civilization of the same kind as the former belief in the rationality of π, or in the rotation of the firmament about the earth." Misapplication of the "Law of the Excluded Middle" is the great evil in mathematics. Classical "true" and "false" should be replaced by "constructively true," "constructively false," and "neither."

Before Brouwer, the French analysts Henri Poincaré, Émile Borel, and Henri Lebesgue had misgivings and disagreements with nonconstructive methods and free use of infinite sets. (Brouwer called them pre-intuitionists.) But Brouwer's demand to restructure analysis from the ground up went much further. To most mathematicians it seemed excessive.

There is also the SECOND ACT OF INTUITIONISM: "Admitting two ways of creating new mathematical entities: firstly in the shape of more or less freely proceeding infinite sequences of mathematical entities previously acquired (so that, for example, infinite decimal fractions having neither exact values nor any guarantee of ever getting exact values are admitted; secondly, in the shape of mathematical species, i.e. properties supposable for mathematical entities previously acquired, satisfying the condition that if they hold for a certain mathematical entity, they also hold for all mathematical entities which have been defined to be 'equal' to it, definitions of equality having to satisfy the conditions of symmetry, reflexivity and transitivity."

Brouwer's most famous contribution to topology was his "fixed point theorem."** It's a powerful tool in classical and applied branches such as differential equations. But it isn't constructive. Nor is the rest of his great work in topology. Ultimately he decided his own best work was wrong.

His interests went beyond mathematics. In youth he wrote a strange book called *Life, Art and Mysticism*. His biographer, van Stigt, calls it "the manifesto of an 'angry young man' rejecting and attacking all he sees at the surface of human society." Some quotes from the book:

"Intellect has done mankind a devil's service by linking the two phantasies of means and end . . . there are others (scientists) who do not know when to stop, who keep on and on until they go mad. They grow bald, short-sighted and fat, their stomachs stop working, and moaning with asthma and gastric trouble they fancy that in this way equilibrium is within reach and almost reached. . . . So much for science, the last flower and ossification of culture."

We present a famous example of Brouwer's intuitionism. It's about the law of trichotomy, which says: "Every real number is either positive, negative, or zero." Brouwer says the Law is false, and gives a counter-example—a real number that is neither positive, negative, nor zero! Most mathematicians vehemently reject this claim. His number *is* either zero, negative or positive, we say! We simply don't know how to determine which it is.

Since there's hardly any mathematics that doesn't depend on the real numbers, the example shows that Platonism is intimately associated with the practice of mathematics today.

To give the example, start with π, the ratio of the circumference of a circle to its diameter. From its decimal expansion we will shortly define a second number, π^ (read: "pi-hat"). The use of π in this construction is arbitrary. We could start with almost any other irrational number. We need two properties: (1) as with π, the capacity to (in principle!) compute its decimal expansion as far out as we wish; (2) some property of the expansion—for instance, appearance somewhere in it of a sequence of 100 successive zeros—that is "accidental" in the sense that we know no reason why this property is excluded or required by the definition of π. To determine whether somewhere in the decimal expansion of π there is a row of 100 successive zeros, we have no recourse but to generate the decimal expansion of π. If there is such a row, eventually (!!!) we'll find it. If there isn't, we won't know that's the case until we look at the complete infinite expansion—that is, never!

Let P be the statement, "In the decimal expansion of π, somewhere occurs a sequence of 100 successive zeros." Let –P be the negation, "In the decimal expansion of π, there's no sequence of 100 successive zeros."

What about the statement "Either P or –P" ? True or false? Is it true or false that P is either true or false? Most folks say "True." The law of the excluded middle (L.E.M.) says "True."

Brouwer says no! The law of the excluded middle doesn't apply. "The expansion of π" is a mythical beast. There's no such thing. The mistaken belief that either P or $-P$ must be true comes from the delusion that the expansion of π exists as a completed object. All that exists, however, all we know how to construct, is a finite piece of this expansion.

The argument may seem a bit theological. Why does it matter?

In fact, if we give up our belief in the expansion of π, in the truth of either P or $-P$, we must restructure all analysis.

We show how Brouwer disproved the law of trichotomy, one of the fundamental properties of the real numbers.

We define the new number $\pi^\wedge$, by a rule that successively determines the first thousand, million, or hundred billion digits of the decimal expansion of $\pi^\wedge$. That's all that's meant by "defining" a real number.

$\pi^\wedge$ looks a lot like π. In fact, it's the same as π in the first hundred, thousand, even the first hundred thousand places. The rule is: expand π until you find a row of 100 successive zeros (or until you reach the desired precision for $\pi^\wedge$, whichever comes first). Up to the first run of 100 successive zeros, the expansion of $\pi^\wedge$ is identical to that of π. Suppose the first run of 100 successive zeros starts in the 93d place. Then $\pi^\wedge$ terminates with its 93d digit. This makes $\pi^\wedge$ less than π. If the first run of 100 successive zeroes starts in the 94th place, put a 1 in the next, the 95th place, and then terminate. This makes $\pi^\wedge$ greater than π. If the first 100-zero run starts in any other place, it has to be either an even-numbered or an odd-numbered place. If in an odd-numbered place, $\pi^\wedge$ terminates there. If in an even-numbered place, $\pi^\wedge$ stops with a 1 in the next, odd-numbered place.

We don't know, and quite possibly never will know, if there is a place where a 100-zero row starts. There may not be any. Nevertheless, our recipe for constructing $\pi^\wedge$ is perfectly definite; we know it to as many decimal places as we know π. If π doesn't include 100 successive zeros anywhere, $\pi^\wedge = \pi$. If it does include such a sequence, and that sequence starts at an even-numbered place, $\pi^\wedge > \pi$. If it starts at an odd-numbered place, $\pi^\wedge < \pi$.

Now consider the difference, $\pi^\wedge - \pi$. Call it Q. Is Q positive, negative, or zero? Try to find out by calculating the expansion of π on a computer. You won't get an answer until the computer finds 100 successive zeros in π. If the computer runs a million years and doesn't find 100 successive zeroes, you still don't know if Q is positive, negative, or zero. If there actually is no hundred-zero sequence, you'll never know.

π has been expanded to billions of places by the brothers Gregory and David Chudnovsky in New York and Tanaka in Tokyo (Preston). No sequence of 100 zeroes has occurred so far. We know nothing about the next hundred billion digits.

Even if 100 successive zeros turn up tomorrow, we can ask instead about 1,000 successive 9s (for example) and again have an open question. There are plenty of questions like this that we'll never answer.

So what about the "law of trichotomy"? It says Q is either positive, negative, or zero, regardless of the fact that we may never know which.

The constructivist says, "None of the three is true! Q *will* be zero, positive or negative *when* someone determines which of the three is the case. Until then, it's none of the three. Any conclusion based on the compound statement

'Either $Q > O$, $Q = O$, or $Q < O$'

is unjustified. Any conclusion about an infinite set is defective if it relies on the law of the excluded middle. As the example shows, a statement may be (in the constructive sense) neither true nor false."

The standard mathematician finds the argument annoying. He has no intention of giving up classical mathematics for a more restricted version. Neither does he admit that his mathematical practice depends on a Platonist ontology. He neither defends Platonism nor reconsiders it. He just pretends nothing happened.

Some aspects of the intuitionist viewpoint are attractive to mathematicians who want to escape Platonism and formalism. The intuitionists insist that mathematics is meaningful, that it's a kind of human mental activity. You can accept these ideas, without saying that classical mathematics is lacking in meaning.

Errett Bishop

À bas with L.E.M.

The U.S. analyst Errett Bishop revised intuitionism, and created a cleaned up, streamlined version he called "constructivism." Constructivism is concerned above all with throwing out the law of the excluded middle for infinite sets. It's closer to normal mathematical practice than Brouwer. It's not tainted with mysticism. Bishop's book *Constructive Analysis* goes a long way toward reconstructing analysis constructively. Here and there in the mathematical community, a few cells of constructivists are still active. But their dream of converting the rest of us is dead. The overwhelming majority long since rejected intuitionism and constructivism, or never even heard of them.

Like Brouwer, Bishop said a lot of standard mathematics is meaningless. He went far beyond Brouwer in remaking it constructively. Some quotes:

"One gets the impression that some of the model-builders are no longer interested in reality. Their models have become autonomous. This has clearly happened in mathematical philosophy: the models (formal systems) are accepted as the preferred tools for investigating the nature of mathematics, and even as the fount of meaning" (p. 2).

"One of the hardest concepts to communicate to the undergraduate is the concept of a proof. With good reason, the concept *is* esoteric. Most mathematicians, when pressed to say what they mean by a proof, will have recourse to formal criteria. The constructive notion of proof by contrast is very simple, as we shall see in due course. Equally esoteric, and perhaps more troublesome, is the concept of existence. Some of the problems associated with this concept have already been mentioned, and we shall return to the subject again. Finally, I wish to point to the esoteric nature of the classical concept of truth. As we shall see later, truth is not a source of trouble to the contructivist, because of his emphasis on meaning."

"One could probably make a long list of schizophrenic attributes of contemporary mathematics, but I think the following short list covers most of the ground: rejection of common sense in favor of formalism, debasement of meaning by the willful refusal to accommodate certain aspects of reality; inappropriateness of means to ends; the esoteric quality of the communication; and fragmentation" (p. 1).

"The codification of insight is commendable only to the extent that the resulting methodology is not elevated to dogma and thereby allowed to impede the formation of new insight. Contemporary mathematics has witnessed the triumph of formalist dogma, which had its inception in the important insight that most arguments of modern mathematics can be broken down and presented as successive applications of a few basic schemes. The experts now routinely equate the panorama of mathematics with productions of this or that formal system. Proofs are thought of as manipulations of strings of symbols. Mathematical philosophy consists of the creation, comparison and investigation of formal systems. Consistency is the goal. In consequence meaning is debased, and even ceases to exist at a primary level.

"The debasement of meaning has yet another source, the wilful refusal of the contemporary mathematician to examine the content of certain of his terms, such as the phrase, 'there exists.' He refuses to distinguish among the different meanings that might be ascribed to this phrase. Moreover he is vague about what meaning it has for him. When pressed he is apt to take refuge in formalism, declaring that the meaning of the phrase and the statement of which it forms a part can only be understood in the context of the entire set of assumptions and techniques at his command. Thus he inverts the natural order, which would be to develop meaning first, and then to base his assumptions and techniques on the rock of meaning. Concern about this debasement of meaning is a principal force behind constructivism."

While other mathematicians say "there exists" as if existence were a clear, unproblematical notion, Bishop says "meaning" and "meaningful" as if those were clear, unproblematical notions. Has he simply shifted the fundamental ambiguity from one place to another?

Most mathematicians responded to Bishop's work with indifference or hostility. We nonconstructivists should do better. We should try to state our philosophies as clearly as the constructivists state theirs. We have a right to our viewpoint, but we ought to be able to say what it is.

The account of constructivism here is given from the viewpoint of classical mathematics. That means it's unacceptable from the constructivist viewpoint.

From that point of view, classical mathematics is the aberration, a jumble of myth and reality. Constructivism is just refusing to accept a myth.

Hilbert, Formalism, Gödel
Beautiful Idea; Didn't Work

Formalism is credited to David Hilbert, the outstanding mathematician of the first half of the twentieth century. It's said that his dive into philosophy of mathematics was a response to the flirtation of his favorite pupil, Hermann Weyl, with Brouwer's intuitionism. Hilbert was alarmed. He said that depriving the mathematician of proof by contradiction was like tying a boxer's hands behind his back.

"What Weyl and Brouwer do comes to the same thing as to follow in the footsteps of Kronecker! They seek to save mathematics by throwing overboard all that which is troublesome. . . . They would chop up and mangle the science. If we would follow such a reform as the one they suggest, we would run the risk of losing a great part of our most valuable treasure!" (C. Reid, p. 155).

Hilbert met the crisis in foundations by inventing proof theory. He proposed to prove that mathematics is *consistent*. To do this, he had a brilliant idea: work with formulas, not content. He intended to do so by purely finitistic, combinatorial arguments—arguments Brouwer couldn't reject! His program had three steps:

1. Introduce a formal language and formal rules of inference, so every classical proof could be replaced by a formal derivation from formal axioms by mechanically checkable steps. This had already been accomplished in large part by Frege, Russell, and Whitehead. Once this was done, the axioms of mathematics could be treated as strings of meaningless symbols. The theorems would be other meaningless strings. The transformation from axioms to theorems—the *proof*—could be treated as a rearrangement of symbols.

2. Develop a combinatorial theory of these "proof" rearrangements. The rules of inference now will be regarded as rules for rearranging formulas. This theory was called "meta-mathematics."

3. Permutations of symbols are finite mathematical objects, studied by "combinatorics" or "combinatorial analysis." To prove mathematics is consistent, Hilbert had to use finite combinatorial arguments to prove that the permutations allowed in mathematical proof, starting with the axioms, could never yield a falsehood, such as

$$1 = 0.$$

That is, to prove by purely finite arguments that a contradiction, for example, $1 = 0$, cannot be derived within the system.

In this way, mathematics would be given a guarantee of consistency. As a foundation this would have been weaker than one known to be *true* (as geometry was once believed to be true) or impossible to doubt (like, possibly, the laws of elementary logic.)

Hilbert's formalism, like logicism, offered certainty and reliability for a price. The logicist would save mathematics by turning it into a tautology. The formalist would save it by turning it into a meaningless game. After mathematics is coded in a formal language and its proofs written in a way checkable by machine, the meaning of the symbols becomes extramathematical.

It's very instructive that Hilbert's writing and conversation displayed full conviction that mathematical problems are about real objects, and have answers that are true in the same sense that any statement about reality is true. He advocated a formalist interpretation of mathematics only as the price of obtaining certainty.

"The goal of my theory is to establish once and for all the certitude of mathematical methods. . . . The present state of affairs where we run up against the paradoxes is intolerable. Just think, the definitions and deductive methods which everyone learns, teaches and uses in mathematics, the paragon of truth and certitude, lead to absurdities! If mathematical thinking is defective, where are we to find truth and certitude?" (D. Hilbert, "On the Infinite," in *Philosophy of Mathematics* by Benacerraf and Putnam).

As it happened, certainty could not be had, even at this price. A few years later, Kurt Gödel proved consistency could never be proved by the methods of proof Hilbert allowed. Gödel's incompleteness theorems showed that Hilbert's program was hopeless. Any formal system strong enough to contain arithmetic could never prove its own consistency. This theorem of Gödel's is usually cited as the death blow to Hilbert's program, and to formalism as a philosophy of mathematics.

(A simple new proof of Gödel's theorem by George Boolos is given in the mathematical Notes and Comments.)

The search for secure foundations has never recovered from this defeat.

John von Neumann tells how "working mathematicians" responded to Brouwer, Hilbert, and Gödel:

1. "Only very few mathematicians were willing to accept the new, exigent standards for their own daily use. Very many, however, admitted that Weyl and Brouwer were *prima facie* right, but they themselves continued to trespass, that is, to do their own mathematics in the old, 'easy' fashion-probably in the hope that somebody else, at some other time, might find the answer to the intuitionistic critique and thereby justify them *a posteriori*.

2. "Hilbert came forward with the following ingenious idea to justify 'classical' (i.e., pre-intuitionist) mathematics: Even in the intuitionistic system it is possible to give a rigorous account of how classical mathematics operates, that is, one can describe how the classical system works, although one cannot justify its workings. It might therefore be possible to demonstrate intuitionistically that classical procedures can never lead into contradiction—into conflicts with each

other. It was clear that such a proof would be very difficult, but there were certain indications how it might be attempted. Had this scheme worked, it would have provided a most remarkable justification of classical mathematics on the basis of the opposing intuitionistic system itself! At least, this interpretation would have been legitimate in a system of the philosophy of mathematics which most mathematicians were willing to accept.

3. "After about a decade of attempts to carry out this program, Gödel produced a most remarkable result. This result cannot be stated absolutely precisely without several clauses and caveats which are too technical to be formulated here. Its essential import, however, was this: If a system of mathematics does not lead into contradiction, then this fact cannot be demonstrated with the procedures of that system. Gödel's proof satisfied the strictest criterion of mathematical rigor—the intuitionistic one. Its influence on Hilbert's program is somewhat controversial, for reasons which again are too technical for this occasion. My personal opinion, which is shared by many others, is, that Gödel has shown that Hilbert's program is essentially hopeless.

4. "The main hope of justification of classical mathematics—in the sense of Hilbert or of Brouwer and Weyl—being gone, most mathematicians decided to use that system anyway. After all, classical mathematics was producing results which were both elegant and useful, and, even though one could never again be absolutely certain of its reliability, it stood on at least as sound a foundation as, for example, the existence of the electron. Hence, as one was willing to accept the sciences, one might as well accept the classical system of mathematics. Such views turned out to be acceptable even to some of the original protagonists of the intuitionistic system. At present the controversy about the 'foundations' is certainly not closed, but it seems most unlikely that the classical system should be abandoned by any but a small minority.

"I have told this story of this controversy in such detail, because I think that it constitutes the best caution against taking the immovable rigor of mathematics too much for granted. This happened in our own lifetime, and I know myself how humiliatingly easily my own views regarding the absolute mathematical truth changed during the episode, and how they changed three times in succession!" (pp. 2058–59).

Instead of providing foundations for mathematics, Russell's logic and Hilbert's proof theory became starting points for new mathematics. Model theory and proof theory became integral parts of contemporary mathematics. They need foundations as much or little as the rest of mathematics.

Hilbert's program rested on two unexamined premises. First, the Kantian premise: *Something* in mathematics—at least the purely "finitary part"—is a solid foundation, is indubitable. Second, the formalist premise: A solid theory of formal sentences could validate the mathematical activity of real life, where the possibility of formalization is in the remote background, if present at all.

The first premise was shared by the constructivists; the second, of course, they rejected. Formalization amounts to mapping set theory and analysis into a part of itself—finite combinatorics. Even if Hilbert had been able to carry out his program, the best he could have claimed would have been that all mathematics is consistent *if* the "finitistic" principle allowed in "metamathematics" is reliable. Still looking for the last tortoise under the last elephant!

The bottom tortoise is the Kantian synthetic a priori, the intuition. Although Hilbert doesn't explicitly refer to Kant, his conviction that mathematics must provide truth and certainty is in the Platonic heritage transmitted through the rationalists to Kant, and thereby to intellectual nineteenth-century western Europe. In this respect, Hilbert is as much a Kantian as Brouwer, whose label of intuitionism avows his Kantian descent.

To Brouwer, the Hilbert program was misconceived at Step 1, because it rested on identifying mathematics itself with formulas used to represent or express it. But it was only by this transition to languages and formulas that Hilbert was able to envision the possibility of a *mathematical* justification of mathematics.

Like Hilbert, Brouwer was sure that mathematics had to be established on a sound and firm foundation. He took the other road, insisting that mathematics must start from the intuitively given, the finite, and must contain only what is obtained in a constructive way from this intuitively given starting point. Intuition here means the intuition of *counting* and that alone. For both Brouwer and Hilbert, the acceptance of geometric intuition as a basic or fundamental "given" on a par with arithmetic would have seemed utterly retrograde and unacceptable *within the context of foundational discussions.* Like Brouwer, Hilbert, the formalist, regarded the finitistic part of mathematics as indubitable. His way of securing mathematics, making it free of doubt, was to reduce the infinitistic part—analysis and set theory—to the finite part by use of the finite *formulas*, which described these nonfinite structures.

In the mid-twentieth century, formalism became the predominant philosophical attitude in textbooks and other official writing on mathematics. Constructivism remained a heresy with a few adherents. Platonism was and is believed by nearly all mathematicians. Like an underground religion, it's observed in private, rarely mentioned in public.

Contemporary formalism is descended from Hilbert's formalism, but it's not the same thing. Hilbert believed in the reality of finite mathematics. He invented metamathematics to justify the mathematics of the infinite. Today's formalist doesn't bother with this distinction. For him, all mathematics, from arithmetic on up, is a game of logical deduction.

He defines mathematics as the science of rigorous proof. In other fields some theory may be advocated on the basis of experience or plausibility, but in mathematics, he says, we have a proof or we have nothing.

Any proof has a starting point. So a mathematician must start with some undefined terms, and some unproved statements. These are "assumptions" or "axioms." In geometry we have undefined terms "point" and "line" and the axiom "Through any two distinct points passes exactly one straight line." The formalist points out that the logical import of this statement doesn't depend on the mental picture we associate with it. Nothing keeps us from using other words—"Any two distinct bleeps ook exactly one bloop." If we give interpretations to the terms bleep, ook, and bloop, or the terms point, pass, and line, the axioms may become true or false. To pure mathematics, any such interpretation is irrelevant. It's concerned only with logical deductions from them.

Results deduced in this way are called theorems. You can't say a theorem is true, any more than you can say an axiom is true. As a statement in pure mathematics, it's neither true nor false, since it talks about undefined terms. All mathematics can say is whether the theorem follows logically from the axioms. Mathematical theorems have no content; they're not *about* anything. On the other hand, they're absolutely free of doubt or error, because a rigorous proof has no gaps or loopholes.

In some textbooks the formalist viewpoint is stated as simple matter of fact. The unwary student may swallow it as the official view. It's no simple matter of fact, but a matter of controversial interpretation. The reader has the right to be skeptical, and to demand evidence to justify this view.

Indeed, formalism contradicts ordinary mathematical experience. Every school teacher talks about "facts of arithmetic" or "facts of geometry." In high school the Pythagorean theorem and the prime factorization theorem are learned as true statements about right triangles or about natural numbers. Yet the formalist says any talk of facts or truth is incorrect.

One argument for formalism comes from the dethronement of Euclidean geometry.

For Euclid, the axioms of geometry were not assumptions but self-evident truths. The formalist view results, in part, from rejecting the idea of self-evident truths.

In Chapter 6 we saw how the attempt to prove Euclid's fifth postulate led to discovery of non-Euclidean geometries, in which Euclid's parallel postulate is assumed to be false.

Can we claim that Euclid's parallel postulate and its negation are *both* true? The formalist concludes that to keep our freedom to study both Euclidean and non-Euclidean geometries, we must give up the idea that either is true. They need only be consistent.

But Euclidean and non-Euclidean geometry conflict only if we believe in an objective physical space, which obeys a single set of laws, which both theories attempt to describe. If we give up this belief, Euclidean and non-Euclidean geometry are no longer rival candidates for solving the same problem, but two different mathematical theories. The parallel postulate is true for the Euclidean straight line, false for the non-Euclidean.

Are the theorems of geometry meaningful apart from physical interpretation? May we still use the words "true" and "false" about statements in geometry? The formalist says no, the statements aren't true or false, they aren't about anything and don't mean anything. The Platonist says yes, since mathematical objects exist in their own world, apart from the physical world. The humanist says yes, they exist in the shared conceptual world of mathematical ideas and practices.

The formalist makes a distinction between geometry as a deductive structure and geometry as a descriptive science. Only the first is mathematical. The use of pictures or diagrams or mental imagery is nonmathematical. In principle, they are unnecessary. He may even regard them as inappropriate in a mathematics text or a mathematics class.

Why give *this* definition and not another?

Why *these* axioms and not others?

To the formalist, such questions are premathematical. If they're in his text or his course, they'll be in parentheses, and in brief.

What examples or applications come from the general theory he develops? This is not really relevant. It may be a parenthetical remark, or left as a problem.

For the formalist, you don't get started doing mathematics until you state hypotheses and begin a proof. Once you reach your conclusion, the mathematics is over. Any more is superfluous. You measure the progress of your class by how much you prove in your lectures. What was understood and retained is a nonmathematical question.

One reason for the past dominance of formalism was its link with logical positivism. This was the dominant trend in philosophy of science during the 1940s and 1950s. Its aftereffects linger on, for nothing definitive has replaced it. (See the section on the "Vienna circle" in Chapter 9.) Logical positivists advocated a unified science coded in a formal logical calculus with a single deductive method. Formalization was the goal for all science. Formalization meant choosing a basic vocabulary of terms, stating fundamental laws, and logically developing a theory from fundamental laws. Classical and quantum mechanics were the models.

The most influential formalists in mathematical exposition was the group "Nicolas Bourbaki." Under this pseudonym they produced graduate texts that had worldwide influence in the 1950s and 1960s. The formalist style dripped down into undergraduate teaching and even reached kindergarten, with preschool texts on set theory. A game called "WFF and Proof" was used to help grade-school children learn about "well-formed formulas" (WFF's) according to formal logic.

In recent years, a reaction against formalism has grown. There's a turn toward the concrete and the applicable. There's more respect for examples, less strictness in formal exposition. The formalist philosophy of mathematics is the source of the formalist style of mathematical work. The signs say that the formalist philosophy is losing its privileged status.

n i n e

Foundationism Dies/ Mainstream Lives

This chapter attempts to bring the story of the Mainstream up to date, which largely means surveying recent analytic philosophy of mathematics. Husserl isn't in the analytic mainstream, but his international stature and his mathematical qualifications justify including him. Carnap and Quine are the two unavoidable analytic philosophers. Both are "icons" among academic philosophers of science and mathematics.

Among "younger" philosophers who seem to me to be mainstream, I have benefited from reading Charles Castonguay, Charles Chihara, Hartry Field, Juliet Floyd, Penelope Maddy, Michael Resnik, Stuart Shapiro, David Sherry, and Mark Steiner.

I report briefly on structuralism and fictionalism. These alternatives to Platonism are attracting attention as I write. I regret the limitations that keep me from describing the work of other interesting authors. For this I recommend Aspray and Kitcher. I reserve for Chapter 12 the philosophers writing today whom I regard as fellow travelers in the humanist direction. There also I've been unable to pay due attention to most of them.

Edmund Husserl (1859–1938)

Phenomenologist/Weierstrass Student

Husserl is the creator of phenomenology, intellectual father of Karl Heidegger and grandfather of Jean-Paul Sartre. He wrote his doctoral dissertation on the calculus of variations under Karl Weierstrass, one of the greatest nineteenth-century mathematicians, and took lifelong pride in being Weierstrass's pupil. Husserl's first philosophical work was in philosophy of mathematics. Yet he's never mentioned today in talk about the philosophy of mathematics, which has been monopolized by analytic philosophers, descendants of Bertrand Russell.

In a posthumous essay, Gödel expressed high hopes for Husserl's phenomenology. ". . . the certainty of mathematics is to be secured . . . by cultivating (deepening) knowledge of the abstract concepts themselves. . . . Now in fact, there exists today the beginning of a science which claims to possess a systematic method for such a clarification of meaning, and that is the phenomenology founded by Husserl. Here clarification of meaning consists in focusing more sharply on the concepts concerned by directing our attention in a certain way, namely onto our own acts in the use of these concepts, onto our powers in carrying out our acts, etc. . . . I believe there is no reason at all to reject such a procedure at the outset as hopeless . . . quite divergent directions have developed out of Kant's thought—none of which, however, really did justice to the core of Kant's thought. This requirement seems to me to be met for the first time by phenomenology, which, entirely as intended by Kant, avoids both the death-defying leaps of idealism into a new metaphysics as well as the positivistic rejection of all metaphysics." (Gödel, 1995, p. 383, 387)

In a review of Husserl's first book, Frege charged him with psychologism. Husserl faithfully avoided psychologism ever after.

Husserl's early works on philosophy of mathematics came before he developed his major ideas on phenomenology. They are influenced by logicism and formalism. I present some of his mature thinking about mathematics, the well-known essay, "The Origin of Geometry."

There he argues that since geometry has a historic origin, someone "must have" made the first geometric discovery. For that primal geometer, geometric terms and concepts "must have" had clear, unmistakable meaning. Centuries passed. New generations enlarged geometry. We inherit it as a technology and a logical structure, but we've lost the meaning of the subject.

We must recover this meaning.

"Our interest shall be the inquiry back into the most original sense in which geometry once arose, was present as the tradition of millennia. . . . The progress of deduction follows formal-logical self-evidence, but without the actually developed capacity for reactivating the original activities contained within its fundamental concepts, i.e. without the "what" and the "how" of its prescientific materials, geometry would be a tradition empty of meaning; and if we ourselves did not have this capacity, we could never even know whether geometry had or ever did have a genuine meaning, one that could really be 'cashed in.' This is our situation, and that of the whole modern age.

"By exhibiting the essential presuppositions upon which rests the historical possibility of a genuine tradition, true to its origins, of sciences like geometry, we can understand how such sciences can vitally develop through the centuries and still not be genuine. The inheritance of propositions and of the method of logically constructing new propositions and idealities can continue without interruption from one period to the next, while the capacity for reactivating the primal beginnings, i.e. the sources of meaning for everything that comes later, has not been handed down with it. What is lacking is this, precisely what had given and had to give meaning to all propositions and theories, a meaning arising from the primal sources which can be made self-evident again and again. . . ."

Husserl isn't asking for the usual fact-obsessed historical research. Nor for the usual theorem-obsessed geometrical research. Unfortunately, he doesn't give an example of what he is asking for.

Yet he ends on a transcendent note:

"Do we not stand here before the great and profound problem-horizon of reason, the same reason that functions in every man, the *animal rationale*, no matter how primitive he is?"

Rota presents a remarkably readable expert account of phenomenology and mathematics.

Vienna Circle. Carnap, etc.

In the late 1920s the logicist tradition was picked up by the Vienna Circle of logical positivists. They tacked together philosophy of language from Wittgenstein's *Tractatus* and logicism from Frege and Russell to fabricate what they considered "scientific philosophy." For them, thinking about mathematics meant thinking about logic and axiomatic set theory. The proper model for all science was mechanics. Ernst Mach had arranged classical mechanics in a deductive system. Mass, length, and time are his undefined terms. His axioms are Newton's laws. Classical mechanics has rules to interpret measurements as values of mass, length, or time.

From the axioms and interpretation rules, everything else must be deduced.

The Vienna Circle ordered all science to conform to that model. To each science, its own axioms and undefined terms. (The undefined terms are [informally] empirical measurements.) To each science, its own interpretation rules to connect theory and data. To do science, you should:

1. Choose basic observables.
2. Find formulas for their relationships (called "axioms").
3. Express all other observables as functions of the basic observables.
4. From (1,2,3) derive the rest of the subject by mathematics. For this school of philosophy, mathematics is a language, and a tool for formulating and developing physical theory. The fundamental laws of science are equations and inequalities. (In mechanics, they're differential equations.) The scientist does mathematical calculations to derive consequences of the fundamental laws. But mathematics has no content of its own. Indeed, mathematics has no empirical observations to which to apply interpretation rules! For logical positivism, mathematics is *nothing but* a language for science, a *contentless* formal structure.

Rudolf Carnap wrote, "Thus we arrived at the conception that all valid statements of mathematics are analytic in the specific sense, that they hold in all possible cases and therefore do not have any factual content" (*Autobiography*).

So logical positivism in philosophy of science matches formalism in philosophy of mathematics. (This is so, even though Carnap's philosophy of mathematics was logicist, not formalist.) As an account of the nature of mathematics, formalism is incompatible with the thinking of working mathematicians. But this was no problem for the positivists! Entirely oriented on theoretical physics, they saw mathematics only as a tool, not a living, growing subject. For a physicist or other user it may be convenient to identify mathematics itself with a particular axiomatic presentation of it. For the producer of mathematics, quite the contrary. Axiomatics is an embellishment added after the main work is done. But this was irrelevant to philosophers whose idea of mathematics came from logic and foundationist philosophy.

With this philosophy came a test of meaningfulness. If a statement isn't "in principle" refutable by the senses, it's meaningless. That kind of statement is no more than a grunt or a groan. In particular, esthetic and ethical judgments have no *factual* content. I say "The *Emperor Concerto* is beautiful. Hitler is evil." I'm just saying, "I like the *Emperor Concerto*. I don't like Hitler."

In retribution, an embarrassment dogged logical positivism. Its own philosophical edicts can't be empirically refuted—not even in principle So by its own test, its own edicts were—mere grunts and groans!

Despite this glitch, logical positivism reigned over American philosophy of science in the 1930s and 1940s, under a group of brilliant refugees from Hitler. Foremost was Rudolph Carnap. W. V. O. Quine wrote, "Carnap more than anyone else was the embodiment of logical positivism, logical empiricism, the Vienna Circle" (*The Ways of Paradox, and Other Essays*, 1966–76, pp. 40–41).

Look at his influential *Introduction to Symbolic Logic*.

Part One describes three formal languages, A, B, and C. Some simple properties of these languages are proved. No nonobvious property or nontrivial problem is stated.

Part Two, "Applications of Symbolic Logic," has a chapter on theory languages and a chapter on coordinate languages. It presents axiom systems for geometry, physics, biology, and set theory/arithmetic. The "applications" are specializing one of his three languages to one of his four subjects.

He doesn't discuss whether these formalizations should interest mathematicians, physicists, or biologists. Such formalizations are rarely seen in biology, physics, or even mainstream mathematics (analysis, algebra, number theory, geometry). Carnap does cite Woodger and some of his own papers.

On p. 21 he says, "if certain scientific elements—concepts, theories, assertions, derivations, and the like—are to be analyzed logically, often the best procedure is to translate them in symbolic language." He provides no support for this claim. In 1957 it evidently was possible to believe in ever-increasing interest among mathematicians, philosophers, and "those working in quite specialized fields who give attention to the analysis of the concepts of their discipline." There's no longer such a belief.

"Symbolic logic" (now called "formal logic") did turn out to be vitally useful in designing and programming digital computers. Carnap doesn't mention any such application. Contrary to his statement, we rarely use formal languages to analyze scientific concepts. We use natural language and mathematics. Carnap's identification of philosophy of mathematics with formalization of mathematics was a dead end. Language A, Language B, and Language C are dead.

On page 49 of his *Autobiography* he wrote: "According to my principle of tolerance, I emphasized that, whereas it is important to make distinctions between constructivist and non-constructivist definitions and proofs, it seems advisable not to prohibit certain forms of procedure but to investigate all practically useful forms. It is true that certain procedures, e.g., those admitted by constructivism or intuitionism, are safer than others. Therefore it is advisable to apply these procedures as far as possible. However, there are other forms and methods which, though less safe because we do not have a proof of their consistency, appear to be practically indispensable for physics. In such a case there seems to be no good reason for prohibiting these procedures so long as no contradictions have been found."

As if Carnap ever had the authority to "prohibit these procedures"!

By the 1950s the sway of logical positivism was shaky. Physicists never accepted its description of their work. Few physicists are interested in axiomatics. Physicists take more pleasure in tearing down axioms. They look for provocative conjectures, and experiments to disprove them.

Hao Wang quotes Carnap's *Intellectual Autobiography*. "From 1952 to 1954 he was at the Princeton Institute and had separate talks 'with John von Neumann, Wolfgang Pauli and some specialists in statistical mechanics on some questions of theoretical physics with which I was concerned.' He had expected that 'we would reach, if not agreement, then at least mutual understanding.' But they failed despite their serious efforts. One physicist said, 'Physics is not like geometry; in physics there are no definitions and no axioms'" (*Intellectual Autobiography*, in Schilpp, pp. 36–37). Wang provides a meticulous criticism of Carnap.

Russell, Frege, and Wittgenstein brought philosophy of mathematics under the sway of analytic philosophy: The central problem is meaning, the essential tool is logic. Since mathematics is the branch of knowledge whose logical structure is best understood, some philosophers think philosophy of mathematics is a model for all philosophy. As the dominant style of Anglo-American philosophy, analytic philosophy perpetuates identification of philosophy of mathematics with logic and the study of formal systems. Central problems for the mathematician become invisible—the development of pre-formal mathematics, and how pre-formal mathematics relates to formalization.

Willard Van Ormond Quine

Most Influential Living Philosopher

Quine is "the most distinguished and influential of living philosophers" says the eminent English philosopher P. F. Strawson (on the jacket of *Quiddities*, Quine's latest book as of 1994).

Quine proved that the real numbers exist—exist philosophically, not just mathematically. He proved that you're guilty of bad faith if you say the real numbers are fictions. We will present and refute Quine's argument. First we sketch a few of his other contributions.

Quine makes no separation between philosophy and logic. Formalization—presentation in a formal language—makes a philosophical theory legitimate. Apart from what can be said in a formal language, it makes no sense to talk philosophy.

Quine granted an interview to *Harvard Magazine*. "'Someone who was a student here many years ago recently sent me a copy of *Methods of Logic* and asked me to inscribe it for him and to write something about my philosophy of life.' (The last three words spoken in gravelly disbelief.)

'And what did you write?'

'Life is agid. Life is fulgid. Life is what the least of us make most of us feel the least of us make the most of. Life is a burgeoning, a quickening of the dim primordial urge in the murky wastes of time.'

'Agid?'

'Yes, it's a made-up word.'

'What you're saying is it's not a serious question.'

'That's right, it's not a serious question. Not a question you can make adequate sense of.'

"For Quine it is important for philosophy to be a technical, specialized discipline (with subdisciplines) and give up contact with people" (Hao Wang, p. 205. Wang provides an infinitely detailed and complete report on Quine's publications, with fascinating critiques.).

Quine's most famous bon mot is his definition of existence: "To be is to be the value of a variable."

This has the merit of shock value.

In the *Old Testament*, Yahweh roars "I am that I am." Must we construe this as: "I, the value of a variable, am the value of a variable!"

Or Hamlet's "To be the value of a variable or not to be the value of a variable?"

Or Descartes's *Meditations*: "I think, therefore I am the value of a variable."

To all this, Quine would have a quick reply. Yahweh, Shakespeare, and Descartes are like the ex-student who asked about his philosophy of life. They all talk nonsense.

The only "existence" of philosophical interest is the existence associated with the existential quantifier of formal logic.

Quine's definition loses its charm when you see he has simply "conflated" the domain of formal logic with the whole material and spiritual universe. His definition could be paraphrased: "To someone interested in existence only as a term in formal logic, to be is. . . ."

"To be" is "to be visible through W. V. O. Quine's personal filter, which is formal logic."

Like a monomaniac photographer saying, "To be is to be recorded on my film," or Geraldo Rivera saying, "To be is to be seen on the Geraldo Rivera show."

Professor Quine also "famously" discovered that translation doesn't exist. (They say he's fluent in six languages.) The insult to common sense is what gets attention. Someone not seeking to shock would say, "*Perfect* or *precise* translation is impossible." That would be banal. Better say something shocking and false.

A real question is being overlooked. Why does the impossibility of perfect translation make no difference in practice? Such an investigation would be empirical, particular, detailed—not Quine's cup of tea.

Our concern with Quine is his new, original argument for mathematical Platonism—for actual existence of real numbers and the set structure logicists erect under them.

Quine calls his idea "ontological commitment." Physics, he tells us, is inextricably interwoven with the real numbers, to such a pitch that it's impossible to make sense of physics without believing real numbers exist. Anyone who turns on a VCR or tests a "nuclear device" believes in physics. It's "bad faith" to drive a car or switch off an electric light without accepting the reality of the real numbers.

In "The Scope and Language of Science" Quine writes: "Certain things we want to say in science may compel us to admit into the range of values of the variables of quantification not only physical objects but also classes and relations of them; also numbers, functions, and other objects of pure mathematics. For, mathematics—not uninterpreted mathematics, but genuine set theory, logic, number theory, algebra of real and complex numbers, differential and integral calculus, and so on—is best looked upon as an integral part of science, on a par with the physics, economics, etc., in which mathematics is said to receive its applications.

"Researches in the foundations of mathematics have made it clear that *all* of mathematics in the above sense can be got down to logic and set theory, and that the objects needed for mathematics in this sense can be got down to a single category; that of *classes*—including classes of classes, classes of classes of classes, and so on." (His "class" is virtually our "set." A real number** is a set of sets of rational numbers, each of which is a set of pairs of natural numbers, each of which is a set of sets.**)

"Our tentative ontology for science, our tentative range of values for the variables of quantification, comes therefore to this: physical objects, classes of them, classes in turn of the elements of this combined domain, and so on up."

This argument of Professsor Quine's is taken seriously. In *Science Without Numbers* Hartry Field says it's the only proof of existence of the real numbers worthy of attention. Field is a nominalist. He denies that numbers exist, so he has to knock down Professor Quine. He writes, "This objection to fictionalism about mathematics can be undercut by showing that there is an alternative formulation of science that does not require the use of any part of mathematics that refers to quantifiers over abstract entities. I believe that such a formulation is possible; consequently, without intellectual doublethink, I can deny that there are abstract entities."

Field says the best exposition of Quine's existence argument is in Putnam's *Philosophy of Logic* (Chapter 5). But after publishing *Philosophy of Logic* Putnam reconsidered Quinism. In *Realism with a Human Face*, in a chapter titled "The Greatest Logical Positivist," Putnam wrote: [In Quine's theory] "mathematical statements, for example, are only justified insofar as they help to make successful predictions in physics, engineering, and so forth." He finds "this claim almost totally unsupported by actual mathematical practice." I agree.

Because we use phone and TV, Quine says we have an "ontological commitment" to the reality of the real numbers (pun intended) and therefore to uncountable sets. In view of his definition of "is," is he saying that the *set* of real numbers is the value of some variable? Which variable?

To Quine it's irrelevant that almost all mathematicians would say they're working on something real, with or without any connection with physics. As far as I can see, he has three options. All are unpleasant and unacceptable. (A) All mathematics is used in physics. (B) The part of mathematics not used in physics doesn't matter. (C) The part of mathematics used in physics not only exists, but somehow causes the unphysical part also to exist. I don't know if Professor Quine upheld any of these absurdities, or if he still thinks about the matter.

What's important is that Professor Quine's leading "insight"—the reality of physics implies the reality of math—is wrong.

The following paragraphs expound a simple remark that shows Quine's argument is without merit (which is what Field wants to do).

Think about digital computers. They're ubiquitous in physics. Physical calculations are either short enough to do by hand, or too long. The ones too long are done on a calculator or a computer. The short ones could also be done on a calculator or computer, if one wished to do so.

To go into a digital calculator or computer, information must be discretized and finitized. Digital computers only accept finite amounts of discrete information.

(A Turing machine is defined to have an infinite tape. But a Turing machine isn't a real machine, it's a mathematical construct. In the whole world, now and to the end of humanity, there's only going to be finitely many miles of tape.)

"Discretized" means there's a smallest increment that the machine and the program read. Anything smaller is read as zero. If a machine and program have a smallest increment of 2^{-100}, and the largest number they accept is 2^{100}, then they work in a number system of 2^{200} numbers. (The number of steps of size 2^{-100} to climb from 2^{-100} to 2^{100}.) 2^{200} is large, but finite.

Real numbers are written as infinite decimals. Nearly all of them need infinitely many digits for their complete description. A computer can't accept such a real number. The biggest computer ever built doesn't have space for even *one* infinite nonterminating nonrepeating decimal number. How does it cope? It *truncates*—keeps the first 100 or the first 1,000 digits, drops the rest.

So physics is dependent on machines that accept only finite decimals. Physics dispenses with real numbers!

Here's an objection. The calculations are done after a theory is formulated. Formation of theory is done by humans, not machines. In formulating a theory, physicists use classical mathematics with real numbers. Doesn't that save Quine's argument?

No.

Quantum mechanics is set in an infinite-dimensional blow-up of Euclidean space known as "Hilbert space." Any coordinate system for Hilbert space has infinitely many "basis vectors." A point in Hilbert space represents a "state" of a quantum-mechanical system. A typical "state" has infinitely many coordinates, and each coordinate is a real number, that is, an infinite decimal. However, not every infinite sequence of real numbers defines a vector in the Hilbert space. The vector must have "finite norm." That means the sum of the squares of the coordinates must be finite.

$$(1, 1, 1, \ldots)$$

isn't a Hilbert space vector, but

$$(1, .5, .25, .125, .0625 \ldots)$$

is a Hilbert space vector. The condition of finite norm means that the "tail," the last part of the coordinate expansion, is negligibly small. For some large finite number N we need only look at the first N terms in the expansion of our vector. In geometrical language, the N-dimensional projection of our infinite-dimensional vector is so close to the vector itself that the distance between them is negligible. Each infinite-dimensional Hilbert-space vector is approximated, as closely as we wish, by N-dimensional vectors, finite-dimensional vectors.

But aren't the component-coordinates of this finite-dimensional projection each separately a real number—an infinite decimal? Yes, but we can choose any N we like, and truncate that infinite decimal after N digits, making an error of only 10^{-N}. If N is large, this is physically undetectable. So the state vector is represented, to arbitrarily high accuracy, by an N-tuple of finite decimals, each containing N digits.

We are interested not only in vectors *in* Hilbert space—(creatures with infinitely many coordinates that I just introduced)—but especially in linear transformations or operators *on* the space. If we choose convenient coordinates, such an operator is represented by an infinite-by-infinite matrix, all of whose infinitely many entries or elements are real numbers. To be a legitimate operator, the rows and columns of the matrix must satisfy a requirement similar to the one satisfied by vectors. If you go far out along a row or column of this infinite matrix, eventually the elements there will be so small they are physically undetectable. There you can truncate the matrix, make it finite. Then you can store the truncated matrix, whose elements are truncated real numbers, in your computer.

The sophisticated reader may ask if this *finite* mathematical system is "isometrically isomorphic" to Quine's (equivalent in a precise mathematical sense.) Isomorphic structures differ only in the names of their elements. But my proposed alternative is certainly *not* isomorphic to the reals. The reals are uncountably infinite; my substitute is finite.

Another objection may come from the Quinist. "You say the computer is a finite-state machine. But the machine is made of silicon, copper, and plastic. It's a physical object and it obeys the laws of physics. It's subject to infinitely many different states of temperature, electrostatic field, kinetic and dynamical variables. There's no such thing as a finite-state machine."

Agreed, the real computer in the computer room is a physical object. We think of it as a finite state machine for simplicity and convenience. But to claim that we can't describe this piece of metal and plastic without using the full system of real numbers is just repeating Quine's original claim that physics requires the real number system. It doesn't. That's true of the physics of digital computers. To consider a Cray or a Sun as a physical system, not an ideal computing machine, we need a much more detailed description of it. The much more detailed description will still be finite. Any description we give of anything is finite.

Another defender of Quine might say, "The real number system developed out of necessity. Mathematical analysis and mathematical physics are impossible without the completeness property—the ability to define or construct a number as the limit of a convergent sequence. How can you do anything without π or $\sqrt{2}$?"

Answer: π and $\sqrt{2}$ exist conceptually, not physically or computationally. Computationally,

$$3.141592653589793238462643383279502884197169399375105820$$

is the circumference of the unit circle, and 1.414213562 is the length of the diagonal of the unit square (the square root of 2.) These finite decimals have errors smaller than we can detect by any physical measurement. Such error is of mathematical interest, not physical interest.

Mathematicians defined an infinite decimal as a "real number" to get a theory in which these negligible errors actually vanish. To compute, we go back to finite decimals or rational numbers.

The same discretization/finitization works for infinite-dimensional manifolds, Lie groups, Lie algebras, and so forth. If some infinite mathematical structure couldn't be approximated by a finite structure, then in general and in principle it would be impossible to carry out physical computations in it. Such a structure might fascinate mathematicians or logicians, but it wouldn't interest physicists. The uncountability of the reals fascinates mathematicians and philosophers. Physically it's meaningless. Physicists don't need infinite sets, and they don't need to compare infinities.

We use real numbers in physical theory out of convenience, tradition, and habit. For physical purposes we could start and end with finite, discrete models. Physical measurements are discrete, and finite in size and accuracy. To compute with them, we have discretized, finitized models physically indistinguishable from the real number model. The mesh size (increment size) must be small enough, the upper bound (maximum admitted number) must be big enough, and our computing algorithm must be stable. Real numbers make calculus convenient. Mathematics is smoother and more pleasant in the garden of real numbers. But they aren't essential for theoretical physics, and they aren't used for real calculations.

It's strange that Quine offers this argument about the real numbers, for it makes the same error he attacked in philosophy of science. In Quine's famous contribution to philosophy of science, "Two Dogmas of Empiricism," he showed that physical theory isn't completely determined by data. Physical theory is a loose-hanging network connected to data along its boundary. For any experimental finding, several explanations are possible. No experiment by itself can establish or refute a physical theory. The data do not determine the physical theory uniquely. If a prediction of a theory is refuted, it may not thereby determine which axiom in the theory needs to be revised.

The traditional view of Bacon, Mill, and Popper said each statement in a physical theory can be tested by experiment or measurement. If the measurement obeys the claim, the statement is confirmed (Mill) or at least not disconfirmed (Popper) and can be retained in our world picture. If measurement contradicts claim, we revise the claim.

However, Poincaré "famously" pointed out that no observation could compel us to consider physical space as Euclidean or non-Euclidean. Anomalous behavior by light rays could instead be explained by anomalous light-transmitting properties of the medium. Rejected theories like Ptolemy's theory of planetary motions, the phlogiston theory of fire, and the ether theory of light propagation all could continue to explain the phenomena by ad-hoc adjustments ("finagling"). Their successors—Copernicus's heliocentric solar system, Dalton's oxidation,

Einstein's relativistic light propagation—weren't the only possible explanation of observations. They were the simplest, most convenient, and therefore most credible.

Yet Quine, having shown the nonuniqueness of physical theory, takes for granted the uniqueness of mathematical theory. The mathematical part of the theory is determined uniquely, he thinks, and it must be the real number system and the set theory that logicians prop underneath the real number system. This is just as wrong as the idea that data determine a physical theory.

Hilary Putnam

Somewhat Influential Living Philosopher

In "What is Mathematical Truth?" Putnam parted company from Quine. Mathematical statements are true, not about objects, but about possibilities. Mathematics has conditional objects, not absolute objects. Putnam refers to Kreisel's remark that mathematics needs objectivity, not objects.

Putnam cites the use in mathematics of nondemonstrative, heuristic reasoning and the difficulty of believing in unspecified, unearthly abstract objects. Implicitly, he excludes physical or mental objects as mathematical objects.

Kreisel was right. In the first instance, mathematics needs objectivity rather than objects. Mathematical truths are objective, in the sense that they're accepted by all qualified persons, regardless of race, color, age, gender, or political or religious belief. What's correct in Seoul is correct in Winnipeg. This "invariance" of mathematics is its very essence. Since Pythagoras and Plato, philosophers have used it to support religion. Putnam's objectivity without objects, like standard Platonism, can be regarded as another form of mathematical spiritualism.

But need we really settle for objectivity without objects? "Not only are the 'objects' of pure mathematics conditional upon material objects," writes Putnam; "they are in a sense merely abstract possibilities. Studying how mathematical objects behave might be better described as studying what structures are abstractly possible and what structures are not abstractly possible. The important thing is that the mathematician is studying something objective, even if he is not studying an unconditional 'reality' of nonmaterial things, . . . mathematical knowledge resembles *empirical* knowledge—that is, the criterion of truth in mathematics just as much as in physics it is success of our ideas in practice and that mathematical knowledge is corrigible and not absolute. . . . What he asserts is that certain things are *possible* and certain things are *impossible*—in a strong and uniquely mathematical sense of 'possible' and 'impossible.' In short, mathematics is essentially *modal* rather than existential."

Possible in what sense? Perhaps logically possible—noncontradictory. If so, he's agreeing with Poincaré from a hundred years ago, and Parmenides from two millennia ago. The mathematician is studying something "real"—the consistency or inconsistency of his ideas. This is close to Frege.

On the other hand, maybe Putnam doesn't mean *logically* possible. Maybe he means *physically* possible. Is an infinite-dimensional infinitely smooth manifold of infinite connectivity "physically possible"? Does he mean there "could be" physical objects modeled by such manifolds? He seems to be running up against Lesson 1 in Applied Mathematics, cited above against Professor Quine: *No real phenomenon (physical, biological, or social) is perfectly described by any mathematical model. There's usually a choice among several incompatible models, each more or less suitable.*

Putnam doesn't think a possibility is an object. He doesn't explain what he means by "object," so it's hard to know if he's right or wrong. Is an object something that can affect human life or consciousness? If so, some probabilities are objects.

"The main burden of this paper is that one does not have to buy Platonist epistemology to be a realist in the philosophy of mathematics. The theory of mathematics as the study of special objects has a certain implausibility which, in my view, the theory of mathematics as the study of ordinary objects with the aid of a special concept does not. . . ."

The argument claims that the consistency and fertility of classical mathematics is evidence that it or most of it is true under *some interpretation*. The interpretation might not be a realist one.

The doctrine of objectivity without objects is not easy to understand or believe. It's proposed because of inability to find appropriate objects to correspond to numbers and spaces. Abstract objects are vacuous. Mental or physical objects are ruled out. So Kreisel and Putnam think no kind of object can be a mathematical object. They overlook the kind of object that works. Social-historic objects.

Patterns/Structuralism

Shades of Bourbaki

The dichotomy between neo-Fregean and humanist maverick must be applied with a light touch. There are neo-Fregeans, there are humanist mavericks, and there are others. In this and the following section, I present two influential recent trends in the philosophy of mathematics. I don't classify them one way or the other.

Structuralism—defining mathematics as "the science of patterns"—may be new to some philosophers, but not to mathematicians. Bourbaki said as much, and called it structuralism. Before Bourbaki there was Hardy: "A mathematician, like a painter or a poet, is a maker of patterns . . . the mathematician's patterns, like the painter's or the poet's, must be beautiful. . . . There is no permanent place in the world for ugly mathematics" (Hardy, 1940). Structuralism is the core of Saunders MacLane's (1986).

It has been adopted by philosophers Michael Resnik and Stuart Shapiro. The definition, "science of patterns," is appealing. It's closer to the mark than "the science that draws necessary conclusions" (Benjamin Peirce) or "the study of form and quantity" (*Webster's Unabridged Dictionary*). Unlike formalism, structuralism allows mathematics a subject matter. Unlike Platonism, is doesn't rely on a transcendental abstract reality. Structuralism grants mathematics unlimited generality and applicability. Watch a mathematician working, and you indeed see her studying patterns.

Structuralism is valid as a partial description of mathematics—an illuminating comment. As a complete description, it's unsatisfactory. Saunders MacLane in *Philosophia Mathematica* has pointed out that elementary and analytic number theory, for example, are understood much more plausibly as about objects than about patterns.

Not everyone who studies patterns is a mathematician. What about a dressmaker's patterns? What about "pattern makers" in machine factories? Resnik and Shapiro don't mean to call machinists and dressmakers mathematicians. By "pattern" they mean, not a piece of paper or sheet metal, but a nonmaterial pattern. (Though mathematicians do use physical models on occasion!)

Then can we say "Mathematics is the science of nonmaterial patterns"? No.

There are physicists, astronomers, chemists, biologists, geologists, historians, ethnographers, sociologists, psychologists, literary critics, journalists of the better class, novelists, and poets who also study "nonmaterial patterns."

The cure of this over-inclusiveness is simple. Resnik and Shapiro mean to define mathematics as the study of *mathematical* patterns. But it's no easier to explain "mathematical pattern" than to explain "mathematics."

And that would still leave a difficulty. "Mathematics is the study of mathematical patterns" would no longer be over-inclusive in subject matter, but still would be over-inclusive in methodology. Some people are studying mathematical patterns—geometric patterns, for example—by computer graphics or by physical models or by statistical sampling. By any of various empirical models, without major use of demonstrative reasoning.

People who do this *are* studying mathematical patterns. They aren't mathematicians. The definition of mathematics has to involve methodology, the "mathematical way of thinking." I believe it would be accurate finally to say "Mathematics is the mathematical study of mathematical patterns." Accurate, but not exciting.

In *Realism in Mathematics* Penelope Maddy said that structuralism differs only verbally from her set-theoretic realism. I have difficulty seeing structuralism and set-theoretic realism as essentially the same. We have a much more definite idea of "set" than of "pattern."

It's easy to give examples of pattern. Less easy to give a coherent, inclusive definition. Maclane quotes and rejects Bourbaki's "precise" definition of "structure."

Maybe Resnik and Shapiro mean that a pattern is a structure; they do call their pattern doctrine "structuralism." "Patternsism" is a bit awkward.

"Pattern" is like "game," Wittgenstein's example whose referents can't be isolated by any explicit definition, only by "family resemblance." Solitaire can be connected by a chain of intermediate games to soccer. In the same way, very likely, the pattern of continuity-discontinuity in mathematical analysis could be connected by a chain of intermediate patterns to the pattern of quotient rings and ideals in algebra.

The structuralist definition fits mathematical practice, because it's all-inclusive. All mathematics easily falls under its scope. Almost everything else falls under it too. The set-theoretic picture of mathematics is unconvincing because set theory is irrelevant to the bulk of mainstream mathematics. Structuralism suffers the opposite shortcoming. It recognizes something present not only in mathematics, but in all analytical thinking. The set-theoretic picture is restrictive; the structuralist picture, over-inclusive.

Fictionalism

Hamlet/Hypotenuse

Recall the three familiar number systems: natural, rational, and real.

Natural numbers are for counting. Most everybody seems to think the natural numbers exist in some sense, though they are infinitely many, and some people gag at anything infinite.

Rational numbers are fractions, positive and negative. They're just pairs of natural numbers. (The rational number $1/2$ is the pair $[1, 2]$). No special difficulty.

The irrational real numbers are the headache. They can be defined as nonrepeating infinite decimals. But no one has ever seen an infinite decimal written out all the way to infinity. Does the irrational number π exist? Its first *billion* decimal places were computed by the Chudnovskys and Tanaka, as I mentioned above. But even after two and three billion decimal places, the Chudnovsky's will only see a finite piece of π. The unseen piece will still be infinite. Wherever we quit, we'll be in the dark about infinitely many digits in the decimal expansion of π.

Cantor proved there are uncountably many real numbers.** That's too many to grasp. Any list of real numbers leaves out nearly all of them! Do so many real numbers really exist, or are they a fairy tale?

Ordinary mathematicians say "They exist!" Fictionalists say "They don't!" They try to show that science doesn't require (as W. V. O. Quine claimed) actual existence of mathematical entities. You can do science, they say, while regarding mathematical entities as fictional—not actually existing. Among philosophers of mathematics who can be called fictionalists are Charles Chihara, Hartry Field, and Charles Castonguay.

What does either side mean by "exist"? They don't say! In today's Platonist-fictionalist argument, as in yesterday's foundationist controversies, this question is ducked. But what you mean by "exist" determines what you believe exists. If only physical entities exist, then numbers don't exist. If *relations* between physical entities exist, then small positive whole numbers exist. If exist means "having its own properties independent of what anybody thinks," then real numbers exist.

Could this difference of opinion affect mathematical practice? What evidence could settle it? I know no practical consequence of this dispute, nor any way to settle it. Argument about it usually comes down to, "To me, this opinion is more palatable than that one."

Some fictionalists are materialists. They notice that mathematics is imponderable, without location or size. Since, as they think, only material objects, ponderable and volume-occupying, are real, mathematics isn't real. They state this unreality by saying mathematics is a fiction. What it means to be a "fiction" isn't further explained.

"Fiction" means a "made-up story." And mathematics *is* a kind of made-up story. But its uniqueness, its difference from other made-up stories, is what we care about. Can the literary notion of fiction assist the philosophy of mathematics?

Aristotle wrote: "The distinction between non-fiction author and fiction author consists really in this, that the one describes the thing that has been, and the other a kind of thing that might be. Hence fiction is something more philosophic and of graver import than non-fiction, since its statements are of the nature rather of universals, whereas those of non-fiction are singular" (*Poetics*, p. 1451). (Modernized translation, replacing "poetry" with "fiction," "history" with "non-fiction.")

Charles Chihara makes an analogy between mathematics and Shakespeare's Prince Hamlet. Hamlet is a fiction, yet we know a lot about him. A stage director putting on *Hamlet* knows more about Hamlet than Shakespeare wrote down. Hamlet probably ate grapes. Certainly didn't eat fried armadillo. In mathematics we also reach conclusions about things that don't exist. Numbers are fictions like Hamlet.

Nonfiction corresponds to empirical science; fiction corresponds to mathematics!

Fictionalism rejects Platonism. In that sense, I'm a fictionalist. But Chihara, Field and I aren't in the same boat.

What do you mean by real? By fictional? Only if that's made clear can we say mathematics is a fiction. How is it distinguished from other fictions? Is Huck Finn the same kind of thing as the hypotenuse of a right triangle?

It's good to reject Platonism. To replace it, we need to understand its powerful hold. Then it becomes possible to let it go.

Our conviction when we work with mathematics that we're working with something real isn't a mass delusion. To each of us, mathematics is an external reality. Working with it demands we submit to its objective character. It's what it is, not what we want it to be. It's ineffective to deny the reality of mathematics without confronting this objectivity.

Fictionalism is refreshingly disrespectful to that holy of holies—mathematical truth. It puts human creativity at center stage. But it's a metaphor, not a theory.

The difficulty is failing to recognize *different levels* of existence. Numbers aren't physical objects. Yet they exist outside our individual consciousness. We encounter them as external entities. They're as real as homework grades or speeding tickets. They're real in the sense of social-cultural constructs. Their existence is as palpable as that of other social constructs that we must recognize or get our heads banged. That's why it's wrong to call numbers fiction, even though they possess neither physical nor transcendental reality, even though they are, like Hamlet, creations of human mind/brains.

We need to start by recognizing nonmaterial realities—mental reality and social-cultural-historical reality. Then it becomes apparent that mathematics is a social-cultural-historical reality with mental and physical aspects. Does that make it a fiction? Sure, if the U.S. Supreme Court and the prime interest rate and the baseball pennant race are all fiction. But they're not. Mathematics *is* at once a fictional reality and a realistic fiction. The interesting question is the intertwining of real and fictitious that make it what it is.

ten

Humanists and Mavericks of Old

About the Humanist Trend in Philosophy of Mathematics

The idea of mathematics as a human creation has been advocated many times, by Aristotle, by the empiricists John Locke, David Hume, and John Stuart Mill, and by many others.

I use "humanist and maverick" to include all these writers. I call my own slant on humanist philosophy of mathematics "social-cultural-historic" or just "social-historic."

Some humanist mavericks weren't primarily philosophers: Jean Piaget, psychologist; Leslie White, anthropologist; Michael Polányi, chemist; Paul Ernest, educationist; and Alfréd Rényi, George Pólya, Raymond Wilder, mathematicians.

Those who were mainly philosophers were nonstandard or off-beat: the pragmatist, Charles S. Peirce; the Hegelian mystical historicist, Oswald Spengler; the critical realist, Roy Sellars; the quasi-empiricist or fallibilist, Imre Lakatos; the scientific materialist, Mario Bunge; the objectivist, Karl Popper; the naturalist, Philip Kitcher; and the quasi-empiricist, Thomas Tymoczko.

Aristotle (384–322 B.C.)

The First Scientist

The first humanist in the philosophy of mathematics is modern compared to Pythagoras or Plato. Aristotle's first concern is careful logical analysis of terms and concepts. His next concern is whether speculative theories conform to known facts. There's no mysticism in his dry reports of the mystics Pythagoras, Plato, and Plato's successors Speusippus and Xenocrates.

H. G. Apostle collected Aristotle's writings on mathematics. I made it my main source on Aristotle. Apostle writes, "Of Aristotle's extant works no one treats of mathematics systematically. . . . However, numerous passages on mathematics are distributed throughout the works we possess and indicate a definite philosophy of mathematics."

In Aristotle's philosophy of mathematics, the key concept is abstraction. Numbers and geometrical figures are abstracted from physical objects by setting aside irrelevant properties—color, location, price, etc.—until nothing's left but size and shape (in the case of geometric figures) or "numerosity" (in the case of finite sets). As an account of elementary mathematics, this is not bad. Today it's inadequate, because mathematics includes much more than circles, triangles, and the counting numbers. His account of abstraction is clear and reasonable. But by twentieth-century standards it's not precise. It would be difficult to give a formal definition of abstraction.

Aristotle gets bad press in the survey course on Western Civilization where many people meet him. We learn there that modern science was born in the struggle of Galileo, Copernicus, and Descartes against the followers of Aquinas and Aristotle. But history is more complicated than that. Much of European philosophical thought developed as a contest between Platonists and Aristotelians. From the time of Augustine (fifth century), Plato was Church dogma. For centuries, Aristotle's writings were lost from Western Europe. They were retrieved in the twelfth century, thanks to Arab scholars of North Africa and Spain.

The recovery of Aristotle's writings led to a turn toward scientific realism under Church control. Thomas Aquinas was a leader in bringing Aristotle's philosophy into the Church.

Later there was a revival of Platonism (Vico), as part of a humanist opposition to Descartes's scientific rationalism. Descartes put observation and experiment above authority. By then, Aristotelian scholasticism really had become antiscientific. But in the longer perspective of the centuries-old competition between Aristotle and Plato, Aristotle favored scientific rationalism, Plato, transcendental mysticism.

For a glimpse of a remarkable mind, some samples from Apostle's anthology of Aristotle:

"The infinite cannot be a number or something having a number, for a number is numerable and hence exhaustible. Moreover, if the infinite were an odd number, then by the removal of a unit the resulting number would be even and still infinite; for as finite it could have only finite numbers as parts—and likewise if it were an even number. But it cannot be both odd and even. Further, if the infinite, after the removal of a unit is considered as odd, were divided into two equal parts, then two infinite numbers would result; and, if this were continued, an infinite number could be divided into as many infinite numbers as one pleased . . . (p. 69).

If we bisect the straight line AZ at B, and again the line BZ on the right at C, and continue the bisection of the part which remains on the right, that part exists from the beginning; and the parts which are taken away from it and added to the left still remain. Here, it is also evident that there is both an infinite by division and an infinite by addition at the same time and in the same straight line AZ; and this is true for any finite magnitude. Along with the division there is a corresponding addition taking place, for corresponding to a given division, say at E, there is a magnitude DE added to the magnitudes AD already taken; and just as there is more division (indeed an endless division) to be made in the remaining magnitude EZ, so there are more and more parts in EZ to be taken away and added to AE without the possibility of an end. Yet, as the bisection continues on and on, the sum of the parts taken tend more and more to a certain limit, AZ, which is never reached (p. 73).

"Antiphon, in attempting to square the circle makes the fallacy in thinking that the parts outside to be taken will finally come to an end. He inscribes a square within the circle, erects isosceles triangles on the sides to the square as bases and with the vertices on the circle, continues this process, and concludes that the increasing side of the circumscribed polygon will finally coincide with the points on the circle. Thus, since a square can be erected equal in area to each set of isosceles triangles added to the previous inscribed regular polygon, the circle itself will ultimately be squared. But this is impossible, for the diminishing sides of the inscribed regular polygon will never become points, and there will always be isosceles triangles outside yet to be taken" (p. 76).

This critique of Antiphon is followed by an analysis of Zeno's paradoxes against motion. This is so clear, one wonders why anyone after Aristotle ever bothered with those paradoxes.

"A difficulty in connection with the infinite concerns the mathematician. If the infinite is not actual and the magnitude of the universe is finite, his theorems concerning numbers will not be true for all numbers but only for a finite number of them; and he will not be able to extend his straight lines and planes indefinitely to demonstrate certain theorems in geometry. . . . If at a certain time the number one million does not exist, then it is false to say that the theorem is true for the number one million at that time; and the theorem is false, not simply, but in a qualified way, at such-and-such a time and for such-and-such a number. The theorem is stated in universal terms and has a potential nature; it is true for any number, not at this or that time or place but whenever and wherever a number exists. The fact that numbers exist or can exist shows that arithmetic does not deal with not-being" (p. 78).

He produced pages criticizing Plato's "Ideas" on grounds of vagueness and inconsistency. From p. 182: "There is also a difficulty in defining an Idea, if we are to predicate definitions of them as we predicate definitions of the corresponding species and genera of things. A definition is a predicate of many individuals and not of only one, but an Idea is one individual. If a man is defined as a rational animal, will the definition of the Idea of Man be 'Absolute Rational

Absolute Animal' or 'Absolute Rational Animal' or 'Absolute Rationality and Absolute Animal'? Moreover, can we truly say that Absolute Man is Rational or Absolute Rational if Absolute Man and Absolute Rationality are two different individual Ideas? This will be equivalent to saying that Plato is Socrates. If Absolute Triangle is defined as "three-sided figure," and "three-sided figure" is a predicate of Absolute Triangle, then "three-sided figure" will also be a predicate of Absolute Isosceles Triangle; and Absolute Triangle will be a part of Absolute Isosceles Triangle and will not exist separately. Further, Absolute Triangle is numerically one, and, since one of two contradictories is true, either it is isosceles (or Isosceles or else Absolute Isosceles) or it is not. If it is isosceles, then it does not differ from Absolute Isosceles Triangle, and the two Ideas will be one Idea; besides, it has just as much reason to be isosceles as to be equilateral or scalene, and it cannot be all of them. If it is not isosceles, then for the same reason it is not equilateral or scalene; but it must be one of them, for the three sides must be related in one of the three ways. Also, whatever is a predicate of an isosceles is also a predicate of an isosceles triangle and conversely. Hence, Absolute Isosceles and Absolute Isosceles Triangle turn out to be one and the same Idea. Again, if the One is ultimately the formal cause of an Idea and "one" is a predicate of the Idea, perhaps the One should be in the Idea. But the One is unique, and so is each Idea of its form. Hence, it would be absurd to try to show that all things have One as form."

In reading this passage, I am reminded of Frege. The direction of their arguments is opposite, for Aristotle is tearing up the Platonic Idea of number, which Frege upholds. But the tone—the cat-like delight in chewing up a philosophical mouse! Across 2,000 years, the two could be cousins.

Euclid

Axioms and Diagrams

The philosophy of mathematics of the Greeks ought to include not only the Greek philosophers but the great mathematicians Archimedes, Eudoxus, Euclid, and Apollonius. Unfortunately, the fragments left to us are insufficient to draw firm conclusions. We may suspect that the philosophy of mathematics accepted by the Greek mathematicians may not have been identical with the teachings of the philosophers.

We are told that the Greeks despised applications. Plutarch says Archimedes didn't think his great military engineering in defense of Syracuse was worth being preserved in writing. Astronomy, of which Ptolemy was the preeminent practitioner, must have been put on a higher plane than earthly calculations.

It's said that Euclid's axiomatics was "material axiomatics"—statements of true facts about real objects—unlike modern "formal axiomatics," which isn't about anything. It's risky to say much more, with so little evidence.

In Chapter 3 I wrote, "In the middle half of the twentieth century formalism was the philosophy of mathematics most advocated in public. In that period, the style in mathematical journals, texts, and treatises was: Insist on details of definitions and proofs; tell little or nothing of why a problem is interesting or why a method of proof is used."

Does that sound like Euclid? Was Euclid a formalist?

Yes, it sounds like Euclid if we look at Euclid with a formalist eye—if we see his text as essential, and his diagrams as unfortunate breaches of rigor.

Without the diagrams, the *Elements* could pass for a formalist text. But Euclid comes with diagrams, not without!

A recent book called *Proof without Words* is instructive. This is a charming collection of proofs using only pictures and diagrams, not a single word. There are theorems on plane geometry, finite and infinite sums and integrals, algebraic inequalities, and more.

The introduction contains a depressing disclaimer. The editor warns us that the proofs in his book aren't really proofs. Only the usual verbal or symbolic proof is really a proof. A sad testimony to the grip of formalism.

A proof can be words only, of course. It can be, as in Euclid, words and diagrams. Or it can be, as in *Proofs without Words*, diagram only. There's no textual or historical evidence that Euclid's diagrams were thought to be unimportant or unnecessary. They supply the motivation and insight that are lacking in the text. They free Euclid from suspicion of formalism.

John Locke (1632–1704)

Tabula Rasa

Locke doesn't have an inclusive, comprehensive philosophy of mathematics. But he understands that mathematics is a creation and an activity of the human mind. Unlike Berkeley and Hume, he has no criticism to make of mathematicians.

In freshman philosophy "Locke-Berkeley-Hume" are presented as opposites of "Descartes-Spinoza-Leibniz"—English empiricists versus continental rationalists. But among the empiricists, Berkeley's goal was utterly different from that of Locke or Hume—closer to that of the rationalist Leibniz, which was not to uphold science against scholasticism, but to prove that matter exists only in the Mind of God.

What did Locke-Berkeley-Hume think about mathematics? No two were alike. Take them in chronological order. Locke insisted everything in common knowledge and in natural science comes from observation. There's no innate or nonempirical knowledge of the exterior world. One might think this position would have led him to attempt an empiricist explanation of mathematics. But he didn't try to reduce mathematical knowledge to the empirical, as Mill did 200 years later. Instead, he saw it virtually as introspective.

"The knowledge we have of mathematical truths is not only certain but real knowledge; and not the bare empty vision of vain, insignificant chimeras of the brain: and yet, if we will consider, we shall find that it is only of our own ideas. The mathematician considers the truth and properties belonging to a rectangle or circle only as they are an idea in his own mind. For it is possible he never found either of them existing mathematically, i.e., precisely true, in his life. But yet the knowledge he has of any truths or properties belonging to a circle, or any other mathematical figure, are never the less true and certain even of real things existing; because real things are no further concerned, nor intended to be meant, by any such propositions, than as things really agree to those archetypes in his mind. Is it true of the idea of a triangle, that its three angles are equal to two right ones? It is true also of a triangle wherever it really exists" *(Essay,* IV, 6).

"All the discourses of the mathematicians about the squaring of a circle, conic sections, or any other part of mathematics, concern not the existence of any of those figures, but their demonstrations, which depend on their ideas, are the same, whether there be any square or circle existing in the world or no" (para. 8).

This sounds like rationalism; triangle and circle as archetypes in our minds. The next sentence makes an analogy between moral knowledge and mathematical knowledge, and says that both types of knowledge "abstract" from observation. "In the same manner, the truth and certainty of moral discourses abstracts from the lives of men, and the existence of those virtues in the world whereof they treat: nor are Tully's *Offices* less true, because there is nobody in the world that exactly practises his rules." In this sentence "abstract" means "idealized." Moral truths, like mathematical truths, refer to ideal, not to empirical reality. Locke equivocates between internal and external as the source of mathematical and moral conceptions. He doesn't, in the manner of the rationalists, try to use mathematical certainty to justify religious certainty. He does use mathematical knowledge as an analogue to moral knowledge.

From Baum, p. 126: "When we nicely reflect upon them, we shall find that general ideas are fictions and contrivances of the mind, that carry difficulty with them, and do not so easily offer themselves as we are apt to imagine. For example, does it not require some pains and skill to form the general idea of a triangle (which is yet none of the most abstract, comprehensive, and difficult), for it must be neither oblique nor rectangle, neither equilateral, equicrural, nor scalenon; but all and one of these at once. (Echo of Aristotle!) In effect, it is something imperfect, that cannot exist; an idea wherein some parts of several different and inconsistent ideas are put together.

"Amongst all the ideas we have, there is none more simple, than that of unity, or one; . . . by repeating this idea in our minds, and adding the repetitions together, we come by the complex ideas of the modes of it. Thus, by adding one to one, we have the complex idea of a couple; by putting twelve units together, we have the complex idea of a dozen, and of a score, or a million, or any other number. . . . Because the ideas of numbers are more precise and distinguishable than in extension, where every equality and excess are not so easy to be observed or measured, because our thoughts cannot in space carrie at any determined smallness beyond which it cannot go, as an unit; and therefore the quantity or proportion of any the least excess cannot be discovered. . . . This I think to be the reason why some American [Indians] I have spoken with (who were otherwise of quick and rational parts enough), could not, as we do, by any means count to 1000, nor had any distinct idea of that number, though they could reckon very well to 20. Because their language being scanty, and accommodated only to the few necessaries of a needy, simple life, unacquainted either with trade or mathematics, had no words in it to stand for 1000. . . . Let a man collect into one sum as great a number as he please, this multitude, how great soever, lessens not one jot the power of adding to it, or brings him any nearer the end of the inexhaustible stock of number, where still there remains as much to be added as if none were taken out. And this *endless addition* or *addibility* of numbers, so apparent to the mind, is that, I think, which gives us the clearest and most distinct idea of infinity." (Anticipating Poincaré by two centuries!)

Isaiah Berlin thinks that "About mathematical knowledge Locke shows great acumen. He sees that, for example, geometrical propositions are true of certain ideal constructions of the human mind and not of, e.g., chalk marks or surveyor's chains in the real world" (Berlin, p. 108). The circle and the line are ideas in the mathematician's head, which is why the mathematician can discover facts about them by mere contemplation. Locke doesn't seem to mind letting mathematics be an exception to his empiricist doctrine. Nor is he troubled by the objection Frege would make later: if a circle is in the mathematician's head, then there must be many different circles, one in each mathematician's head.

Locke also has his proof of the existence of God, based on "intuitively clear" ideas, such as "I exist," "I came from somewhere," "there had to be a first cause," and so forth.

Berkeley is conventionally presented in chronological order, between Locke and Hume. But in the history of the philosophy of mathematics, he cannot be counted as a forerunner of the social-cultural tendency. We treated Berkeley in our previous historical section, the history of Mainstream formalism and mysticism.

David Hume (1711–1776)

Commit It to the Flames!

Locke and Berkeley's successor David Hume was the only well-known philosopher before the twentieth century to recognize that mathematics, like the other sciences, gives only probable knowledge. He explicitly describes the role of the community of mathematicians in the process of mathematical growth and discovery. "In all demonstrative sciences the rules are certain and infallible, but when we apply them, our fallible and incertain faculties are very apt to depart from them, and fall to error. . . . By this means all knowledge degenerates into probability; . . . There is no Algebraist nor Mathematician so expert in his science, as to place entire confidence in any truth immediately upon his discovery of it, or regard it as any thing, but a mere probability. Every time he runs over his proofs, his confidence encreases; but still more by the approbation of his friends; and is rais'd to its utmost perfection by the universal assent and applauses of the learned world. Now 'tis evident that this gradual encrease of assurance is nothing but the addition of new probabilities" (*Treatise*, p. 231).

I respond, "How true!" One of the rare statements about mathematics in philosophy texts to which I so respond.

Hume came under Berkeley's influence early in life. "Perhaps the most important formative influence in Hume's formative years was his membership in the Rankenian Club of Edinburgh during the early 1720's. . . . The club carried on a correspondence with Berkeley, whose philosophy was apparently one of the central topics for discussion. . . . Nowhere, according to [Berkeley], was he better understood" (Turbayne).

Nearly all of Part 2, Book 1, of Hume's *Treatise of Human Nature* is devoted to refuting the infinite divisibility of the line. This doctrine has been universally accepted in geometry at least since Euclid. Euclid's simple construction to bisect a line segment, when repeated enough times, produces a line segment as short as you please. But Hume argues that time and space must consist of indivisible atoms. "For the same reason that the year 1737 cannot concur with the present year 1738, every moment must be distinct from and posterior or antecedent to another. 'Tis certain then, that time, as it exists, must be composed of indivisible moments. For if in time we could never arrive at an end of division, and if each moment, as it succeeds another, were not perfectly single and indivisible, there would be an infinite number of co-existent moments, or parts of time; which I believe will be allowed to be an arrant contradiction. The infinite divisibility of space implies that of time, as is evident from the nature of motion. If the latter, therefor be impossible, the former must be equally so. . . . Tis an establish'd maxim in metaphysics, That whatever the mind clearly conceives includes the idea of possible existence, or in other words, that nothing we imagine is absolutely impossible."

Hume and Berkeley are sure there must be shortest, indivisible units of length, because, they think, "the Mind" cannot conceive of a finite interval being composed of infinitely many parts. By saying so, they imply an attack on the differential calculus of Newton and Leibniz, where arbitrarily short intervals play a central role, and which at this very time in the hands of Huygens, the Bernoullis, and Euler, was making wonderful progress, establishing the dynamics of particles, rigid bodies, fluids, and solids. Did Hume realize he was challenging the best mathematics and science of his time? To Hume it's axiomatic that an event is possible if and only if "the Mind" can conceive it clearly. This thinking goes back to Descartes.

Hume likes to say an infinitely divisible line segment is as inconceivable (and therefore as impossible) as a square circle.** More than geometry, it was modern physics that wiped out the doctrine that an event is possible if and only if the Mind finds it conceivable. The quantum-mechanical uncertainty principle, the complementarity of particle and wave, the relativization of simultaneity by Einstein, the cosmos as *curved four-dimensional* space-time—all were *inconceivable*! They're incompatible with our deepest convictions about the world. Nevertheless, they are effective explanatory theories. Grandpa's rule of possibility says, "If it happens, it's possible!" To understand the world, we have to stretch the limits of what our minds can conceive.

"Thus it appears that the definitions of mathematics destroy the pretended demonstrations, and that if we have the idea of indivisible points, lines and surfaces conformable to the definition, their existence is certainly possible; but if we have no such idea, 'tis impossible we can ever conceive the termination of any figure, without which conception there can be no geometrical demonstration. The first principles are founded on the imagination and senses: The conclusion, therefor, can never go beyond, much less contradict these faculties. No geometrical demonstration for the infinite divisibility of extension can have so much force as what we naturally attribute to every argument, which is supported by such magnificent pretensions. . . . For tis evident, that as no idea of quantity is infinitely divisible, there cannot be imagin'd a more glaring absurdity than to endeavor to prove that quantity itself admits of such a division."

One fallacy here is "conflation" of mathematical space and physical space. He uses an intuitive *physical* argument to prove that infinite divisibility is *mathematically* impossible. We cannot fault him for this. The distinction between physical space and mathematical space was yet unborn (until the discovery of non-Euclidean geometry).

Even if *physical* space were granular, that wouldn't stop us from using infinite divisibility in a *mathematical* theory, just as we use such exotica as infinite-dimensional spaces, the set of all subsets of an uncountable set, or a well-ordering of the real numbers.

As a matter of fact, Hume is wrong both physically and mathematically. Physically, he claims to prove that there's a minimum possible size of a material particle—without benefit of any physical measurement, based only on his inability to conceive. But you can't limit physical reality by the limits of your imagination.

Mathematically, he's saying the notion of an infinitely divisible interval is nonsense because an interval can't be divided into infinitely many pieces of *equal* length. But division into an arbitrarily large finite number of pieces is enough to refute his idea of a shortest possible length. And can he not have noticed the possibility of infinitely many pieces of decreasing length? The success of differential calculus refutes his claim that infinite divisibility is incoherent.

He thinks that points are indivisible, and line segments are made of finitely many points, so the minimum line segment must have positive length. This puzzle comes up in modern measure theory. How can a segment of positive length be composed of points of zero length? Hume is tripping over a consequential matter—the uncountability of the continuum, a path-breaking discovery of George Cantor, in the late nineteenth century.

When Hume says that instants of time are linearly ordered, few would disagree. But then he says that since they're linearly ordered, they're discrete. This is a mistake. We mustn't blame him for that. Unlike Berkeley, whose arguments on this matter are insubstantial, *Hume is grappling with a real mathematical problem.* The example of an ordered set that is dense, not discrete, was there for all to see, in the set of rational numbers. But it wasn't noticed until Cantor did so, a century later. Hume has not received credit for raising the problem, though of course he couldn't solve it.

Hume's major contribution to epistemology was to show that the "law of cause and effect" is not deductively valid. Our faith in it rests only on habit, so we can't have certain knowledge about anything external or material.

This corrosive skepticism attacks science as well as religion. Science doesn't hope for absolute certainty, but it assumes that knowledge is possible, if only partial knowledge or tentative knowledge. Hume destructively discounts the possibility of fundamental scientific advance. Elasticity, gravity, and other observed physical phenomena should be taken at face value, he says; to seek for deeper explanation would be in vain.

"As to the causes of these general causes, we should in vain attempt their discovery, nor shall we ever be able to satisfy ourselves by any particular explication of them. These ultimate springs and principles are totally shut up from human curiosity and inquiry. Elasticity, gravity, cohesion of parts, communication of motion by impulse—these are probably the ultimate causes and principles which we shall ever discover in nature; and we may esteem ourselves sufficiently happy if, by accurate inquiry and reasoning, we can trace up the particular phenomena to, or near to, these general principles . . . the observation of human blindness and weakness is the result of all philosophy, and meets us at every turn. Nor is geometry . . . able to remedy this defect . . . by all the accuracy of reasoning for which it is so justly celebrated" (*Inquiry*, p. 45).

Still, he does acknowledge that mathematical knowledge is different from empirical knowledge. "From (cause and effect reasoning) is derived all philosophy excepting only geometry and arithmetic" (*Abstract*, p. 187).

So, in his famous peroration, he spares us from the bonfire:

"If we take in our hand any volume—of divinity or school metaphysics, for instance—let us ask, Does it contain any abstract reasoning concerning quantity or number? No. Does it contain any experimental reasoning concerning matter of fact and existence? No. Commit it then to the flames, for it can contain nothing but sophistry and illusion" (*Inquiry*, p. 173).

Jean Le Rond D'Alembert (1717–1783)
At Last, Enlightenment

D'Alembert, Diderot, Rousseau, and Condillac were the leading "Encyclopedists" who played a major role in forming the modern mind.

"D'Alembert was the natural son of a soldier aristocrat, the chevalier Destouches, and Madame de Tencin, one of the most notorious and fascinating aristocratic women of the century. A renegade nun, she acquired a fortune as mistress to the powerful minister, Cardinal Dubois, and after a successful career of political scheming, she rounded out her life by establishing a salon which attracted the most brilliant writers and philosophers of France. It is reported that d'Alembert was not the first of the inconvenient offspring she abandoned. In any case, he was found shortly after his birth on the steps of the Parisian church of Saint-Jean Lerond [note the name]. He was raised by a humble nurse, Madame Rousseau, whom he treated as his mother and with whom he lived until long after he achieved international fame."

A self-taught physicist and mathematician, he published his *Treatise on Dynamics* in 1743, at age 26. His formula for the one-dimensional linear wave equation with constant density is beloved by every student of applied mathematics. "A combination of virtuosity, ambition, aggressiveness, and personal charm eventually won him a most honored position in the intellectual community of Europe, including the lasting friendship of both Voltaire and Frederick the Great. . . . His entry in the Académie Française in 1754 marked a major victory for the encyclopedic party. Eventually he became the perpetual secretary of that academy. . . ."

"He never married. However, after 1754 he became the intimate friend of an aristocratic lady, likewise of illegitimate birth, the famous Julie de Lespinasse, with whom he was ever more closely bound until her death left him desolate in 1776. Their strictly spiritual and intellectual relationship was something exceptional among the *philosophes*" (Schwab in D'Alembert, 1963, p. 15 ff.).

D'Alembert is on the short list of notable contributors to both mathematics and philosophy. His word of encouragement to a beginner in calculus is often quoted: "Continue. Eventually, faith will come."

In his *Preliminary Discourse to the Encyclopedia of Diderot,* he stated his opinions on the nature of mathematics and logic:

"We will note two limits within which almost all of the certain knowledge that is accorded to our natural intelligence is concentrated, so to speak. One of those limits, our point of departure, is the idea of ourselves, which leads to that of the Omnipotent Being, and of our principal duties. [Echo of Descartes!] The other is that part of mathematics whose object is the general properties of bodies, of extension and magnitude. Between these two boundaries is an immense gap where the Supreme Intelligence seems to have tried to tantalize the human curiosity, as much by the innumerable clouds it has spread there as by the rays of light that seem to break out at intervals to attract us. . . ."

"With respect to the mathematical sciences, which constitute the second of the limits of which we have spoken, their nature and their number should not overawe us. It is principally to the simplicity of their object that they owe their certitude. Indeed, one must confess that, since all the parts of mathematics do not have an equally simple aim, so also certainty, which is founded, properly speaking, on necessarily true and self-evident principles, does not belong equally or in the same way to all these parts. . . . Only those that deal with the calculation of magnitudes and with the general properties of extension, that is, Algebra, Geometry, and Mechanics, can be regarded as stamped by the seal of evidence. . . . The broader the object they embrace and the more it is considered in a general and abstract manner, the more also their principles are exempt from obscurities. It is for this reason that geometry is simpler than mechanics, and both are less simple than Algebra. . . . Thus one can hardly avoid admitting that the mind is not satisfied to the same degree by all the parts of mathematical knowledge."

[Quite different from the earlier views of Plato, and the later views of foundationism, that all mathematics must and should be absolutely certain.]

"Let us go further and examine without bias the essentials to which this knowledge may be reduced. Viewed at first glance, the information of mathematics is very considerable, and even in a way inexhaustible. But when, after having gathered it together, we make a philosophical enumeration of it, we perceive that we are far less rich than we believed ourselves to be. I am not speaking here of the meager application and usage to which a number of these mathematical truths lend themselves; that would perhaps be a rather feeble argument against them. I speak of these truths considered in themselves. What indeed are most of those axioms of which Geometry is so proud, if not the expression of a single simple idea by means of two different signs or words? Does he who says two and two equals four have more knowledge than the person who would be content to say two and two equals two and two? Are not the ideas of 'all,' of 'part,' of 'larger,' and of 'smaller,' strictly speaking, the same simple and individual idea, since we cannot have the one without all the others presenting themselves at the same time? As some philosophers have observed, we owe many errors to the abuse of words. It is perhaps to this same abuse that we owe axioms. My intention is not, however, to condemn their use; I wish only to point out that their true purpose is merely to render simple ideas more familiar to us by usage, and more suitable for the different uses to which we can apply them. I say virtually the same thing of the use of mathematical theorems, although with the appropriate qualifications. Viewed without prejudice, they are reducible to a rather small number of primary truths. If one examines a succession of geometrical propositions, deduced one from the other so that two neighboring propositions are immediately contiguous without any interval between them, it will be observed that they are all only the first proposition which is successively and gradually reshaped, so to speak, as it passes from one consequence to the next, but which, nevertheless, has not really been multiplied by this chain of connections; it has merely received different forms. It is almost as if one were trying to express this proposition by means of a language whose nature was being imperceptibly altered, so that the proposition was successively expressed in different ways representing the different states through which the language had passed. Each of these states would be recognized in the one immediately neighboring it; but in a more remote state we would no longer make it out, although it would still be dependent upon those states which preceded it, and designed to transmit the same ideas. Thus, the chain of connection of several geometrical truths can be regarded as more or less different and more or less complicated translations of the same proposition and often of the same hypothesis. These translations are, to be sure, highly advantageous in that they put us in a position to make various uses of the theorem they express—uses more estimable or less, in proportion to their import and consequence. But, while conceding the substantial merit of the mathematical translation of a proposition, we must recognize also that this merit resides originally in the proposition itself. This should make us realize how much we owe to the inventive geniuses who have substantially enriched Geometry and extended its domain by discovering one of these fundamental truths which are the source, and, so to speak, the original of a large number of others . . . (p. 25 ff.).

"The advantage men found in enlarging the sphere of their ideas, whether by their own efforts or by the aid of their fellows, made them think that it would be useful to reduce to an art the very manner of acquiring information and of reciprocally communicating their own ideas. This art was found and named Logic. . . . It teaches how to arrange ideas in the most natural order, how to link them together in the most direct sequence, how to break up those which include too large a number of simple ideas, how to view ideas in all their facets, and finally how to present them to others in a form that makes them easy to grasp. . . . This is what constitutes this science of reasoning, which is rightly considered the key to all our knowledge. [This looks like a veiled reference to the great *Port-Royal Logic* of the Jansenist Antoine Arnauld.] However, it should not be thought that it [the formal discipline of Logic] belongs among the first in the order of discovery. . . . The art of reasoning is a gift which Nature bestows of her own accord upon men of intelligence, and it can be said that the books which treat this subject are hardly useful except to those who can get along without them. People reasoned validly long before Logic, reduced to principles, taught them how to recognize false reasoning, and sometimes even how to cloak them in a subtle and deceiving form" (p. 30).

John Stuart Mill (1808–1873)
Classic Liberal

Mill was a father-created child prodigy, like William Rowen Hamilton and Norbert Wiener. He performed wonders as an infantile multilingual classical scholar. He's remembered today for courageous writings in defense of liberty and against the subjection of women. He and Harriet Taylor achieved the first famous collaboration between male and female authors.

His well-known *Utilitarianism* has this provocative remark: "Confusion and uncertainty exist respecting the principles of all the sciences, not excepting that which is deemed the most certain of them—mathematics, without much impairing, generally indeed without impairing at all, the trustworthiness of the conclusions of those sciences. An apparent anomaly, the explanation of which is that the detailed doctrines of a science are not usually deduced from, nor depend for their evidence upon, what are called its first principles. Were it not so, there would be no science more precarious, or whose conclusions were more insufficiently made out, than algebra, which derives none of its certainty from what are commonly taught to learners as its elements, since these, as laid down by some of its most eminent teachers, are as full of fictions as English law, and of mysteries as theology. The truths which are ultimately accepted as the first principles of a science are really the last results of metaphysical analysis practiced on the elementary notions with which the science is conversant; and their relation to the science is not that of foundations to an edifice, but of roots to a tree, which perform their office equally well though they be never dug down to and exposed to light."

A rejection of foundationism before foundationism came into flower!

Mill's major contribution to philosophy of mathematics is the *System of logic* . . . , finished in 1841. Today's student knows of this book only by Frege's merciless attack in the *Grundlagen*. But one can't rely on Frege for a fair account. Frege wrote, claiming to paraphrase Mill: "From this we can see that it is really incorrect to speak of three strokes when the clock strikes three." But on page 189 Mill writes, "*Ten* must mean ten bodies, or ten sounds, or ten beatings of the pulse." Today's revival of empiricism in philosophy of mathematics calls for a new look at Mill.

Mill's main thesis was that laws of mathematics are objective truths about physical reality. They are derived from elementary principles or laws (*not* arbitrary axioms) that we learn by observing the world. To Mill, the number 3 is defined independently of 1 and 2, by generalization from observed triples.

"2 + 1 = 3" is a truth of observation, not essentially different from "all swans are white." It turned out that in Australia there are black swans. It could turn out that 2 + 1 sometimes isn't three. But since we have a tremendous amount of confirmation of this law, we rightly have tremendous confidence in it. Mill's idea is close to Aristotle's. But in Mill it seems radical or paradoxical, because by Mill's time rationalist-idealist philosophy of mathematics had become dominant.

Mill fought against two different theories of arithmetic. The nominalists (Hobbes, Condillac, and J. S. Mill's father James Mill) said $2 + 1$ is the *definition* of 3; therefore (foreshadowing Bertrand Russell) the equation $2 + 1 = 3$ is an empty tautology. To Mill the formula $2 + 1 = 3$ is not tautologous but informative; it says a triple can be separated into a pair and a singleton, or a pair and a singleton united to become a triple.

The other philosophy that Mill opposed was called intuitionist (no connection with Brouwer!). Its leading advocate was William Whewell, a famous philosopher of science who was then tremendously influential in university mathematics. Whewell said anything inconceivable must be false. (Echoes of Berkeley and Hume!) Therefore, the negation of anything inconceivable must be true! It's inconceivable that

> $2 + 1$ isn't equal to 3;
> therefore, $2 + 1 = 3$.

In reply, Mill recalled "inconceivables" of previous times: a spherical earth, the earth revolving round the sun, instantaneous action at a distance by gravitation. "There was a time when men of the most cultivated intellects, and the most emancipated from the domain of early prejudice, could not credit the existence of antipodes; were unable to conceive, in opposition to old association, the force of gravity acting upward instead of downward. The Cartesians long rejected the Newtonian doctrine of the gravitation of all bodies toward one another, on the faith of a general proposition, the reverse of which seemed to them to be inconceivable—the proposition that a body can not act where it is not. . . . And they no doubt found it as impossible to conceive that a body should act upon the earth from the distance of the sun or moon, as we find it to conceive an end to space or time, or two straight lines enclosing a space" (Mill, pp. 178–79).

What's inconceivable last century is common sense in this.

Mill even quoted Whewell against himself. In Whewell's *Philosophy of the Inductive Sciences*, Mill found these lines: "We now despise those who, in the Copernican controversy, could not conceive the apparent motion of the sun on the heliocentric hypothesis. . . . The very essence of these triumphs is that they lead us to regard the views we reject as not only false but inconceivable."

Mill didn't confront the difficulties of the empiricism he advocated. It's believed that there are finitely many physical objects, but there are infinitely many numbers. Already in Mill's time mathematics included non-Euclidean geometry, which had (as yet) no physical interpretation.

From Kubitz, pp. 267–68: "The axioms which the *Logic* examines are the laws of causation and the axioms of mathematics. Mill gave an empirical explanation of these because "*the notion that truths external to the mind may be known by intuition or consciousness, independently of observation and experience,"* was the *"great intellectual support of false doctrines and bad institutions."* [My emphasis.] By the aid of this theory, every inveterate belief and every intense feeling of which the origin is not remembered, is enabled to dispense with the obligation of justifying itself by reason, and is erected into its own all-sufficient voucher and justification. There never was such an instrument devised for consecrating all deep-seated prejudices. And the chief strength of this false philosophy in morals, politics, and religion, lies in the appeal which it is accustomed to make to the evidence of mathematics and of the cognate branches of physical science."

The *System of Logic* explained the character of necessary truths on the basis of experience and education. The existence of "necessary truths, which is adduced as proof that their evidence must come from a deeper source than experience," was explained in such a way as to help justify the need for political reform. More on this point in Chapter 13.

eleven

Modern Humanists and Mavericks

Charles Sanders Peirce (1839–1914)

American Tragedy

Charles Sanders Peirce was a great logician, a great philosopher, and a respectable mathematician.[1]

He was a life-long freelancer in logic, mathematics, physics, and philosophy. Independently of Frege, Peirce's student, O. H. Mitchell, introduced the famous quantifiers "there exists" and "for all." Peirce was the founder of semiotics—the abstract, general study of signs and meaning.

I quote Christopher Hookway in the *Companion to Epistemology.* "From his earliest writings Peirce was critical of Cartesian approaches to epistemology. He charged that the method of doubt encouraged people to pretend to doubt what they did not doubt in their hearts, and criticized its individualist insistence that 'the ultimate test of certainty is to be found in the individual consciousness.' We should rather begin from what we cannot in fact doubt, progressing towards the truth as part of a community of inquirers trusting to the multitude and variety of our reasoning rather than to the strength of any one. He claimed to be a contrite Fallibilist and urged that our reasoning should not form a chain that is not stronger than its weakest link, but a cable whose fibres may be ever so slender, provided they are sufficiently numerous and intimately connected."

Peirce's view of mathematics is radically different from the foundationist sects. His surprising separation of mathematics and logic, and his full acceptance of the role of the mathematics community and of intuition in mathematics, put him on the humanist side. I quote from his essay "The Essence Of Mathematics." His remarks on the nature of mathematics could have been written yesterday.

[1] The first chapter of Corrington's book tells the tragic and embarrassing story of Peirce's life.

"Mathematics is distinguished from all other sciences, except only ethics, in standing in no need of ethics" (p. 267).

"Mathematics, along with ethics and logic alone of the sciences, stands in no need of logic. Make of logic what the majority of treatises in the past have made of it—that is to say, mainly formal logic, and the formal logic represented as an art of reasoning—and in my opinion, this objection is more than sound, for such logic is a great hindrance to right reasoning. True mathematical reasoning is so much more evident than it is possible to render any doctrine of logic proper—without just such reasoning—that an appeal in mathematics to logic could only embroil a situation. On the contrary, such difficulties as may arise concerning necessary reasoning have to be solved by the logician by reducing them to questions of mathematics."

"One singular consequence of the notion, which prevailed during the greater part of the history of philosophy, that metaphysical reasoning ought to be similar to that of mathematics, only more so, has been that sundry mathematicians have thought themselves, as mathematicians, qualified to discuss philosophy; and no worse metaphysics than theirs is to be found."

"In the major theorems it will not do to confine oneself to general terms. It is necessary to set down some individual and definite schema, or diagram—in geometry, a figure composed of lines with letters attached; in algebra an array of letters of which some are repeated. After the schema has been constructed according to the precept virtually contained in these, the assertion of the theorem is evidently true. Thinking in general terms is not enough. It is necessary that something should be *done*" (pp. 260–61).

From p. 267: "It is a remarkable historical fact that there is a branch of science in which there has never been a prolonged dispute concerning the proper objects of that science. It is the mathematics. Mistakes in mathematics occur not infrequently, and not being detected give rise to false doctrine, which may continue a long time. Thus, a mistake in the evaluation of a definite integral by Laplace in his *Mécanique céleste*, led to an erroneous doctrine about the motion of the moon which remained undetected for nearly half a century. But after the question had once been raised, all dispute was brought to a close within a year" (1960, para. 3,426; reprinted in the *American Mathematical Monthly*, 1978, p. 275).

Besides his path-breaking contributions to semiotics and logic, Peirce was the founder of pragmatism. The great American pragmatists William James and John Dewey were his followers. All three of them thought of mathematics as a human activity rather than a transcendental hyper-reality. Martin Gardner writes in his book *The Meaning of Truth*, that "James argues for a mind-dependent view of mathematics very close to that of Davis, Hersh, and Kline." Gardner thinks that "James makes a good case for a cultural approach to mathematics that was shared by F. C. S. Schiller and (he thinks) by John Dewey." Gardner himself, I should say, thinks otherwise.

Henri Poincaré (1854–1912)

Mozart of Mathematics

Poincaré was one of the supreme mathematicians, rivaled or surpassed in his time only by David Hilbert. Like other French masters of the turn of the century, he was a virtuoso at complex function theory. His qualitative study of the three-body problem was a fountainhead of algebraic topology, one of the great achievements of twentieth-century mathematics.

Poincaré's work on the Lorentz transformation and Maxwell's equations could have led him to special relativity. But in physics Poincaré was a conventionalist. He thought it a matter of convenience which mathematical model one uses to describe a physical situation. For example, he said, it makes no sense to ask if physical space is Euclidean or non-Euclidean. Whichever is convenient is best. His conventionalism hid from him the deep physical meaning of his mathematical results on relativity.

Einstein was not a conventionalist but a realist. He opposed Ostwald and Mach, who thought atoms and molecules were only convenient fictions. His 1905 paper on Brownian motion proved that molecules actually have definite volumes—they are real! As a philosophical realist, Einstein could more readily appreciate the physical consequences of mathematical relativity.

Poincaré disliked Peano's work on a formal language for mathematics, then called "logistic." He wrote of Russell's paradox, with evident satisfaction, "Logistic has finally proved that it is not completely sterile. At last it has given birth—to a contradiction."

With his colleagues Emile Borel and Henri Lebesgue, he opposed uninhibited proliferation of infinite sets. Arithmetic should be the common starting point of all mathematics. Mathematical induction is the fundamental source of novelty in mathematics. In this he was a forerunner of Brouwer's intuitionism. But he never joined Brouwer in condemning indirect proof (the law of the excluded middle.)

His brilliant, penetrating articles on philosophy of science and mathematics still find admiring readers.

Ludwig Wittgenstein (1889–1951)

Vienna in Cambridge

Ludwig Wittgenstein is one of the remarkable personalities of the century. His father was a top capitalist in Austrian steel. He had two brothers who committed suicide, and he spent a lifetime with self-hatred and religious guilt. In philosophy he created two revolutions.

He went to England to work on aeronautics. Bertrand Russell's *Philosophy of Mathematics* drew him to philosophy. Russell told Wittgenstein's sister that Wittgenstein was the philosopher of the future.

In 1921 he published the *Tractatus Logico-Philosophicus.* Russell wrote an admiring introduction to it. Wittgenstein said Russell didn't understand it. Later it became a bible for the Vienna Circle.

The message of the *Tractatus* is: exact correspondence between language, logic, and the world. Language and logic match perfectly everything of which it's possible to speak. The rest must pass in silence.

After the *Tractatus*, no more philosophy was needed. He went to teach school in the Austrian mountains. After a few difficult years, he gave up school teaching and returned to Cambridge as a professor. His opinions had changed. He repudiated the *Tractatus.* The task of philosophy is to show that there are no philosophical problems, "to show the fly the way out of the fly bottle."

In his last years, he gave courses and wrote notes on the philosophy of mathematics. Alan Turing, already famous for his theory of computation, was in the class of 1939. After Wittgenstein died, some of his notes were published as *Remarks on the Foundations of Mathematics.* Then class notes by Bosanquet, Malcolm, Rhees, and Smythies were published as *Wittgenstein's Lectures on the Foundations of Mathematics.*

Wittgenstein's *Remarks* did not meet universal enthusiasm. Logician Georg Kreisel called them "the surprisingly insignificant product of a sparkling mind." Logician Paul Bernays wrote, "He sees himself in the part of the free-thinker combating superstition. The latter's goal, however, is freedom of the mind; whereas it is this very mind which Wittgenstein in many ways restricts, through a mental asceticism for the benefit of an irrationality whose goal is quite undetermined."

Ernest, Klenk, Kielkopf, Shanker, and Wright offer more favorable interpretations of Wittgenstein's philosophy of mathematics. Of course Ernest, Klenk, Kielkopf, Shanker, and Wright disagree with each other. Kielkopf says Wittgenstein was a "strict finitist." Klenk says he wasn't a finitist at all.

Wittgenstein's bête noir was the idea that mathematics exists apart from human activity—Platonism. He rejected the logicist Platonism of his philosophical parents Russell and Frege. In fact, he denied any connection of mathematics with concepts or thought. Mathematics is nothing but calculation, and the rules of calculation are arbitrary. The rules of counting and adding are only custom and habit, "the way we do it." If others do it otherwise, there's no way to say who's right and who's wrong.

The *Remarks* and the *Lectures* are collections of provocative fragments. Many of the *Remarks* are about adding. Not adding big numbers, just adding 1. He doesn't get to subtracting or multiplying.

He doesn't wish to justify adding—quite the contrary! Going to the opposite extreme of Frege and Russell, he denies the logical necessity of the addition table. Even counting is debunked. Wittgenstein says that, given the question

1, 2, 3, 4, ???

"someone" *might* answer

1, 2, 3, 4, 100.

Then "we" would say "100" is wrong.

(I put quotes on "we," because I don't know if "we" means his auditors, or educated twentieth-century Europeans, or all sane adult humans.)

"We" might say, "That person doesn't understand that for all n, the n'th number is n."

But suppose "someone" understands it *differently*, says Wittgenstein.

Then there is an impasse. Explicit formalization will work only if "someone" catches on to it.

Wittgenstein says

$$3 + 5 = 8$$

only because, "That's how we do it."

"We" could say

$$3 + 5 = 9$$

if "we" chose.

"Someone" might get 9 when he adds 3 and 5. He might insist that he's following the rules he learned in school. In that case, according to Wittgenstein, there's nothing more to say. "Someone" does it his way; "we" do it ours.

An alert student might interrupt: "Excuse me, Professor Wittgenstein. The axioms of arithmetic may be arbitrary for abstract logic, but not from a practical viewpoint. And once we fix the axioms, 3 + 5 equals 8. I can show you a proof."

Wittgenstein would not be troubled. He could reply, "Your proof is a proof only if we accept it as a proof. If someone says, 'I don't see that you've proved it,' there's no way to compel him to agree that you have proved it."

He admits that the rules of arithmetic may have been chosen for practical reasons—measuring wood, perhaps. But that wouldn't make them necessary. Only convenient.

Wittgenstein's position can be summarized: "Rules aren't self-enforcing." They're enforced by agreement of the people concerned.

He's not questioning just the axioms of arithmetic. He's questioning whether *any* axioms compel *any* theorem—except for the reason, "That's how we do it."

Some do a subtler reading of Wittgenstein, based on two plausible premises:

Premise 1: Any fool knows 3 + 5 = 8.
Premise 2: Wittgenstein was no fool.

Then his skepticism about

$$3 + 5 = 8$$

must have been a trick, perhaps to show that we don't know what numbers are, perhaps to teach by example how to be a philosopher. This is sometimes called "the Harvard interpretation." It has the merit of opening the door to a semi-infinite sequence of philosophy dissertations. But, it seems to me, elementary courtesy requires us to accept that what Wittgenstein says is what Wittgenstein means.

The *Remarks* is a list of numbered paragraphs. Here are my favorite Wittgensteinisms, mostly from the *Remarks,* numbered by page and paragraph. I comment in brackets.

"Mathematics consists entirely of calculations. In mathematics everything is algorithm and nothing is meaning, even when it doesn't look like that because we seem to be using words to talk about mathematical things. Even these words are used to construct an algorithm" (*Remarks,* p. 468).

"'Mathematical logic' has completely deformed the thinking of mathematicians and of philosophers, by setting up a superficial interpretation of the forms of our everyday language as an analysis of the structures of facts. Of course, in this it has only continued to build on the Aristotelian logic" (*Remarks,* p. 156, para. 48).

"If in the midst of life we are in death, so in sanity we are surrounded by madness." (157, 53)

"Nothing is hidden, everything is visible."

[But science is the struggle to reveal what's hidden!]

"Imagine someone bewitched so that he calculated

$$\{1, 2, 3; 3, 4, 5; 5, 6, 7; 7, 8, 9; 9, 10\}.$$

Now he is to apply this calculation. He takes 3 nuts four times over, and then 2 more, and he divides them among 10 people and each one gets *one* nut; for he shares them out in a way corresponding to the loops of the calculation, and as often as he gives someone a second nut it disappears" (42, 136). [Classical Wittgenstein. Imagine a fantastic "computation" to prove that computation is arbitrary. Here's an analogy. I insist on crawling instead of walking. I insist that crawling is a natural, proper locomotion. Then, for Wittgenstein, there's no way to prove I'm wrong. Wittgenstein's conclusion: We walk rather than crawl only because "That's the way we do it."

[Another of his "examples" is about people who sell wood. They spread it on the ground, and charge according to the *area* covered, regardless of whether the wood is piled high or low. "What's wrong with that?" asks Wittgenstein.

Easy to answer! Say that in the north woodlot, wood is stacked 2 feet high. In the south woodlot, it's stacked 4 feet high. Customers buy all the high-stacked wood from the south lot and none from the north. A competing wood seller sets price by volume, and Wittgenstein's silly wood seller goes broke. Despite Wittgenstein's off-the-wall "examples" of imaginary "tribes," the universal requirements of buying and selling compel

$3 + 5 = 8$.]

"A proof convinces you that there is a root of an equation (without giving you any idea *where*). How do you know that you understand the proposition that there is a root? How do you know that you are really convinced of anything? You may be convinced that the application of the proved proposition will turn up. But you do not understand the proposition so long as you have not found the application" (146, 25) [A quibble on "understand." I'd be interested to know if there's a hungry lion in the house with me, even if I don't know his exact location.]

"If you know a mathematical proposition, that's not to say you yet know *anything*" (160, 2).

"If mathematics teaches us to count, then why doesn't it also teach us to compare colors?" (187, 38).

"The idea of a (Dedekind) 'cut'** is such a *dangerous* illusion" (148, 29).

"Fractions cannot be arranged in an order of magnitude. At first this sounds extremely interesting and remarkable. One would like to say of it, e.g. 'It introduces us to the mysteries of the mathematical world.' This is the aspect against which I want to give a warning. When . . . I form the picture of an unending row of things, and between each thing and its neighbour new things appear, and more new ones again between each of these things and its neighbour, and so on without end, then certainly there is something here to make one dizzy. But once we see that this picture, though very exciting, is all the same not appropriate, that we ought not to let ourselves be trapped by the words 'series,' 'order,' 'exist,' and others, we shall fall back on the *technique* of calculating fractions, about which there is no longer anything *queer*" (60.11, *Philosophical Grammar*). [What trap? Why dizzy? Fall back from what?]

"Someone makes an addition to mathematics, gives new definitions and discovers new theorems—and in a *certain* respect he can be said to not know what he is doing.—He has a vague imagination of having *discovered* something like a space (at which point he thinks of a room), of having opened up a kingdom, and when asked about it he would talk a great deal of nonsense." [Wittgenstein is mocking his friend G. H. Hardy.] "We are always being told that a mathematician works by instinct (or that he doesn't proceed mechanically like a chess player or the like), but we aren't told what that's supposed to have to do with the nature of mathematics. If such a psychological phenomenon does play a part in mathematics we need to know how far we can speak about mathematics with complete exactitude, and how far we can only speak with the indeterminacy we must use in speaking of instincts etc" (p. 295). [Wittgenstein is unwittingly revealing that to hold onto his desiccated notion of mathematics he refuses to listen to better information.]

"How a proposition is verified is what it says" (p. 458).

[The Pythagorean theorem can be verified in many ways. But what it says is always: If a, b, and c are sides of a right triangle, $a^2 + b^2 = c^2$. According to Wittgenstein there are hundreds of different Pythagorean theorems.

You can arrive in London by land, sea or air. That doesn't mean there are three different Londons.]

"The only point there can be to elegance in a mathematical proof is to reveal certain analogies in a particularly striking manner when that is what is wanted; otherwise it is a product of stupidity and its only effect is to obscure what ought to be clear and manifest. The stupid pursuit of elegance is a principal cause of the mathematicians' failure to understand their own operations, or perhaps the lack of understanding and the pursuit of elegance have a common origin" (top, p. 462). [Dislike of mathematical elegance disqualifies anyone from talking about mathematics. In his preface to the *Tractatus*, Wittgenstein says its propositions are obviously true, he doesn't care if they're original. In fact, most of them are blatantly false. Their charm is originality and *elegance*—qualities he despises.]

Wittgenstein confuses what's logically undetermined and what's arbitrary. A mathematical rule may be determined by *convenience*. Such a determination isn't arbitrary, even though it's not compelled logically.

Besides their usual rules, arithmetic and mathematics have an unstated metarule:

Preserve the old rules!

Hermann Hankel called it "the law of preservation of forms." It's a principle of maximal convenience.

In the natural numbers, there's a rule that everything except 1 is greater than 1. When we introduce 0 and negative numbers, we give up that rule. Giving it up leads to a useful extension, the negative numbers and the integers—*if we preserve the other old rules.*

When we introduce negative numbers, we have to decide, what is

$$-1 \times -1 \,?$$

-1 is new, not previously defined or regulated.

So logically, we could choose any value we please, for example,

$$-1 \times -1 = -95.$$

But then -1 would violate the associative and distributive laws. If we want to keep the associative and distributive laws for negative numbers as well as positive, we find**

$$-1 \times -1 = 1.$$

In a sense this formula is just a convention. In another sense it's a theorem. It's *not* arbitrary.

Wittgenstein detected conventions in mathematics. He didn't know or didn't care that important convenience dictated those conventions. He acknowledged kitchen convenience—measuring wood, counting potatoes. He didn't recognize *mathematical* convenience.

He says:

1. Mathematics is just something we do.
2. We could just as well do it any other way.

1. Is correct.
2. Is nonsense.

The absurd (2) obscures his important insight (1).

(1) Can be restated: "Mathematics is an activity of the community. It doesn't exist apart from people." That's right. It's a courageous corrective to Frege's Platonism.

(2) Can be restated: "We can do mathematics any way we please." That's so wrong that it obscures the merit of the first statement.

Wittgenstein may have been misled by an analogy between mathematics and language. Language and mathematics both have rules. In language, the surface forms of grammar are indeed conventional, or in a sense arbitrary. They're not determined by intrinsic necessity. Mathematics is different. It's more than a semitransparent transmission medium. It has content. Its rules are not arbitrary. They're determined by mathematical convenience and mathematical necessity.

If Wittgenstein were right, why do we never hear of anyone who thinks

$$3 + 5 = 9?$$

It's true, school and society tell us

$$3 + 5 = 8.$$

But in politics, in music, or in sexual orientation, some people reject the dictate of school and society. Some people dare question the Holy Trinity, the American flag, whether God should save Our Gracious Queen, and so on. But *nobody* questions elementary arithmetic. A few poor souls trisect angles by compass and straight-edge, despite a famous proof that it can't be done. But in that problem the discoverer easily confuses himself. We *never* get letters claiming that

$$3 + 5 = 9.$$

If arithmetic can be whatever you like, why has no one in recorded history written

$$3 + 5 = 9?$$

Anthropologists in the Sepik Valley of New Guinea find surprising practices and beliefs about medicine, about rain, about gods and devils. Not about arithmetic.

"Mathematics is a practice," Wittgenstein said. Right.

He says that since it's a practice, it's not a body of knowledge.

Non sequitur!

The practice of carpentry has a body of knowledge. The practice of swimming has a body of knowledge. The practice of flamenco dancing has a body of knowledge. Mathematics is the example par excellence of a practice inseparable from its theory, its body of knowledge. Wittgenstein's claim that theory is just an adjunct secondary to calculation betrays embarrassing unfamiliarity with mathematics.

I will present my own example of Wittgenstein arithmetic.[2]

It's cousin to an example of Saul Kripke, with the merit of coming from real life. In this example,

$$12 + 1 = 14$$

and

$$12 + 2 = 15.$$

I explain.

I live on the twelfth floor. There's no thirteenth floor in my building. (13 is unlucky, so it's hard to rent apartments on 13.) Carlos is one flight above me, on 14.

Veronka took the elevator to Carlos on 14. Then, looking for me on 12, she walked down two flights, to arrive—at 11!

Some tenants say the floors from 14 up are misnumbered. "What's marked 14 is really 13."

Others say, "The name of a floor is whatever the people who live there call it." If it's marked 14, it is 14.

Veronka's experience shows that in this building,

$$12 + 1 = 14.$$

The management provides an addition table for the convenience of delivery boys. The "eights" row reads:

$8 + 1 = 9$	$8 + 2 = 10$	$8 + 3 = 11$
$8 + 4 = 12$	$8 + 5 = 14$	$8 + 6 = 15$ etc.

This example proves Wittgenstein's theory. The standard addition table isn't the only one possible!

Some say this is a weird addition table.

Some say it isn't real addition. It's some other funny operation.

To a mathematician, there's no argument. We have many functions of two variables at our disposal. The one called "addition" or "+" is useful. At times

[2] After the fact, I found a connected example on page 83, Lecture VIII.

another may be more appropriate. It doesn't matter if we call the other function "alternative addition" or just a different function. The two peacefully coexist.

Imre Lakatos (1922–1973)

Hegel Comes In

If we want to separate the history of humanistic philosophy of mathematics from its prehistory, we could say that Aristotle, Locke, Hume, Mill, Peirce, and Wittgenstein are our prehistory. Our history starts not with Frege, but with Lakatos.

From the *London Times*, Wednesday, February 6, 1974:

> Professor Imre Lakatos died suddenly on February 3, at the age of 51. He was the foremost philosopher of mathematics in his generation, and a gifted and original philosopher of empirical science, and a forceful and colorful personality.
>
> He was born in Hungary on November 9, 1922. After a brilliant school and university career he graduated from Debrecen in Mathematics, Physics and Philosophy in 1944. Under the Nazi occupation he joined the underground resistance. He avoided capture, but his mother and grandmother, who had brought him up, were deported and perished in Auschwitz.
>
> After the war he became a research student at Budapest University. He was briefly associated with Lukacs. At this period he was a convinced communist. In 1947 he had the post of "Secretary" in the Ministry of Education, and was virtually in charge of the democratic reform of higher education in Hungary. He spent 1949 at Moscow University.
>
> His political prominence soon got him into trouble. He was arrested in the spring of 1950. He used to say afterward that two factors helped him to survive: his unwavering communist faith and his resolve not to fabricate evidence. (He also said, and one believes it, that the strain of interrogation proved too much—for one of his interrogators!)
>
> He was released late in 1952. He had no job and had been deprived of every material possession (with the exception of his watch, which was returned to him and which he wore until his death.) In 1954, the mathematician Rényi got him a job in the Mathematical Research Institute of the Hungarian Academy of Science translating mathematical works. One of them was Pólya's *How to Solve It*, which introduced him to the subject in which he later became preeminent, the logic of mathematical discovery. He now

had access to a library containing books, not publicly available, by western thinkers, including Hayek and Popper. This opened his eyes to the possibility of an approach to social and political questions that was non-Marxist yet scientific. His communist certainties began to dissolve.

After the Hungarian uprising he escaped to Vienna. On Victor Kraft's advice, and with the help of a Rockefeller fellowship, he went to Cambridge to study under Braithwaite and Smiley. . . .

In 1958 he met Pólya, who put him on to the history of the "Descartes-Euler conjecture" for his doctorate. This grew into his *Proofs and Refutations* (1963–64), a brilliant imaginary dialogue that recapitulates the historical development. It is full of originality, wit, and scholarship. It founded a new, quasi-empiricist philosophy of mathematics.

In England the man whose ideas came to attract him most was Professor (now Sir Karl) Popper, whom he joined at LSE in 1960. (There he rose rapidly, becoming Professor of Logic in 1969.). . .

When he lectured, the room would be crowded, the atmosphere electric, and from time to time there would be a gale of laughter. He inspired a group of young scholars to do original research; he would often spend days with them on their manuscripts before publication. With his sharp tongue and strong opinions he sometimes seemed authoritarian; but he was "Imre" to everyone; and he invited searching criticism of his ideas and of his writings over which he took endless trouble before they were finally allowed to appear in print. . . .

He was not without enemies; for he was a fighter and went for the things he believed in fearlessly and tirelessly. But he had friends all over the world who will be deeply shocked by his untimely death.

Foundationism, the attempt to establish a basis for mathematical indubitability, dominated the philosophy of mathematics in the twentieth century. Imre Lakatos offered a radically different alternative. It grew out of new trends in philosophy of science.

In science, the search for foundations leads to the classical problem of inductive logic. How can you derive general laws from a few experiments and observations? In 1934 Karl Popper revolutionized philosophy of science when he said that justifying inductive reasoning is neither possible nor necessary. Popper said scientific theories aren't derived inductively from facts. They're invented as hypotheses, speculations, guesses, then subjected to experimental test in which critics try to refute them. A theory is scientific, said Popper, if it's in principle capable of being tested, risking refutation. If a theory survives such testing, it gains credibility, and may be tentatively established; but it's never proved. Even if a scientific theory is objectively true, we can never know it to be so with certainty.

Popper's ideas are sometimes considered one-sided or incomplete, but his criticism of the inductivist dogma made a fundamental change in how people think about scientific knowledge.

While Popper and others transformed philosophy of science, philosophy of mathematics stagnated. This is the aftermath of the foundationist controversies of the early twentieth century. Formalism, intuitionism, and logicism each left its trace as a mathematical research program that made its contribution to mathematics. As *philosophical* programs, as attempts at a secure foundation for mathematical knowledge, they all ran their course, petered out, and dried up. Yet there remains a residue, an unstated consensus that philosophy of mathematics *is* foundations. If I find foundations uninteresting, I conclude I'm not interested in philosophy—and lose the chance to confront my uncertainties about the meaning or significance of mathematical work.

The introduction to *Proofs and Refutations* is a blistering attack on formalism, which Lakatos defines as the school "which tends to identify mathematics with its formal axiomatic abstraction and the philosophy of mathematics with metamathematics. Formalism disconnects the history of mathematics from the philosophy of mathematics. Formalism denies the status of mathematics to most of what has been commonly understood to be mathematics, and can say nothing about its growth.

"Under the present dominance of formalism, one is tempted to paraphrase Kant: the history of mathematics, lacking the guidance of philosophy, has become blind, while the philosophy of mathematics, turning its back on the most intriguing phenomena in the history of mathematics, has become empty. . . . The formalist philosophy of mathematics has very deep roots. It is the latest link in the long chain of dogmatist philosophies of mathematics. For more than 2,000 years there has been an argument between dogmatists and sceptics. In this great debate, mathematics has been the proud fortress of dogmatism. . . . A challenge is now overdue."

Lakatos doesn't make that overdue challenge. He writes,

"The core of this case-study will challenge mathematical formalism, but will not challenge directly the ultimate positions of mathematical dogmatism. Its modest aim is to elaborate the point that informal, quasi-empirical, mathematics does not grow through a monotonous increase of the number of indubitably established theorems, but through the incessant improvement of guesses by speculation and criticism, by the logic of proofs and refutations."

Instead of symbols and rules of combination, Lakatos presents human beings, a teacher and students. *Proofs and Refutations* is a classroom dialogue, continuing one in Pólya's *Induction and Analogy in Mathematics.*

Pólya considers various polyhedra: prisms, pyramids, double pyramids, roofed prisms, and so on. If the number of faces of a polyhedron is F, the number of vertices is V, and the number of edges is E, he shows that in "all" cases

$$V - E + F = 2 \text{ (the Descartes-Euler formula).}$$

Lakatos continues from this "inductive " introduction by Pólya. The teacher presents Cauchy's proof of Euler's formula:

Stretch the edges of your polyhedron onto a plane. Then simplify the resulting network, removing one edge or two edges at a time. At each removal, check that although V, E, and F are reduced, the expression $V - E + F$ is unchanged. After enough reductions, the network becomes a single triangle. In this simple case $V = 3$, $E = 3$, and $F = 2$. (One face inside the triangle, one face outside.) And

$$3 - 3 + 2 = 2.$$

This completes the "proof."

No sooner is the proof finished than the class produces a whole menagerie of counter-examples. The battle is on. What did the proof prove? What do we know in mathematics, and how do we know it? The discussion reaches ever greater sophistication, mathematical and logical. There are always several viewpoints in contest, and occasional about-faces when one student takes up a position abandoned by his antagonist.

Instead of a general system starting from first principles, Lakatos presents clashing views, arguments, and counter-arguments; instead of fossilized mathematics, mathematics growing unpredictably out of a problem and a conjecture. In the heat of debate and disagreement, a theory takes shape. Doubt gives way to certainty, then to new doubt.

In counterpoint with these dialectical fireworks, footnotes tell the genuine history of the Euler-Descartes conjecture in amazing complexity. The main text is in part a "rational reconstruction" of the actual history. Lakatos once said that the actual history is a parody of its rational reconstruction. *Proofs and Refutations* is overwhelming in its historical learning, complex argument, and self-conscious intellectual sophistication. Its polemical brilliance dazzles.

For fifteen years *Proofs and Refutations* was an underground classic, read by the venturesome few who browse in the *British Journal for Philosophy of Science*. In 1976, three years after Lakatos's death at age 51, it was published by Cambridge University Press.

Proofs and Refutations takes history as the text for its sermon: Mathematics, like natural science, is fallible. It too grows by criticism and correction of theories, which may always be subject to ambiguity, error, or oversight. Starting from a problem or a conjecture, there's a search for *both* proof and counter-examples.

Proof explains counter-examples, counter examples undermine proof. Proof isn't a mechanical procedure that carries an unbreakable chain of truth from assumption to conclusion. It's explanation, justification, and elaboration, which make the conjecture convincing, while the counter-examples make it detailed and accurate. Each step of the proof is subject to criticism, which may be mere skepticism or may be a counter-example to a particular argument. Lakatos calls a counter-example that challenges one step in the argument a "local counter-example"; one that violates the conclusion itself, he calls a "global counter-example."

Lakatos does epistemological analysis of *informal* mathematics, mathematics in process of growth and discovery, the mathematics known to mathematicians and mathematics students. Formalized mathematics, to which philosophy has been devoted, is hardly found on earth outside texts of symbolic logic.

Lakatos *argues* that dogmatic philosophies of mathematics (logicist or formalist) are unacceptable. He does not argue, but *shows* that a Popperian philosophy of mathematics is possible. But he doesn't discover an ontology to go with his fallibilist epistemology. In the main text of *Proofs and Refutations* we hear the author's puppets, not the author himself. He shows us mathematics as it's lived. But he doesn't tell the import of what he is showing us. Or rather, he states its import in the critical sense, in all-out tooth-and-nail attack on formalism. But what is its import in the positive sense? We need to know what mathematics is *about*. The Platonist says it's about objectively existing ideal entities, which a certain intellectual faculty lets us perceive, as eyesight lets us perceive physical objects. But few modern readers, certainly not Lakatos, are prepared to contemplate seriously the existence, objectively, timelessly, and spacelessly, of all the entities in set theory, let alone in future theories yet to be revealed.

The formalist says mathematics isn't about anything, it just *is*. A mathematical formula is just a formula. Our belief that it has content is an illusion. This position is tenable if you forget that informal mathematics *is* mathematics. Formalization is merely an abstract possibility, which one rarely wants or is able to carry out.

Lakatos thinks informal mathematics is science in the sense of Popper. It grows by successive criticism and refinement of theories and the introduction of new, competing theories—*not* the deductive pattern of formalized mathematics. But in natural science Popper's doctrine depends on the objective existence of the world of nature. Singular spatio-temporal statements such as "The voltmeter showed a reading of 2.6" provide the test whereby scientific theories are criticized and sometimes refuted. Popper calls these "basic statements" the "potential falsifiers." If informal mathematics is like natural science, it needs its objects. What are the data, the "basic statements," which provide potential falsifiers to theories in informal mathematics? This question is not even posed in *Proofs and Refutations*. Yet it's the main question if you want a fallibilist philosophy of mathematics.

After *Proofs and Refutations* Lakatos fought about philosophy of science with Rudolf Carnap, Karl Popper, Thomas Kuhn, Michael Polányi, Stephen Toulmin, and Paul Feyerabend. He never returned to philosophy of mathematics. His posthumous papers contain an article, "A renaissance of empiricism in the philosophy of mathematics?" with quotations from eminent mathematicians and logicians, both logicist and formalist, all agreeing that the search for foundations has been given up. The only reason to believe in mathematics is—it works! John von Neumann said modern mathematics is no worse than modern physics, which many people believe in. Having cut the ground under his opponents by showing that his "heretical" view isn't opposed to the mathematical establishment, Lakatos contrasts "Euclidean" theories like the traditional foundationist philosophies of mathematics, with "quasi-empiricist" theories that regard mathematics as conjectural and fallible. His own theory is quasi-empiricist, not empiricist *tout court*, because the potential falsifiers or basic statements of mathematics, unlike those of natural science, are not singular spatio-temporal statements (e.g., "the reading on the volt-meter was 6.2"). For formalized mathematical theories, he said, the potential falsifiers are informal theories. To decide whether to accept some axiom for set theory, for example, we would investigate how it conforms to the informal set theory we know. We may decide, as Lakatos acknowledges, not to fit the formal theory to the informal one, but instead to modify the informal theory. The choice may be complex and controversial. At this point, he's facing the main problem. What are the "objects" of *informal* mathematical theories?

When we talk about numbers or triangles apart from axioms or definitions, what kinds of entities are we talking about? There are many answers. Some go back to Aristotle and Plato. All have difficulties, and long attempts to evade the difficulties. The fallibilist position should lead to a critique of the old answers, perhaps to a new answer that would bring the philosophy of mathematics into the mainstream of contemporary philosophy of science. Lakatos doesn't commit himself. "The answer will scarcely be a monolithic one. Careful historico-critical case-studies will probably lead to a sophisticated and composite solution." A reasonable answer. But a disappointing one.

Except for a summary by Paul Ernest in *Mathematical Reviews*, I know no public response to *Proofs and Refutations* before its publication in 1976 by Cambridge University Press.

The response is in the book itself, in notes by editors Elie Zahar and John Worrall. On p. 138 they correct Lakatos's statement that to revise the infallibilist philosophy of mathematics "one had to give up the idea that our deductive inferential intuition is infallible." They write, "This passage seems to us mistaken and we have no doubt that Lakatos, who came to have the highest regard for formal deductive logic, would himself have changed it. First-order logic has arrived at a characterization of the validity of an inference which (relative to a characterization of the "logical" terms of a language) does make valid inference essentially infallible." The claim is repeated elsewhere. Lakatos "underplays a little the achievements of the mathematical 'rigorists'." "There is no serious sense in which such proofs are fallible." They correct him wherever he questions the achievement of a complete solution of the problem of mathematical rigor. A naive reader might conclude that today there's no doubt whether a rigorous mathematical proof is valid. A modern formal deductive proof is said to be infallible; doubt can come only from doubting the premises. If you think of a theorem as a conditional statement: "If the hypotheses are true, the conclusion is true," then in this conditional form the achievements of logic make it indubitable. Thus Lakatos's fallibilism is incorrect.

But Lakatos is right, Zahar and Worral wrong. It is remarkable that their objections repeat the very same error Lakatos attacked in his introduction—the error of identifying mathematics itself (what real mathematicians do in real life) with its model or representation, metamathematics, first-order logic.

Worrall and Zahar say a formal derivation in first-order logic isn't fallible in any serious sense. But such formal derivations don't exist, except for toy problems—homework exercises in logic courses.

On one side we have mathematics, with proofs established by "consensus of the qualified." Real proofs aren't checkable by machine, or by live mathematicians not privy to the mode of thinking of the appropriate field of mathematics. Even qualified readers may differ whether a real proof (one that's actually spoken or written down) is complete and correct. Such doubts are resolved by communication and explanation.

Once a well-known analyst at lunch told us how as a graduate student he was caught reading *Logic for Mathematicians*, by Paul Rosenbloom. His major professor ordered him to get rid of the book, saying, "You'll have time for that stuff when you're too old and tired for real mathematics." The group at lunch laughed. We weren't shocked. In fact, studying logic at that earlier time wouldn't have helped an analyst, and might have interfered with him. Today this is no longer true. Mathematical logic has produced tools that have been used by analyst or algebraist. But that has nothing to do with justifying proofs by rewriting them in first-order logic. Once a proof is accepted, its conclusion is regarded as true with high probability. To detect an error in a proof can take generations. If a theorem is widely used, if alternate proofs are found, if it is applied and generalized, if it's analogous to other accepted results, it can become "rock bottom." Arithmetic and Euclidean geometry are rock bottom.

In contrast to the role of proof in mainstream mathematics, there's "metamathematics" or "first-order logic." It's about mathematics in a certain sense, and at the same time it's part of mathematics. It lets us study mathematically the consequences of an imagined ability to construct infallible proofs. For example, the consequences of constructivist variations on classical deduction.

How does this metamathematical picture of mathematics affect our understanding and practice of real mathematics? Worrall and Zahar think the problem of fallibility in real proofs, which Lakatos is talking about, has been settled by the notion of infallible proof in metamathematics. (This term of Hilbert's is old-fashioned, but convenient.) How would they justify such a claim?

I guess they would say a real proof is merely an abbreviated or incomplete formal proof. This raises several difficulties. In mathematical practice we distinguish between a complete informal proof and an incomplete informal proof. In a complete informal proof, every step is convincing to the intended reader. In an incomplete informal proof, some step fails to convince. As *formal* proofs, *both* are incomplete. We considered this question in Chapter 4.

So what does it mean to say a real mathematical proof is an abbreviation of a formal one, since the same can be said whether an informal "proof" is correct or incorrect?

"Formalists," as Lakatos called people like Zahar and Worrall who "conflate" mathematics with its metathematical model, don't explain *in what sense* formal systems are a model of mathematics. Normatively, mathematics *should* be like a formal system? Or descriptively, mathematics *is* like a formal system?

If descriptive, then it has to be judged by how faithful it is to what it purports to describe. If normative, it has to explain why mathematics is prospering so well while paying so little heed to its norms.

Logicians today claim that their work is descriptive. Sol Feferman writes that the aim of a logical theory is "to model the reasoning of an idealized Platonistic or an idealized constructivistic mathematician." Comparison is sometimes made between logicians' use of formal systems to study mathematical reasoning and physicists' use of differential equations to study physical problems. But he points out that the analogy breaks down, since there's no analogue of the physicists' experimental method for the logicians to use in testing their model against experience.

"We have no such tests of logical theories. Rather, it is primarily a matter of individual judgment how well these square with ordinary experience. The accumulation of favorable judgment by many individuals is of course significant."

There is no infallibility.

Proofs and Refutations presents a picture of mathematics utterly at variance with the one presented by logic and metamathematics. It's clear which is truer to life. Feferman writes, "Clearly logic as it stands fails to give a direct account of either the historical growth of mathematics or the day-to-day experience of its practitioners. It is also clear that the search for ultimate foundations via formal systems has failed to arrive at any convincing conclusion."

Feferman has reservations about Lakatos's work. Lakatos's scheme of proofs and refutations is not adequate to explain the growth of all branches of mathematics. Other principles, such as the drive toward the unification of diverse topics, seem to provide the best explanation for the development of abstract group theory or of point-set topology. But Feferman acknowledges Lakatos's achievement. "Many of those who are interested in the practice, teaching and/or history of mathematics will respond with larger sympathy to Lakatos' program. It fits well with the increasingly critical and anti-authoritarian temper of these times. Personally, I have found much to agree with both in his general approach and in his detailed analysis."

Hungarian Style

In addition to Imre Lakatos, four other brilliant Hungarian Jews belong in our story. George Pólya and John von Neumann became Americans; Michael Polányi escaped to England. Only Alfréd Rényi lived and died in Budapest.

Pólya, like Wilder, had a second career after years of mathematical research. Pólya's late-life specialty was heuristics. How do we solve problems? How do we teach others to solve problems? (See the 4-cube in Chapter 1.)

In expounding the heuristic side of mathematics—what I called "the back" in Chapter 3—he was implicitly exposing the incompleteness of the formalist and logicist pictures of mathematics. Like Wilder and Gauss, Pólya disliked philosophical controversy. Perhaps, like Gauss, he was impatient with "Boeotians." (See non-Euclidean geometry.**) Pólya once said he became a mathematician because he wasn't good enough for physics, but too good for philosophy.

In the preface to *Mathematics and Plausible Reasoning*, Pólya wrote: "Finished mathematics presented in a finished form appears as purely demonstrative, consisting of proofs only. Yet mathematics in the making resembles any other human knowledge in the making. You have to guess a mathematical theorem before you prove it; you have to have the idea of the proof before you carry through the details. You have to combine observations and follow analogies; you have to try and try again."

But a few pages later, "I do not know whether the contents of these four chapters deserve to be called philosophy. If this is philosophy, it is certainly a pretty low-brow kind of philosophy, more concerned with understanding concrete examples and the concrete behavior of people than with expounding generalities."

Polányi was a Hungarian-English chemist who in the fullness of his fame became philosopher of science. His opinions came out of the laboratory and lecture hall, out of collaboration and clash with other chemists. Like Lakatos and Pólya, he paid little mind to the disputations of philosophers of science. His books were read by scientists and the public, widely ignored by philosophers.

He contributed the notion of "tacit knowledge." We know more than we can say, and this tacit knowledge is essential to scientific discovery. This doctrine is indigestible to people who equate knowledge with written text. It clashes with the logicist and formalist pictures of mathematical knowledge as explicit formulas and sentences. It also contradicts Wittgenstein's much-quoted mot: "What we cannot speak about, we must pass over in silence."

I must quote these paragraphs of Polányi's (1962, pp. 187–88): "Even supposing mathematics were wholly consistent, the criterion of consistency, which the tautology doctrine is intended to support, would still be ludicrously inadequate for defining mathematics. One might as well regard a machine which goes on printing letters and typographical signs at random as producing the text of all future scientific discoveries, poems, laws, speeches, editorials, etc. for just as only a tiny fraction of true statements about matters of fact constitute science and only a tiny fraction of conceivable operational principles constitute technology, so also only a tiny fraction of statements believed to be consistent constitute mathematics. Mathematics cannot be properly defined without appeal to the principle which distinguishes this tiny fraction from the overwhelmingly predominant aggregate of other non-self-contradictory statements.

"We may try to supply this criterion by defining mathematics as the totality of theorems derived from certain axioms according to certain operations which will assure their self-consistency, provided the axioms themselves are mutually consistent. But this is still inadequate. First, because it leaves completely unaccounted for the choice of axioms, which hence must appear arbitrary—which it is not; second, because not all mathematics considered to be well established has ever been completely formalized according to strict procedure; and third—as K. R. Popper has pointed out—among the propositions that can be derived from some accepted set of axioms there are still, for every single one that represents a significant mathematical theorem, an infinite number that are trivial.

"All these difficulties are but consequences of our refusal to see that mathematics cannot be defined without acknowledging its most obvious feature: namely, that it is interesting. Nowhere is intellectual beauty so deeply felt and fastidiously appreciated in its various grades and qualities, as in mathematics, and only the informal appreciation of mathematical value can distinguish what is mathematics from a welter of formally similar, yet altogether trivial statements and operations."

Mathematics is created and pursued by people. It must have meaning or value to them or else they would not create or pursue it. Whatever mathematics we look at, out of the huge pile that humanity has created, one thing can always be said. It was interesting to someone, some time, and somewhere.

The next Hungarian on our list is Alfréd Rényi.

Rényi was the son of an engineer of wide learning and grandson of Bernat Alexander, a most influential professor of philosophy and aesthetics at Budapest. His uncle was the psychoanalyst, Franz Alexander. He went to a humanistic gymnasium, and maintained life-long interest in classical Greece. He had the rare ability to be equally at home in pure and applied mathematics.

His *Dialogues on Mathematics* are beautiful examinations of mathematical truth and meaning. They deal in profound and original ways with fundamental philosophical issues, yet their light touch and dramatic flair make them readable by anyone. There are dialogues with Socrates, Archimedes, and Galileo. "For Zeus's sake," asks Socrates, "is it not mysterious that one can know more about things which do not exist than about things which do exist?" Socrates not only asks this penetrating question, he answers it. It's astonishing that to this day (to my knowledge) no philosopher of mathematics has responded to Rényi's *Dialogues.*

Finally, among the great Hungarian Jewish mathematicians, we remember Neumann Janos—in the United States, John von Neumann (1903–1957). In breadth and power, the supreme mathematician of the 1930s, 1940s, and 1950s—along with perhaps Andrei Kolmogorov and Israel Moiseyevich Gel'fand. His contributions to games, economics, quantum mechanics, sets, lattices, linear operators, nuclear weapons, computers, numerical analysis, and fluid dynamics are legendary. His philosophical writings are sparse. In youth he worked on Hilbert's formalism. Later he was philosophically uncommitted. I quote his essay "The Mathematician" in the section on Hilbert in Chapter 8.

One can ask, what did these Hungarians have in common, besides Budapest? Widely as their specialties and their politics differed, there is a "Hungarian style." It's hard to describe, but not hard to recognize. Lively affection for the concrete, specific, and human. Quiet avoidance of blown-up pretension, vapid generality, words for words' sake. Intellectual and cultural breadth, encompassing history, literature and philosophy. Understanding that mathematics flowers as part of human culture.

Leslie White/Raymond Wilder
Anthropology Comes In

Wilder was a topologist at Ann Arbor. White was an anthropologist there. They became friends. In his later years, retired at Santa Barbara, Wilder had a second career as philosopher/historian of mathematics. He credited White for the insight that mathematics is a cultural phenomenon, which can be studied with the methods of anthropology. Wilder knew this viewpoint was opposed to Platonism, formalism, and intuitionism. But in his writings he tried to avoid philosophical controversy. Like Gauss who suppressed his discovery of non-Euclidean geometry, he usually preferred not to stir up the "Boeotians" (Aristophanes's name for ignorant hicks).

White's essay is a beautiful statement of the locus of mathematical reality. Its weakness is failure to confront the uniqueness of mathematics—what makes mathematics different from other social phenomena. He wrote: "We can see how the belief that mathematical truths and realities lie outside the human mind arose and flourished. They *do* lie outside the mind of each individual organism. They enter the individual mind as Durkheim says from the outside. . . . Mathematics is not something that is secreted, like bile; it is something drunk, like wine. . . . Heinrich Hertz, the discoverer of wireless waves, once said: 'One cannot escape the feeling that these mathematical formulas have an independent existence and an intelligence of their own, that they are wiser than we are, wiser even than their discoverers, that we get more out of them than was originally put into them.'. . . The concept of culture clarifies the entire situation. Mathematical formulas, like other aspects of culture, do have in a sense 'independent existence and intelligence of their own.' The English language has, in a sense, 'an independent existence of its own.' Not independent of the human species, of course, but independent of any individual or group of individuals, race or nation. It has, in a sense, an 'intelligence of its own.' That is, it behaves, grows and changes in accordance with principles with are inherent in the language itself, not in the human mind. . . . Mathematical concepts are independent of the individual mind but lie wholly within the mind of the species, i.e., culture."

twelve

Contemporary Humanists and Mavericks

Sellars, Popper, Medawar, Leavis, Bunge

The autonomy of the social/cultural/historic isn't a new idea. A school of "critical realism" advocated "levels of reality" and "emergent evolution" 70 and 80 years ago. (The father of the analytic philosopher Wilfred Sellars was the critical realist Roy Sellars.) In the 1930s emergent evolution was buried under analytic philosophy and logical empiricism, and almost forgotten. But it is re-emerging. In 1992 David Blitz wrote, "Despite its temporary eclipse by reductionist and physicalist philosophies in the period from the mid-1930s to the mid-1950's, emergent evolution is an active trend of thought at the interface between philosophy and science." I sketch the ideas of some recent authors.

In Chapter 1 I mentioned Karl Popper's role in dethroning the doctrine of inductive reasoning in empirical science. More recently he introduced a notion of "World 3"—a world of scientific and artistic knowledge, distinct from the physical world (World 1) and the world of mind or thought (World 2). In "Epistemology without a knowing subject" and "On the theory of the Objective Mind" he uses "World 3" to mean a world of intelligibles: ideas in the objective sense, possible objects of thought, theories and their logical relations, arguments, and problem situations (1974, p. 154). It seems Popper wants to put a Platonic world of Ideal Forms alongside the mental and physical worlds.

Peter Medawar was impressed by Popper's World 3. "Popper's new ontology does away with subjectivism in the world of the mind. Human beings, he says, inhabit or interact with three quite distinct worlds; World 1 is the ordinary physical world, or world of physical states; World 2 is the mental world; World 3 is the world of actual or possible objects of thought—the world of concepts, ideas, theories, theorems, arguments, and explanations—the world, let us say, of all artifacts of the mind. The elements of this world interact with each other much like the ordinary objects of the material world; two theories interact and lead to the formulation of a third; Wagner's music influences Strauss's and his in turn all music written since. . . . The existence of World 3, inseparably bound up with human language, is the most distinctly human of all our possessions. The third world is not a fiction, Popper insists, but exists 'in reality.' It is a product of the human mind but yet is in large measure autonomous. This was the conception I had been looking for: The third world is the greater and more important part of human inheritance. It's handing on from generation to generation is what above all else distinguishes man from beast."

But the difficulty of explaining the interaction between these worlds is fatal for Popper, as for his predecessors.

F. R. Leavis in his book *Nor Shall My Sword* calls the physical world "public," the mental world "private." Then he talks about "objects of the third kind," which are neither wholly public not wholly private. This is close to my meaning.

Mario Bunge, a prolific philosopher of physics, attacked Popper's World 3 in *The Mind-Body Problem* (pp. 169–73)—but then developed his own reformulation of it.

Bunge is a materialist. With respect to mathematics, he's a fictionalist. He thinks that claims for autonomy of mind interfere with the scientific study of mind as brain activity.

He writes: Popper's "autonomy of the 'world' of the creations of the human mind is supposed to be relative to the latter as well as to the physical world. Thus the laws of logic are neither psychological nor physiological (granted)—and, moreover, they are supposed to be 'objective' in some unspecified sense. The idea is probably that the formulas of logic (and mathematics), once guessed and proved, hold, come what may. However, this does not prove that conceptual (e.g. mathematical) objects, or any other members of 'world 3,' lead autonomous existences. It only shows that, since they do not represent the real world, their truth does not depend upon it, and, therefore, we can *feign* that they are autonomous objects. However, this is not what Popper claims: he assigns his 'world 3' reality and causal efficacy. . . . We pretend that there are infinitely many integers even though we can think of only finitely many of them—and this because we assign the infinite set of all integers definite properties, such as that of being included in the set of rational numbers. Likewise we make believe that every deductive system contains infinitely many theorems—and this because, if pressed, we can prove any of them. All such fictions are mental creations and, far from being idle, or merely entertaining, are an essential part of modern culture. But it would be sheer animism to endow such fictions with real autonomous existence and causal efficacy. . . . In short, ideas in themselves are fictions and as such have no physical existence; only the brain processes of thinking them up are real. . . ."

"But what about the force of ideas? Do not ideas move mountains? Has it not been in the names of ideas, and particularly ideals, that entire societies have been built or destroyed, nations subjugated or liberated, families formed or wiped out, individuals exalted or crushed? Surely even the crassest of materialists must recognize the power of ideas, particularly those our grandparents used to call idées-forces, such as those of fatherland and freedom. Answer: Certainly, ideation—a brain process—can be powerful. Moreover, ideation is powerful when it consists in imagining and planning courses of action engaging the cooperation of vast masses of individuals. But ideas in themselves, being fictions, are impotent. The power of ideation stems from its materiality, not from its ideality."

This isn't right. The power of ideas stems from their meaning, not their materiality. Someone shouts "Liberté, égalité, fraternité!" That shout inspires someone else to pull a cobblestone from the Rue de la Paix and throw it at a *gendarme*. It wasn't the air vibrating from the shout that budged the cobblestone. It was the *meaning* of the slogan that moved the stone. For that meaning activated a mind-brain that ordered hands to pull up a stone and hurl it. We can't explain the Revolution with neurons and hormones. We explain it with economics, politics, and history.

Yes, a slogan has effect only by actions of people. Yes, people have an effect only by physical action, whether hurling a cobblestone or whispering a word. But the power of the word is its meaning—not its decibels.

Later (p. 214) Bunge has second thoughts: "We have been preaching the reduction of psychology to neurophysiology, but have also warned that such reduction can only be partial or weak, and this for two reasons. One reason is that psychology contains certain concepts and statements that are not to be found in today's neuroscience. Consequently neuroscience must be enriched with some such constructs if it is to yield the known psychological regularities and, a fortiori, the new ones we would like to know. The second reason for the incomplete reducibility of psychology to neurophysiology is that neuroscience does not handle sociological variables, which are essential to account for the behavior and mentation of the social higher vertebrates. For these reasons the reductionistic effort should be supplemented by an integrative one. Let me explain. Behavior and mentation are activities of systems that cross a number of real (not just cognitive) levels, from the physical level to the societal one. Hence they cannot be handled by any one-level science. Whenever the object of study is a multilevel system, only a multi disciplinary approach—one covering all of the intervening levels—holds promise. In such cases pig-headed reductionism is bound to fail for insisting on ab initio procedures that cannot be implemented for want of the necessary cross-level assumptions. (Take into account that it has not been possible to write down, let alone solve, the Schrodinger equation for a biomolecule, let alone for a neuron, even less for a neuronal system.) In such cases, pressing for reduction is quixotic; it is not a fruitful research strategy. In such cases only the opportunistic (or catch-as-catch-can) strategy suggested by systemism and a multilevel world view can bring success, for it is the one that integrates the physical, the chemical, the biological and the sociological approaches, and the one that builds bridges among them. . . (p. 216). Our rejecting psychophysical dualism does not force us to adopt eliminative or vulgar materialism in either of its versions—i.e. the theses that mind and brain are identical, that there is no mind, or that the capacities for perception, imagination and reasoning are inherent in all animals or even in all things. Psychobiology suggests not just psychoneural monism but also emergentism, i.e., the thesis that mentality is an emergent property possessed only by animals endowed with an extremely complex and plastic nervous system. . . . In short, minds do not constitute a supra-organic level, because they form no level at all. But psychosystems do. To repeat the same idea in different words: one can hold that the mental is emergent relative to the merely physical without reifying the former. . . . And so emergentist (or systemic) materialism—unlike eliminative materialism—is seen to be compatible with overall pluralism, or the world view that proclaims the qualitative variety."

What Bunge taketh with one hand, he restoreth with the other. Minds aren't real, but "psychophysical systems" are. Out of habit, we will probably go on calling psychophysical systems "minds."

Still, there is a difference between Bunge's emergent monism and the mind-body dualisms he refutes.

Bunge and others had an easy time attacking Popper. Such attacks would be off the mark with reference to my level 3. I don't claim it's independent of mind and body. On the contrary, a socio-cultural-historical object exists only in some representation, whether physical (books, computer "memories," musical scores and recordings, photographs, drawings) or mental (knowledge or consciousness of people) or both.

There's no thinking without a brain. But the mental is autonomous in the sense that the evolution and interaction of minds has to be understood in terms of thoughts, emotions, habits, desires, and so forth, not just chemistry and electricity. In the same way that the mental exists through the physical, so the social-cultural exists through both the mental and physical.

N.B. In response, Professor Bunge explains that he does not say minds don't exist, only that they don't exist autonomously.

Philip Kitcher

Two important contributions to the humanist philosophy of mathematics are Kitcher's *Nature of Mathematical Knowledge* (NMK) and his edited book with William Aspray, *History and Philosophy of Modern Mathematics* (HP).

The introduction to HP explains that in both the philosophy of mathematics and the history of mathematics, an old tradition persists while a new trend challenges it. The philosophy of mathematics, we're told, was created by Gottlob Frege. (Earlier thinkers about the nature of mathematics were prehistoric.) After Frege came Whitehead and Russell, Hilbert, Brouwer, Wittgenstein, Gödel, the

Vienna Circle, Quine, and so on. In the 1950s philosophy of mathematics arrived at "neo-Fregeanism." This is an orthodoxy that decrees that mathematics is all about sets. For a mathematical statement to be true means it corresponds to the state of affairs in some set. Aspray and Kitcher call their own challenge to neo-Fregean philosophy "maverick."

Kitcher's book, *The Nature of Mathematical Knowledge,* starts with a painstaking critique of "a priorist" epistemologies. Traditional philosophy of mathematics wants to justify mathematical knowledge. To formalism, the justification for a theory is that the theorems follow from the axioms. Neither axioms nor theorems have truth value beyond their logical connections.

To intuitionism, elementary arithmetic is given directly to the intuition; other mathematics gets legitimacy from its connection to arithmetic. Kitcher makes an unsparing critique of both formalism and intuitionism.

His own viewpoint blends empiricism and evolutionism. All mathematical knowledge comes by a rationally explicable process of growth, starting from a basic core, the arithmetic of small numbers and the directly visual properties of simple plane figures. The basic core is experienced in physical acts of collecting, ordering, matching, and counting. The formula

$$2 + 2 = 4$$

can be proved as a theorem in a formal axiomatic system, but it derives its force and conviction from its physical model of collecting coins or pebbles.

Arithmetic is an idealized theory of matching and counting; set theory is an idealized theory of forming collections. An upper bound can be given for all the collections that will ever actually be formed by humans—but in mathematics we allow arbitrarily large sets and numbers, even infinite sets and numbers. This doesn't destroy the empirical nature of arithmetic, any more than the ideal gas hypothesis destroys the empirical character of gas dynamics.

To the question, what is mathematics about? Kitcher answers, it's about collecting. This is a constructivized re-interpretation of the idea that mathematics is about sets—what Kitcher calls neo-Fregeanism. But Kitcher is more generous than the constructivists in what his idealized mathematician can do. For instance, he can collect uncountably many elements into his collections.

Mathematics is a lawful, comprehensible evolution from a basic core. It develops in response to internal strain (here a definition would help) and external pressure. The mathematical knowledge of one generation is rooted in that of its parent's generation. The mathematics of the research journals is validated by the mathematical community through criticism and refereeing. Because most mathematical papers use reasoning too long to survey at a glance, acceptance is tentative. We reconsider our claim if it's disputed by a competent skeptic. Thus, our belief is *not* prior to all experience: it's conditioned on our *social* experience!

This is one of Kitcher's major points. The mathematician described by philosophy of mathematics must be, like the flesh-and-blood mathematician, a social being, not a self-sustaining isolate. This insight suffices to discredit the claim that mathematical knowledge is a priori, for communication with other humans is a precondition to mathematical knowledge.

The classical philosophical account of advanced mathematics relies on the possibility of reducing it all to set theory. Everything in partial differential equations or ergodic theory can be thought of as a set, so the only existence questions one need consider are about existence of sets. This reduction belies the perception of the working mathematician, who sees his own subject as central and autonomous, and set theory as peripheral or irrelevant. Kitcher is with the mathematicians, not the foundationists (philosophical logicians and set theorists). To the question, "How do we acquire knowledge of mathematics?" Kitcher gives the truthful answer: "We learn it at school."

This leads to the next question, "How does mathematical knowledge increase?" Answering this is the historian's job, and Kitcher takes on the historian's responsibility as well as the philosopher's. In a long account of mathematical analysis in the eighteenth and nineteenth centuries, he gives historical reasons for the establishment of real numbers as a foundation for calculus. This account is history at the service of philosophy, concretely answering the question, "What is mathematics?"

Kitcher adopts an important principle of Thomas Kuhn: scientific change means change in practice, not just in theory. He identifies *five components of mathematical practice*: language, metamathematical views, accepted questions, accepted statements, accepted reasonings. The five must be compatible. If one changes, the others must change accordingly.

He identifies *five rational principles* according to which mathematics develops: rigorization, generalization, question-generalization, question-answering, and systematization. These principles yield "rational interpractice transitions." "When these occur in sequence, the mathematical practice may be dramatically changed through a series of rational steps." The validity of today's mathematics comes from its connection, rational step by rational step, to the empirically valid basic mathematics of counting and collecting.

This explanatory scheme is powerful and convincing. It should be a lasting contribution to the historical analysis of mathematics.

Jean Piaget (1896–1982) & Lev Vygotsky (1896–1934)

Piaget was a psychologist, not a philosopher, but he has a respectful entry in the *Encyclopedia of Philosophy*. His painstaking observation of the growth of abstract thinking in children revolutionized cognitive psychology in the decades after the Second World War. In place of statistics on running rats, he used understanding, insight, and open-mindedness in observing children. His books report children's thinking about physics, logic, number, geometry, space, time, and chance, among other things.

Piaget's most popular idea was "stages": The brain can't absorb certain concepts until it attains the right stage of maturation. This idea was harmful to education. "Children in a classroom must be at the same maturational stage. It's useless to try teaching something to a child who hasn't reached the right stage." This idea wasn't supported by later research.

With a Dutch logician, Beth, Piaget wrote a book on the logico-psychological foundations of mathematics. Beth took it for granted that the foundation of mathematics is logic and set theory, so the book is now outdated.

Piaget and the Bourbakiste Jean Dieudonné had an intellectual encounter that gratified them both. Bourbaki had decided that mathematics is built from three fundamental structures: (1) order (in the sense of putting things one after another); (2) algebra (sets with operations, such as addition, multiplication, reflection, etc.); and (3) topology (neighborhoods, closeness). Piaget had independently decided that children's mathematical ideas are built from the same three elements. Naturally, they were happy to encounter each other. This Bourbaki-Piaget philosophy was called "structuralism." Mathematics was a collection of structures built from the three basic structures.

Bourbakisme and structuralism went out of fashion with change of styles. Later Piaget took up category theory. His book *Morphisms and Categories* has a critical introduction by the computer scientist Seymour Papert, inventor of LOGO, who was Piaget's pupil. The "structuralism" now being advanced by Shapiro and Resnik is not the same thing.

From my point of view, what's interesting in Piaget isn't his theory of stages or his encounters with Beth and Dieudonné. It's his epistemology. This has largely been ignored by mainstream philosophy. He presented his epistemology in several books. We look at *Genetic Epistemology*.

Piaget observed that children learn actively: picking up, putting down, moving, using. Presumably the bodily movements associated with two-ness (for example) leave a trace in the mind/brain. That trace would *be* the child's concept of two. Generalizing this idea, Piaget proposed that the concept of counting and natural number are acquired through activity. Real physical activity—not just observation or talk. The child picks up and puts down, manipulates, handles buttons, coins, and pebbles. By these activities the fundamental properties of discrete, permanent objects attain a representation in the child's mind/brain.

This may sound like Aristotle's explanation of number and shape as abstractions. But Aristotle's abstraction was passive observation. We see three apples, three pebbles, three coins, and we abstract "three." We notice that two beans added to two beans makes four beans. But we can't make these observations until we have learned that number is an interesting property. A cat sees two beans added to two beans, but it doesn't discover that $2 + 2 = 4$. A cat doesn't handle, play with, pick up, and put down the beans or the beads. A child does.

From counting, go to motion. As the concept of discreteness comes from playing with beans and beads, the concepts of three-dimensional motion and space come from moving our three-dimensional bodies in three-dimensional space. Raise your arm, turn your head, lower your arm. By doing so, you learn the structure of the three-dimensional continuum. Without that knowledge, you couldn't find your way to breakfast.

This viewpoint is an escape from the Platonist conception of number and space as abstract objects, independent of both mental and physical reality.

If our concepts of number and space are the mental effects of childhood activity, what about more developed mathematics? P-adic analysis, measurable cardinals, square-integrable martingales, the well-ordering theorem, the continuum hypothesis? We can escape from the formalist conception, that each of these entities is nothing more than a formal definition in the context of a certain formal theory. We can recognize the crucial fact that the concept comes before the formal definition. The concept of an abstract group, for example, is the mental effect of calculations and reasoning and mental struggles with concrete groups. The trace or the effect of this activity and struggle on our mind-brain is experienced as an intuitive concept. With effort the mathematician may formalize it, interacting with the history of the subject and the thinking and writing of colleagues and competitors.

Piaget's epistemology is close to Marx's. Marx said our knowledge of natural science comes from interaction with Nature in productive labor. This was part of the doctrine that all classes but the Proletariat could be dispensed with. This Marxism is as dated as Bourbakisme. In rooting mathematical concepts in physical activity, Piaget unlike Marx had no ideological ax to grind.

Kitcher's *The Nature of Mathematical Knowledge*, Ernest's *The Philosophy of Mathematics Education* and Castonguay are among the few books where Piaget receives his due as a philosopher of mathematics.

Among cognitive psychologists today there's increasing interest in Lev Vygotsky, a Russian psychologist active in the 1920s and early 1930s. Vygotsky didn't write about mathematics, but his theory of learning is relevant to mathematics, and several psychologists are developing Vygotskian theories of mathematics education.

Vygotsky insisted that learning and intellectual activity are fundamentally social, not individual. The content of learning and thinking comes from social structures, and is assimilated by the individual learner.

This opinion wouldn't surprise an anthropologist or a sociologist, but psychologists, including Piaget, usually take it for granted that their task is to study the individual mind, detached from other people or from society. Vygotsky was a Marxist, but his intellectual daring and his deep love for western literature made him suspect in Stalin's Russia. (He wrote a book about Shakespeare's *Hamlet*, with a startling new interpretation.) He died of tuberculosis in his 30s, before the worst of the purges and persecutions. His books were banned in the Soviet Union for 30 years.

Paul Ernest

In the young movement to rebuild philosophy of mathematics as part of social reality, Ernest's *Philosophy of Mathematics Education* is one of the most comprehensive and comprehensible. It consists of an introduction and two parts—philosophy of mathematics is Part 1, philosophy of mathematics education is Part 2. Ernest starts with a critique of absolutist philosophies of mathematics. Absolutist here means philosophies that say mathematical truth is or should be absolute (free of all possible manner of doubt). This includes the three standard philosophies Imre Lakatos collected under the label "foundationalists": logicists, formalists, intuitionists. Ernest's critique is in the name of "fallibilism," a term sometimes joined to the name of Lakatos. It simply means the denial of absolutism. Next comes a chapter "reconceptualizing" the philosophy of mathematics, and then three chapters expounding "social constructivism."

This phrase is popular in "behavioral science" and "humanities." Ernest is familiar with and respects social constructivism in sociology and psychology. He provides generous summaries of Piaget and Vygotsky in constructivist psychology, Bloor and Restivo in constructivist sociology. Ernest seems to be the first to speak of social constructivism in philosophy of mathematics.

With each theory he describes he presents both the positive assertions of the theory and the objections that could be made against it. This includes the "conventionalism" of Wittgenstein and the "quasi-empiricism" of Lakatos, which he names as his foundation stones. Of course, it's one thing to state objections, another to deal with them. Once in a while Ernest dismisses with a sentence or two difficulties that deserve many chapters.

From his introductory summary:

"Social constructivism views mathematics as a social construction. It draws on conventionalism [read "Wittgenstein"], in accepting that human language, rules and agreement play a key role in establishing and justifying the truths of mathematics. It takes from quasi-empiricism [read "Lakatos"] its fallibilist epistemology, including the view that mathematical knowledge and concepts develop and change. It also adopts Lakatos' philosophical thesis that mathematical knowledge grows through conjectures and refutations, utilizing a logic of mathematical discovery. . . . A central focus of social constructivism is the genesis of mathematical knowledge, rather than just its justification . . . subjective and objective knowledge of mathematics each contributes to the creation and recreation of the other."

On p. 83, "In summary, the social constructivist thesis is that objective knowledge of mathematics exists in and through the social world of human action, interactions and rules, supported by individuals' subjective knowledge of mathematics (and language and social life), which need constant re-creation. Thus subjective knowledge recreates objective knowledge, without the latter being reducible to the former." Parts of the second half of the book will seem exotic to readers not acquainted with the English school system. Here Ernest classifies philosophies of education by their practice and practical effect as well as their theory. A remarkable table displays five educational ideologies: "industrial trainer," "technological pragmatist," "old humanist," "progressive educator," and "public educator" (in order from worst to best). Listed under each is its political ideology, view of mathematics, moral values, mathematical aims, theory of teaching mathematics, and theories of learning, ability, society, the child, resources, assessment in mathematics, and social diversity.

Of course the construction of the table is partly impressionistic and anecdotal, but it sounds right. Just for a taste, here are the views of maths and of society, listed below each teaching ideology.

Industrial Trainer	*Technological Pragmatist*	*Old Humanist*
Set of truths and rules	Unquestioned body of useful knowledge	Body of structured pure knowledge
Rigid hierarchy Market place	Meritocratic Hierarchy	Elitist class-stratified

Progressive Educator	*Public Educator*
Process view Personalized maths	Social constructivism
Soft hierarchy Welfare state	Inequitable hierarchy Needing reform

The second part of the book advocates the ideology of the public educator. This is a descendant of Deweyan progressivism, politicized and radicalized.

Ernest's forthcoming *Social Constructivism as a Philosophy of Mathematics* (Albany, SUNY) is close to the present book in point of view.

Ethnomathematics

Marcia Ascher has written an instructive book with this title.

Mathematical ideas, like artistic ideas or religious ideas, are a universal part of human culture. This forthright claim isn't made by Ascher, but her book compels me to that conclusion. Mathematics as we know it was invented by the

Greeks. But mathematical ideas involving number and space, probability and logic, even graph theory and group theory—these are present in preliterate societies in North and South America, Africa, the South Pacific, and doubtless many other places if anyone bothers to look.

This is not to say that everybody can do mathematics, any more than everybody can play an instrument or succeed in politics. Many people do not have mathematical or musical or political ability. But every society has its music and its politics; so too, it seems, every society has its mathematics.

Some people count by tens, others by twenties. "There is an oft-repeated idea that numerals involving cycles based on ten are somehow more logical because of human fingers. The Yuki of California are said to believe that their cycles based on eight are most appropriate for *exactly* the same reason. The Yuki, however, are referring to the interfinger spaces." And how about Toba, a language of western South America, in which "the word with value five implies (two plus three), six implies (two times three), and seven implies (two times three) plus one. Then eight implies (two times four), nine implies (two times four) plus one, and ten is (two times four) plus two."

Professor Ascher knows of three cultures that trace patterns in sand—the Bushoong in Zaire, the Tshokwe in Zaire and Angola, and the Malekula in Vanuatu (islands between Fiji and Australia formerly called the New Hebrides). Sand drawings play a different role in each culture. "Among the Malekula, passage to the Land of the Dead is dependent on figures traced in the sand. Generally the entrance is guarded by a ghost or spider-related ogre who is seated on a rock and challenges those trying to enter. There is a figure in the sand in front of the guardian and, as the ghost of the newly dead person approaches, the guardian erases half the figure. The challenge is to complete the figure which should have been learned during life, and failure results in being eaten. . . . The tales emphasize the need to know one's figures *properly* and demonstrate their cultural importance by involving them in the most fundamental of questions—mortality and (survival) beyond death. The figures vary in complexity from simple closed curves to having more than one hundred vertices, some with degrees of 10 or 12."

In all three cultures, there's special concern for Eulerian paths—paths that can be traced through every vertex without tracing any edge more than once. (The seven bridges of Königsberg!) All three seem to know that an Eulerian path is possible if and only if there are zero or two vertices of odd degree.

The Maori of New Zealand play a game of skill called mu torere. The game is played by two players; the "board" is an eight-pointed star. Each player has four markers—pebbles, or bits of broken china. Prof. Ascher shows that with any number of points except eight, the game would be uninteresting. "Mu torere, with four markers per player on an eight-pointed star, is the most enjoyable version of the game."

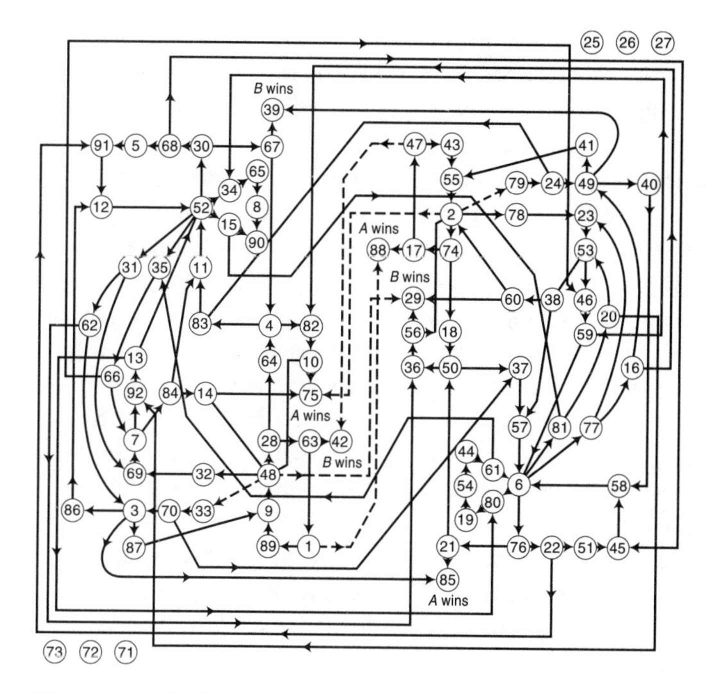

Figure 1. The flow of the game of mu torere.

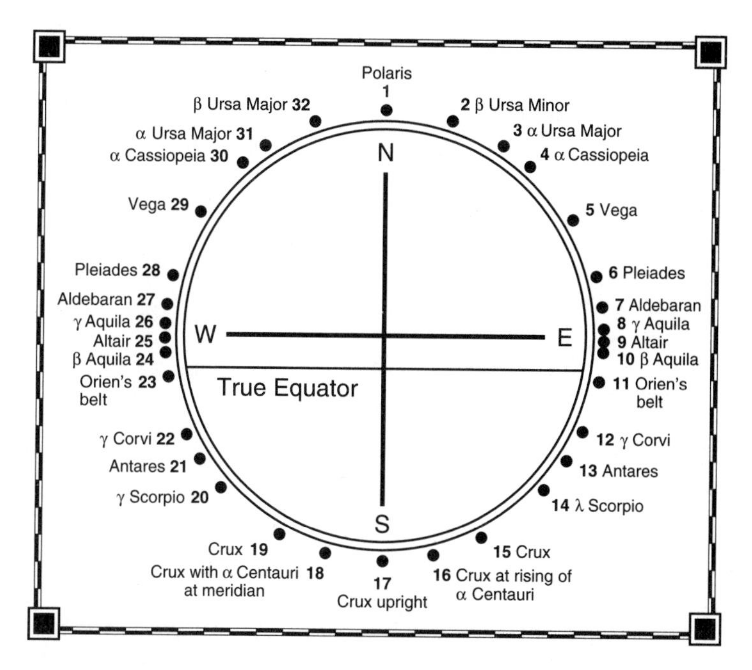

Figure 2. The start compass of the Caroline navigators.

The Caroline Islanders north of New Guinea cross hundreds of miles of empty ocean to Guam or Saipan. "The Caroline navigators do not use any navigational equipment such as our rulers, compasses, and charts; they travel only with what they carry in their minds."

Professor Ascher reminds us that leading anthropologists once taught that preliterate peoples were at an early stage of evolution. (Western society was advanced.) Later it was said that preliterate peoples ("savages") had an utterly different way of thinking from us. They were prelogical. We were logical.

Nowadays anthropologists say there's no objective way to rank societies as more or less advanced, higher or lower. Each is uniquely itself.

Professor Ascher's research is related to ethnomathematics as an educational program. This movement asks schools to respect and use the mathematical skills pupils bring with them—even if they differ from what's taught in school. By increasing understanding and respect for ethnomathematics, this work may benefit education.

There's a lesson for the philosophy of mathematics. Mathematics as an abstract deductive system is associated with our culture. But people created mathematical ideas long before there were abstract deductive systems. Perhaps mathematical ideas will be here after abstract deductive systems have had their day and passed on.

Summary and Recapitulation

thirteen

Mathematics Is a Form of Life

Self-graded Report Card Could Be Worse

Chapter 2 considered a list of criteria for a philosophy of mathematics. How do we look according to our own tests? I reprint the list:

1. Breadth
2. Connected with epistemology and philosophy of science
3. Valid against practice: research, applications, teaching, history, computing, intuition
4. Elegance
5. Economy
6. Comprehensibility
7. Precision
8. Simplicity
9. Consistency
10. Originality/novelty
11. Certitude/indubitability
12. Acceptability

Take them in order.

1. *Breadth.* Philosophy of mathematics should try to take into account all major parts and aspects of mathematics. The neo-Fregean dogma that set theory alone has philosophical interest, is an unacceptable excuse for ignorance. This book reflects an inside view of mathematical life. It's based on 20 years doing research on partial differential equations, stochastic processes, linear operators, and nonstandard analysis, 35 years teaching graduates and undergraduates, and many long hours listening, talking to, and reading philosophers.

2. *Epistemology/philosophy of Science.* Your philosophy of mathematics must fit your theory of knowledge and your philosophy of science. It just doesn't work, to be a Platonist in mathematics and a materialist empiricist in physical science.

Today Platonism is disconnected from *any* epistemology or philosophy of science. In Berkeley and Leibniz it was connected by their over-arching idealism. For them, mathematical knowledge was an aspect of spiritual knowledge, knowledge in the mind of God. Contemporary Platonists just ignore the question of how philosophy of mathematics connects with the rest of philosophy.

Empiricists like Mill and pragmatists like Quine do want to relate philosophy of mathematics to philosophy of science. But they do so without understanding the special character of mathematics. By recognizing mathematics as the study of certain social-cultural-historic objects, humanism connects philosophy of mathematics to the rest of philosophy.

3. *Valid against Practice*

a. *Research.* Taking seriously the actual experience of mathematical research is the main distinguishing feature of this book.
b. *Applications.* From Pythagoras to Russell and beyond, philosophy of mathematics rarely paid serious attention to applied mathematics. (Stephen Körner is one exception.) The present book recognizes the interlinking of the pure and applied viewpoints. The picture of mathematics as a social-cultural-historical phenomenon naturally includes applied mathematics.
c. *Teaching.* The philosophy of mathematics must let it be comprehensible that mathematics is actually taught. This issue is discussed by philosophically concerned educationists, not by philosophers. Tymoczko was an exception. Logicism, formalism, and intuitionism, each in a different way, make the possibility of teaching mathematics a deep mystery. Seeing mathematics as a cultural-social-historical entity makes it obvious that it's taught and learned. I elaborate on this in the section below titled "Teaching."
d. *History.* Like Lakatos and Kitcher, I explicitly incorporate history into the philosophy of mathematics, by identifying mathematics as the study of certain social-historic-cultural objects.
e. *Computing.* Logicist and formalist writers base their theories on an idealized, infallible, nonhuman, nonmaterial notion of computing. Intuitionists see computing as a purely mental activity. I discuss how mathematics is affected by real computing, an activity of real people and real machines. The section on "Proof" in Chapter 4 has more detail on this.
f. *Intuition.* See the section "Intuition" in Chapter 4.

4. *Elegance.* My theory is neither clean, complete, nor self-contained. I have put aside elegance in favor of other criteria.

5. *Economy.* Related to elegance. It means using the smallest number of basic concepts. I fare poorly, for the same reasons as in regard to point 4, elegance.

6. *Comprehensibility.* Some would say this criterion is not philosophical but merely literary. I strive to be clear both conceptually and verbally.

7. *Precision. Mathematics* is precise. *Philosophy* isn't. Mathematicians often mistakenly expect philosophy of mathematics to be part of mathematics. It isn't. It's part of philosophy. If it were part of mathematics, it could be precise. But it's no more necessary or possible for philosophy of mathematics to be precise than for any other branch of philosophy to be precise. (Gian-Carlo Rota brilliantly scores this point in *Indiscreet Thoughts.*) Frege, Russell, Hilbert, Brouwer, and Bishop, in trying to create mathematical foundations for mathematics, actually did contribute to mathematics and logic. In this book I have no intention to do mathematics.

8. *Simplicity.* Related to economy and elegance. Monism, whether materialist or idealist, is simple. Humanism, recognizing three kinds of existence, isn't so simple. If simplicity conflicts with truthfulness, adequacy, or accuracy, it should take second place.

9. *Consistency.* Most philosophers make consistency the chief desideratum, but in mathematics it's a secondary issue. Usually we can patch things up to be consistent. We can't so easily patch them up to be comprehensive or true to life. It's believed that if a set of statements is true, it must be consistent. In that spirit, I claim that what I offer here is consistent.

10. *Originality.* Pólya and Lakatos and White had a big influence on me. Professor Hao Wang of Rockefeller University corrected blunders and gave me courage to persist. Kitcher, Tymoczko, and Ernest wrote related ideas. Everyone in the Acknowledgments made an impact on this book. Nevertheless, this is the only book of its kind. There's no other quite like it.

11. *Certitude or Indubitability.* This book does nothing to establish mathematical certitude or indubitability. On the contrary, it says to forget about them. Move past foundationism.

12. *Acceptability.* Are my proposals acceptable, worthy of criticism by experts and authorities? That remains to be seen.

If you don't like these criteria, throw them out. Put others in. Apply *your* criteria to humanism, Platonism, logicism, intuitionism, constructivism, conventionalism, structuralism, and fictionalism. Find out which one looks good to you.

For Teaching, Philosophy Makes a Difference

What's the connection between philosophy of mathematics and teaching of mathematics? Each influences the other. The teaching of mathematics *should* affect the philosophy of mathematics, in the sense that philosophy of mathematics must be compatible with the fact that mathematics can be taught. A philosophy that obscures the teachability of mathematics is unacceptable. Platonists and formalists ignore this question. If mathematical objects were an other-worldly, nonhuman reality (Platonism), or symbols and formulas whose meaning is irrelevant (formalism), it would be a mystery how we can teach it or learn it. Its teachability is the heart of the humanist conception of mathematics.

In the other direction, the philosophy of mathematics held by the teacher can't help but affect her teaching. The student takes in the teacher's philosophy through her ears and the textbook's philosophy through her eyes. The devastating effect of formalism on teaching has been described by others. (See Khinchin or Ernest.) I haven't seen the effect of Platonism on teaching described in print. But at a teachers' meeting I heard this:

"Teacher thinks she perceives other-worldly mathematics. Student is convinced teacher really does perceive other-worldly mathematics. No way does student believe *he's* about to perceive other-worldly mathematics."

Platonism can justify a student's certainty that it's impossible for her/him to understand mathematics. Platonism can justify the belief that some people can't learn math. Elitism in education and Platonism in philosophy naturally fit together. Humanist philosophy, on the other hand, links mathematics with people, with society, with history. It can't do damage the way formalism and Platonism can. It could even do good. It could narrow the gap between pupil and subject matter.

Such a result would depend on many other factors. But if other factors are compatible, adoption by teachers of a humanist philosophy of mathematics could benefit mathematics education.

This possible educational value is not a warrant for *correctness* of humanist philosophy. In earlier chapters I argued the correctness of humanism. But it's not unexpected that a philosophy epistemologically superior is educationally superior.

Philosophy and Ideology
Politics Makes a Difference

In our half of the twentieth century, it's unacceptable to import ideology into scholarship. We remember the destruction of Russian linguistics and genetics by Stalin's political correctness. Philosophy was mutilated too. Only dialectical materialism was allowed in Moscow.

Nearly forgotten is Hitler's ideology-philosophy. (Think with your blood! When I hear "culture," I reach for my revolver!) Martin Heidegger, deemed by some to be the supreme philosopher of our time, quickly and easily accommodated to Nazism.

Ideological philosophy is twice shameful. Shameful intellectually, by fostering meretricious hacks while crushing genuine intellect. Shameful politically, by complicity in the regime's crimes.

Therefore, one doesn't ask whether a philosophical view is socially harmful or beneficial, or whether it's favored by aristocrats, generals, or coal miners. For intellectual value, none of that should matter. Is it true? Is it interesting? Is it beautiful?

That said, the fact remains, philosophers are human beings. We should care how a philosophy connects with other realms of thought, how it connects with society at large.

In historical Part 2, I described philosophers' religious beliefs in relation to their philosophies of mathematics. Philosophers also hold political beliefs. Can there also be a connection between political position and philosophy of mathematics?

To compare philosophers of different periods and places, we want a uniform terminology for comparative political positions. In classical Athens, democrat/oligarch; during the Enlightenment, clerical/anticlerical, royalist/republican; in the 1930s, Fascist/anti-Fascist. Rather than "progressive/conservative" or "popular/aristocratic," I use "left-wing/right-wing." These terms belong to the French Chamber of Deputies in the nineteenth century, but for convenience I call any politics that restricts popular political rights "right wing"; politics that increases them, "left wing." (Gödel said simply "left" and "right," as quoted in the Preface.)

I start with the Mainstream, as defined in Part 2, with *Pythagoras and the Pythagoreans.* They were progressive in admitting women to their school. Their maxims, such as (Wheelwright) "Abstain from beans" and "Do not urinate in the direction of the sun" seem apolitical. But "it was to the young men of well-to-do families that Pythagoras made his appeal. Pretending to have the power of divination, given at all times to mysticism, and possessed in a remarkable degree of personal magnetism, he gathered about him some 300 of the noble and wealthy young men of Magna Graecia and established a brotherhood that has ever since served as a model for all the secret societies in Europe and America" (Smith, p. 72).

Moreover, "The doctrines of the Pythagoreans and Eleatics may be understood, partly at least, in the light of social patterns which were congenial to the philosophers; these thinkers were not unaffected in their theorizing about eternal values by the actual political structure of which they were a part. The Pythagoreans interpreted the world in terms of order and symmetry, based on fixed mathematical ratios and found similar satisfactory order and symmetry in existing aristocratic schemes of government. . . . As the body must be held in subjection by the soul, so in every society there must be wise and benevolent masters over obedient and grateful inferiors; and of course they had no doubt as to who were qualified to be the masters. Their religious brotherhoods became powerful political influences in Italiot Greece, a training school for aristocratic leadership. . . . Zeno defended the same thesis by a clever series of paradoxes (Agard, p. 42). . . . The philosophy that appealed most to the best families in

Athens was that of the Pythagorean brotherhoods, whose chief intellectual concern was mathematics but whose practical interest lay in a determined defense of aristocratic regimes against the inroads of democracy. They approved especially of the Pythagorean loyalty to ancient laws and customs even if they might be in certain respects inferior to new ones, on the principle that change is in itself a dangerous thing, the greatest sin is anarchy, and in the nature of things some are fitted to rule, others to obey" (Agard, p. 180). I count the Pythagoreans as *right wing.*

Plato is known for his totalitarian politics even more than for his philosophy of mathematics. Popper called the *Republic* a blueprint for fascism. Stone documented Plato's elitism and authoritarianism. Some are offended by Popper and Stone, but nobody claims Plato as a liberal. *Right wing.*

The Platonist Catholic philosophers *Augustine of Hippo* and *Nicolas Cusanus* thought of mathematics mystically. The Aristotelian Catholic *Thomas Aquinas* thought of mathematics more scientifically. But all three took for granted the Church's right and duty to guide society and its morality. *Right wing.*

Descartes's Method was radical in its rejection of authority in scientific work. Nevertheless, he was an obedient son of the Church. "His political views were also extremely orthodox, and closely linked with his religious ones . . . his deep respect for nobility and particularly sovereigns verged on the passionate if not the religious" (Vrooman, p. 42). Descartes's deep respect for royalty proved fatal. He spent his fifty-fourth winter in frigid Stockholm, rising to give philosophy lessons three times a week to Queen Christina at 5 A.M. He caught pneumonia, and died. A *right-winger.*

Spinoza was a subverter of Scripture, cursed by Protestant, Catholic, and Jew. "The *Tractatus Theologico-Politicus* is an eloquent plea for religious liberty. True religion is shown to consist in the practice of simple piety, and to be quite independent of philosophical speculations. The elaborate systems of dogmas framed by theologians are based on superstition, resulting from fear" (Pollock, *Trac.* intro., p. 31).

The *Tractatus* made Spinoza famous in Europe as a dangerous atheist. "Spinozism" was a top-ranking evil. To sell the book to the market created by banning it, booksellers printed it *with a false title page*!

In his *Tractatus Theologico-Politicus* (Chapter xvi), Spinoza wrote: "A Democracy may be defined as a society which wields all its power as a whole. . . . In a democracy, irrational commands are less to be feared: For it is almost impossible that the majority of a people, especially if it be a large one, should agree in an irrational design; and, moreover, the basis and aim of a democracy is to avoid the desires as irrational, and to bring men as far as possible under the control of reason, so that they may live in peace and harmony. . . . I think I have now shown sufficiently clearly the basis of a democracy. I have especially desired to do so, for I believe it to be of all forms of government the most natural, and the most consonant with individual liberty. In it no one transfers his natural right so absolutely that he has no further voice in affairs; he only hands it over to the majority of a society, where he is a unit. Thus all men remain, as they are in the state of Nature, equals.

"This is the only form of government which I have treated at length, for it is the one most akin to my purpose of showing the benefits of freedom in a state" (Ratner, pp. 304–7).

Feuer writes (p. 5), "Spinoza is the early prototype of the European Jewish radical. He was a pioneer in forging methods of scientific study in history and politics. He was cosmopolitan, with scorn for the notion of a privileged people. Above all, Spinoza was attracted to radical political ideas. From his teacher Van den Ende he had learned more than Latin. He had evidently imbibed something of the spirit of that revolutionist whose life was to end on the gallows. . . . In his youth, furthermore, Spinoza's closest friends were Mennonites, members of a sect around which there still hovered the suggestion of an Anabaptist, communistic heritage." A *lefty.*

An Anglican Bishop, *Berkeley* had the right-wing politics expected of his position. Yet when Ireland was wracked by famine, he worked strenuously to get food for the hungry. Still, a *right-winger.*

Leibniz thought he was the man able to save Europe from war and revolution. He wanted to be chief counselor to some principal monarch, to the emperor, or to the Pope (Meyer, pp. 2–4). He wrote, "Those who are not satisfied with what God does seem to me like dissatisfied subjects whose attitude is not very different from that of rebels . . . to act conformably to the love of God it is not sufficient to force oneself to be patient, we must really be satisfied with all that comes to us according to his will" (1992, p. 4). This was the Leibniz caricatured in Voltaire's *Candide. Right wing.*

Kant was a moderate. His life and writings upheld the status quo. Yet in religion he leaned toward free thought, and politically he wasn't out of sympathy with the French Revolution. He was ordered by the King of Prussia not to write about religion. Popper calls him an ardent liberal. *A leftish.*

Frege actually died a Nazi. Sluga reports: "Frege confided in his diary in 1924 that he had once thought of himself as a liberal and was an admirer of Bismarck, but his heroes now were General Ludendorff and Adolf Hitler. This was after the two had tried to topple the elected democratic government in a coup in November 1923. In his diary Frege also used all his analytic skills to devise plans for expelling the Jews from Germany and for suppressing the Social Democrats." Michael Dummett tells of his shock to discover, while reading Frege's diary, that his hero was an outspoken anti-Semite (1973). *Right wing.*

Russell was the "cream" of English society. His grandfather John Russell was a Whig foreign minister. Yet Bertie became a socialist. In World War I he went to jail as a conscientious objector. Then in the 1920s and 1930s he alienated left-wingers by hostility to the Soviet Union. After the explosion of the U.S.-British atom bombs, and before the Soviet atom bomb, he favored a preventive atom-bomb attack on the Soviet Union. But after the Soviet atomic explosion, he became a fervent opponent of atom bombs and an advocate of peace with the Soviet Union. He was on an international tribunal that condemned U.S. war crimes in Vietnam. *Left wing.*

Brouwer was a fanatically antifemale, pro-German eccentric. "There is less difference between a woman in her innermost nature and an animal such as a lioness than between two twin brothers. . . . The usurpation of any work by women will automatically and inexorably debase that work and make it ignoble. . . . When all productive labor has been made dull and ignoble by socialism it will be done exclusively by women. In the meantime men will occupy their time according to their ability and aptitude in sport, gymnastics, fighting, studying philosophy, gardening, wood-carving, traveling, training animals and anything that at the time is regarded as noble work, even gambling away what their wives have earned. For this really is much nobler than building bridges or digging mines" (*Life, Art and Mysticism*). His biographer writes, "At the time, a few months after his wedding, he was supported by his wife, who was herself working for a degree while running her pharmacy. . . . There is a touch of insincerity about most of Brouwer's strong condemnations. He ridicules fashions and many of the human weaknesses which mark his own life, such as ambition, lust for power, jealousy and hypochondria. His condemnation of those seeking security by amassing capital rings rather hollow in a man whose life was so obsessed with money. . . . *Life, Art and Mysticism* cannot be written off as a rash, 'teen-age effort.' . . . He wrote it in 1905, after his doctoral examination. Far from disowning his 'booklet' as a youthful aberration, Brouwer backed it all his life. He discussed the possibility of an English translation as late as 1964. It proudly features in every one of his entries for various biographical dictionaries as the first of his two books. Most important, it is the clearest expression of his philosophy of life, which inspired his intuitionism" (Van Stift).

After World War II his university convicted him of collaboration with the Nazi occupation. Van Stift explains that Brouwer was not so much anti-Semitic as anti-French. "The German occupation seriously affected academic life in Holland during the war years. Brouwer's pragmatic attitude to politics, his concern for the continuance of academic life and his fear of 'becoming involved' laid him open to petty accusations of the many enemies he had made in local government and at the University. In the postwar hysteria these were blown up into serious crimes before a kangaroo court of Amsterdam University. He was reprimanded and suspended from his duties for nine months." *Right wing.*

W. V. O. Quine is a self-identified Republican. *Right wing.*

So among the Mainstream we have Pythagoras, Plato, Augustine, Cusa, Descartes, Berkeley, Leibniz, Frege, Brouwer, and Quine on the right; Spinoza, Kant and Russell on the left. *10 righties, 3 lefties.*

Now the humanists.

Aristotle criticized all forms of government, especially the two competing in Athens—democracy and oligarchy. In the end, he comes out as a cautious, critical democrat. From Barker's translation of *The Politics*, published in 1962 by Oxford:

"It is possible, however, to defend the alternative that the people should be sovereign. The people, when they are assembled, have a combination of qualities which enables them to deliberate wisely and to judge soundly. This suggests that they have a claim to be the sovereign body. . . . It may be argued that experts are better judges than the non-expert, but this objection may be met by reference to (a) the combination of qualities in the assembled people (which makes them collectively better than the expert) and (b) their 'knowing how the shoe pinches' (which enables them to pass judgment on the behavior of magistrates). . . (p. 123).

"Each individual may indeed be a worse judge than the experts; but all, when they meet together, are either better than experts or at any rate no worse. . . . There are a number of arts in which the creative artist is not the only, or even the best judge. . . (p. 126). [This seems to be a cautious suggestion that the art of government is this kind.]

"It is therefore just and proper that the people, from whom the assembly, the council, and the court are constituted, should be sovereign on issues more important than those assigned to the better sort of citizens" p. 127). *A lefty*

John *Locke* was a father of modern democracy. The French Philosophes took him as their teacher. So did Thomas Jefferson, in writing the U.S. Declaration of Independence. A *lefty*.

Hume also belonged to the eighteenth-century enlightenment, which gave intellectual nourishment to the eighteenth-century revolutions, but a Tory!

d'Alembert was a leader of the Philosophes, who laid the ideological groundwork for the French Revolution. On the *left* side.

Mill is better known for liberal politics than for philosophy of mathematics (Kubitz, p. 277). There's a surprising linkage between his philosophy of mathematics and his politics. This is revealed in two book reviews by an anonymous writer who may have been Mill.

"In October, 1830, the *Westminster Review*, the radical periodical with which both Mills were associated, brought a review of *The First Book of Euclid's Elements with Alterations and Familiar Notes. Being an Attempt to get rid of "Axioms" altogether; and to establish the Theory of Parallel Lines, without the introduction of any principle not common to other parts of the Elements.* By a member of the Univ. of Cambridge. 3rd Ed. R. Heward, 1830. The reviewer hails this work with an interesting passage, which, if coming from the hand of Mill, would give us a significant expression about axioms from the time when he was occupied with the question left him by Whately, 'how can the truths of deductive science be all wrapt up in its axioms?' The reviewer, whoever he is, rejoices,

"This is an attempt to carry radicalism into Geometry; always meaning by radicalism, the application of sound reason to tracing consequences to their *roots.* To those who do not happen to be familiar with the facts, it may be useful to be told, that after all the boast of geometricians of possessing an *exact science*, their science has really been founded on taking for granted a number of propositions under the title of Axioms, some of which were only specimens of slovenly acquiescence in assertion where demonstration might easily have been had, but others were in reality the begging of questions which had quite as much need of demonstration, as the generality of those to which demonstration was applied." In July 1833, the reviewer of Whewell's *First Principles of Mechanics* in the *Westminster Review* remarks in the same vein, "Axiom is a word in bad odour, as having been used to signify a lazy sort of *petitio principi* introduced to save the trouble of inquiry into cause. . . ." "These passages make it all the more evident that the group with which Mill was associated were not oppposed to axioms on speculative, but on the practical grounds of political reform" (Kubitz). (Today again, some writers are making a connection between political philosophy and the epistemology and pedagogy of mathematics.) Mill counts on the *left.*

Peirce kept away from politics. He was a notorious elitist and snob. He took great pride in his family connections with Boston "aristocracy." During the Civil War he managed to evade service in the Northern army. Yet his lifelong friend and supporter was the great liberal, William James. For decades he reviewed science and mathematics for the liberal magazine *The Nation.* He wrote a denunciation of "social Darwinism," an ideology supporting the right to rule of the moneyed classes. We class him as borderline.

Rényi, Lakatos, Polányi, Pólya, and Von Neumann were Hungarian Jews. In 1944 Rényi was dragged to a Fascist labor camp, but escaped when his company was sent west. (In the late 1930s the Horthy regime set up labor camps for men "unfit" for military service—Communists, Jews, gypsies. Sometimes they suffered extreme danger and hardship.) For half a year he hid with false papers.

His parents were in the Budapest ghetto. Jews there were being rounded up for annihilation. Rényi stole a soldier's uniform at a Turkish bath, walked into the ghetto, and marched his parents out. One must be familiar with the circumstances to appreciate his courage and skill.

Starting in 1950 he directed the Mathematical Institute of the Hungarian Academy of Science. Under his leadership it became an international research center, and the heart of Hungarian mathematical life. A *lefty.*

Pólya and Polányi were liberal exiles from Horthy's clerical fascism. As a student at Budapest University, Pólya belonged to the liberal Galileo Society. *Two lefties.*

Lakatos was a Communist before his arrest and imprisonment in 1950. By the time he left Hungary after the 1956 uprising, he was an ardent anti-Communist. I classify him as a *right-winger.*

Von Neumann was a hawk in the Cold War with the Soviet Union. He became science adviser to the United States and member of the Atomic Energy Commission. Since we haven't been able to classify him as Mainstream or humanist, he doesn't affect our tabulation.

Hilbert was first and last a mathematician, neither a right nor a left. *Poincaré* may have had political views, but I haven't found them. (His cousin Raymond, with whom he shared a flat in the Latin Quarter while the two were students, became President of the Republic during World War I, and Premier from 1924 to 1929. Raymond was a moderate bourgeois, working between the royalists and the socialists.) Three insufficient informations.

Omitting the inconclusive Peirce, Poincaré and Hilbert, we have among the humanists Aristotle, Locke, d'Alembert, Mill, Rényi, Polányi, and Pólya, on the left; Aquinas and Lakatos on the right (*7 lefties, 2 righties*).

It appears the Mainstream are mostly rightish, humanists mostly leftish.

Following the example of Paul Ernest in *Philosophy of Mathematics*, think of the four entries as proportional to conditional probabilities. There are

$$10 + 2 + 3 + 7 = 22$$

philosophers in the matrix. Pick one at random. If he happens to be a Mainstream, he's "rightish" with probability 10/13, or 77 percent; leftish with probability 3/13, or 23 percent.

If you picked a humanist, the probability he's rightish is much less. Only 2/9, or 22 percent. The probability he's leftish is 7/9, or 78 percent.

Look at it the other way. If the philosopher you pick happens to be rightish, the probability that he's philosophically Mainstream is 10/12, or 83 percent. The probability that he's a humanist is only 2/12, or 17 percent.

If you pick a lefty, the probability he's Mainstream is much less. Only 3/10, or 30 percent. The probability he's a humanist is 7/10, or 70 percent.

Why so?

In Chapter 1 and in a previous section of this chapter I argue that philosophy of mathematics makes a difference for mathematics education. I claim that Platonist philosophy is anti-educational, while humanist philosophy can be pro-educational. I emphasize that the social consequence of a philosophy is *not* the same as its validity. But doesn't make the social consequence unimportant.

Conservative politics and Platonist philosophy of mathematics don't *imply* each other. This is proved by exceptions like Russell and Lakatos. The numbers do suggest a *correlation*.

What can we say that's neither dogmatic nor mere guesswork?

I simply ask: doesn't it make sense?

Political conservatism opposes change. Mathematical Platonism says the world of mathematics never changes.

Political conservatism favors an elite over the lower orders. In mathematics teaching, Platonism suggests that the student either can "see" mathematical reality or she/he can't.

A humanist/social constructivist/social conceptualist/quasi-empiricist/naturalist/maverick philosophy of mathematics pulls mathematics out of the sky and sets it on earth. This fits with left-wing anti-elitism—its historic striving for universal literacy, universal higher education, universal access to knowledge, and culture.

If the Platonist view of number is associated with political conservatism, and the humanist view of number with democratic politics, is that a big surprise?

The Blind Men and the Elephant
(Six Men of Indostan)

J. GODFREY SAXE (1816–1887)

It was six men of Indostan
  To learning much inclined,
Who went to see the Elephant
  (Though all of them were blind).
That each by observation
  Might satisfy his mind.

The First approached the Elephant,
  And happening to fall
Against his broad and sturdy side,
  At once began to bawl:
"God bless me! but the Elephant
  Is very like a wall!"

The Second, feeling of the tusk,
  Cried, "Ho! what have we here
So very round and smooth and sharp?
  To me 'tis mighty clear
This wonder of an Elephant
  Is very like a spear!"

The Third approached the animal,
  And happening to take
The squirming trunk within his hands,
  Thus boldly up and spake:
"I see," quoth he, "the Elephant
  Is very like a snake!"

The Fourth reached out an eager hand,
  And felt about the knee.
"What most this wondrous beast is like
  Is mighty plain," quoth he;
" 'Tis clear enough the Elephant
  Is very like a tree!"

The Fifth who chanced to touch the ear,
Said: "E'en the blindest man
Can tell what this resembles most:
Deny the fact who can,
This marvel of an Elephant
Is very like a fan!"

The Sixth no sooner had begun
About the beast to grope,
Than, seizing on the swinging tail
That fell within his scope,
"I see," quoth he, "the Elephant
Is very like a rope!"

And so these men of Indostan
Disputed loud and long,
Each in his own opinion
Exceeding stiff and strong,
Though each was partly in the right
And all were in the wrong!

This doggerel is a metaphor for the philosophy of mathematics, with its Wise Men groping at the wondrous beast, Mathematics.

What do the six men of Indostan have in common? (We'll call them 6I, to give the conversation a mathematical tinge.)

They obey a common axiom: *Axiom 6I: Cling to an incomprehensible partial truth to avoid a larger, more inclusive truth.*

Let FP denote a Famous Philosopher. (Continuing with the pseudomathematical terminology.) We're ready to formulate our problem in *precise language*:

Problem: Given an FP, does he satisfy axiom 6I?

Poetically speaking, is the Famous Philosopher one of the Six Men of Indostan?

I'll probe this question only with respect to the formalists. I leave the Platonists and intuitionists for the reader's pleasure.

With respect to the formalists, the answer to the *Problem* is YES. David Hilbert proposed that, for philosophical purposes, statements about infinitary objects (including all of calculus and analysis) be regarded as mere formulas, devoid of reference, meaning, or interpretation. This was forgotten whenever Hilbert took off his philosopher hat and resumed his true identity as mathematician. It wasn't taken seriously in real mathematical life. It was just a tactic in Hilbert's campaign against Brouwer's intuitionism.

Because of Hilbert's pre-eminence in mathematics, the formalist philosophy was identified with his name. Formalism was recognized long before; it was one of the bugbears Frege tried to squash. Decades after Hilbert, the Bourbakistes of Paris went deeper into formalism than Hilbert, regarding *all* mathematical statements, finitary or infinitary, as meaning-free formulas. More precisely, any meaning such a formula carries is irrelevant in mathematics, which is concerned with the formulas themselves.

(But Dieudonné shamelessly blurted out, as I quoted in Chapter 1, that actually Bourbaki didn't believe any of this. It was all a trick, to fend off philosophers.)

Whichever formalist represents the tribe, whether David Hilbert, "Nicolas Bourbaki," Haskell Curry, or David Henle, when we ask, "Does he satisfy Axiom 6I?" we must answer, "Yes." Everyone agrees that formalization is *an aspect* of mathematics. On the other hand, formalism doesn't have the whole beast in its view-finder. We need only point to the vital part of mathematics that formalism negates—the intuitive, or informal (see Chapter 4).

To save repetition, I make two declarations:

Declaration 1: The intuitive is an essential aspect of mathematics. Without the intuitive, mathematics wouldn't be mathematics as we have always conceived it.

Declaration 2: The intuitive, by its very conception and definition, is not formalizable.

The formalist can't reject Declaration 1. To do so would be simply incredible.

He can't reject Declaration 2. To do so would be a contradiction in terms.

He can give up any claim to describe mathematics in its totality, and aspire only to describe the formalizable part—*the part that formalism can describe*. This would be correct—but vacuous. The man of Indostan feels the part of the elephant that is like a snake—and calls out, "It's very like a snake!"

It would be good to repeat this discussion, replacing formalists by intuitionists (who see the mental side of the elephant mathematics, but are blind to its social-historic parts) and by logicists and Platonists (who see the dynamically evolving, socially interacting beast of mathematics as a frozen abstraction in the sky). I leave these exercises to the interested reader.

Summary

Mathematics is like money, war, or religion—not physical, not mental, but social. Dealing with mathematics (or money or religion) is impossible in purely physical terms—inches and pounds—or in purely mental terms— thoughts and emotions, habits and reflexes. It can only be done in social-cultural-historic terms. This isn't controversial. It's a fact of life.

Saying that mathematics, like money, war, or religion, is a social-historic phenomenon, is not saying it's the *same* as money or war or religion. Money is different from war, money is different from religion, religion is different from war. But all four are social-historic phenomena. Mathematics is another particular, special social-historical phenomenon. Its most salient special feature is the uniquely high consensus it attains.

War or money don't exist apart from human minds and bodies. Without bodies, no minds; without people's minds and bodies, no society, culture, or history. The emergent social level does not emerge from a vacuum; it emerges from the mental and physical levels. Yet it has qualities and phenomena that can't be understood in terms of the previous levels.

Recognizing that mathematics is a social-cultural-historical entity doesn't automatically solve the big puzzles in the philosophy of mathematics. *It puts those puzzles in the right context, with a new possibility of solving them.*

This is like a standard move in mathematics—*widen the context.* Consider the equation:

$$x^2 + 1 = 0$$

Among the real numbers, it has no solution. If we enlarge the context to the complex numbers, we find two solutions, $+i$ and $-i$. This is the first step to a beautiful and powerful theory.**

Yet from the viewpoint of the original problem, these solutions don't really exist. They aren't fair. They "don't count."

So it is here. Problems intractable from the foundationist or neo-Fregean viewpoint are approachable from the humanist viewpoint. But from the foundationist viewpoint, humanist solutions are no solutions. They're unfair. It's not allowed to give up certainty, indubitability, timelessness, or tenselessness. These restrictions in philosophy of mathematics act like the restriction to the real line in algebra. Dropping the insistence on certainty and indubitability is like moving off the line into the complex plane.

When mathematicians move into the complex plane, we don't throw away all sound sense. We keep the rules of algebra. Dropping indubitability from philosophy of mathematics doesn't mean throwing away all sound sense. The guiding principles remain: intelligibility, consistency with experience, compatibility with philosophy of science and general philosophy. A humanist philosophy of mathematics respects these principles.

Epilogue

The line between humanism and Mainstream is more than a philosophical preference. It's tied to religion and politics.

The humanist philosophy of mathematics has a pedigree as venerable as that of Mainstream. Its advocates are respected thinkers.

The Mainstream continues to dominate. That doesn't sanctify it. Religious obscurantism and propertied self-interest still dominate society. They're not sanctified.

Who's interested in escaping from neo-Fregeanism? Philosophically concerned mathematicians. And people interested in the foundations of mathematics education. Putnam's disconnection from Quinism may have been a straw in the wind. The blossoming of humanism among mathematics educators is another.

Mathematical audiences are impatient with neo-Fregeanism. They show lively interest in alternatives. Yet the typical journal article on philosophy of mathematics still plods after Carnap and Quine. I suspect that game is played out.

Neo-Fregeans will disagree. But neo-Fregeanism no longer reigns by inherited right. If it cares to be taken seriously much longer, it has to face the challenge of humanism.

Mathematical Notes/Comments

Most of these notes keep promises made in previous chapters. "Square circles" is in a light-hearted vein. No one need be offended. David Hume was my favorite philosopher in college.

Except for their use of algebra, the articles on "How Imaginary Becomes Reality" could have come earlier. They show natural, inevitable ways that mathematics grows, with no mystery about invention vs. discovery.

"Calculus refresher" is included because it isn't possible to talk sense about mathematics without an acquaintance with or recollection of calculus. By omitting exercises and formal computations, I present a semester of calculus in a few easy pages.[1]

The last article is a wonderful piece of mathematical artistry. The late George Boolos proves Gödel's great incompleteness theorem in three simple pages![2]

Arithmetic

[1] It's taken, with minor improvements, from the Teacher's Guide which accompanies the study edition of *The Mathematical Experience*, co-authored with Philip J. Davis and Elena Anne Marchisotto, Birkhauser Boston, 1995.

[2] It's reproduced from the *Notices* of the American Mathematical Society to make it available to a larger readership.

Logic

Sets

Geometry

How Imaginary Becomes Reality

Calculus

More Logic

ARITHMETIC

What Are the Dedekind-Peano Axioms?

Instead of constructing the natural numbers out of sets à la Frege-Russell, we can take them as basic, and describe them by axioms from which their other properties can be derived.

The axioms should be consistent, of course; chaos could ensue if they were contradictory (see below). We would like them to be minimal—not include any redundant axioms.

The standard axioms for the natural numbers were given by Richard Dedekind, inventor of the Dedekind cut. Following the usual rule of misattribution in mathematical nomenclature, they are called "Peano's postulates." The undefined terms are "1" and "successor of."

1. 1 is a number.
2. 1 isn't the successor of a number.
3. The successor of any number is a number.
4. No two numbers have the same successor.
5. (Postulate of mathematical induction) If a set contains 1, and if the successor of any number in the set also belongs to the set, then every number belongs to the set.

How to Add 1's

We show that Dedekind's axioms imply

$$2 + 2 = 4.$$

None of the symbols in this equation appears in the axioms, so we must define all four of them.

"=" is defined by the rule that for any x and y, if $x = y$, then in any formula y may be replaced by x and vice versa. This rule is called "substitution."

For present purposes, we need only define addition by 1 and 2, for all n. Let S stand for the successor operation.

Define 2 as S(1), 3 as S(2), 4 as S(3).
Define "n + 1" as S(n) and "n + 2" as S(S(n)).

Then by substitution

(A) 2 + 2 = S(S(2)).

Again by substitution,

(B) 4 = S(S(2)).
Voilà! 2 + 2 = 4.

To define n + k for all n and k would take more work, using recursion on both n and k.

Is 2231 prime?

I know how to find out. 2231 is prime if it's not divisible by any number between 1 and 2231. I could just divide 2231 by all the numbers from 2 to 2230.

This labor can be cut down a lot. If 2231 is factorable, it factors into two numbers, one larger, one smaller or both equal to each other. It's sufficient to find the smaller. Since

$$47^2 = 2249,$$

the smaller factor has to be less than 47. Moreover, it's not necessary to divide by any composite number, because if 2231 has a composite factor, that composite factor has prime factors that also factor 2231.

So we only have to check the prime numbers less than 47—2, 3, 5, 7, 11, 13, 17, 19, 23, 29, 31, 37, 41, 43.

Now 2231 is odd, so 2 isn't a factor. The sum of the digits is 8, which is not divisible by 3, so 3 isn't a factor. It doesn't end in 5 or 0, so 5 isn't a factor. The alternating sum

$$+ 2 - 2 + 3 - 1 \text{ isn't } 0,$$

so 11 isn't a factor. Get your calculator and divide 2231 by 7, 13, 17, 19, 23, 29, 31, 37, 41, and 43. If none of them divides 2231 without remainder, 2231 is prime.

Logic

Zermelo-Fraenkel, Axiom of Choice, and the Unbelievable Banach-Tarski Theorem

"Given any collection of nonempty sets, it is possible to form a set that contains exactly one element from each set in the collection."

Surely a harmless-sounding assumption to make about finite collections of finite sets, and even countably infinite collections of countably infinite sets. But when it's applied to collections and sets of arbitrarily great uncountable cardinality, trouble comes! Consequences follow, which many mathematicians would rather not believe. Zermelo proved, using the axiom of choice, that any set—for instance, the uncountable set of real numbers—can be rearranged to be well-ordered. (But no one can actually do it, and no one expects anyone to be able to do it.) Stefan Banach and Alfred Tarski proved, using the axiom of choice, that it's possible to divide a pea (or a grape or a marshmallow) into 5 pieces such that the pieces can be moved around (translated and rotated) to have volume greater than the sun (see Wagon). As mentioned in Chapter 4 on proof, a transitory movement to avoid the axiom of choice has long been given up.

1/0 Doesn't Work (0 into 1 Doesn't Go)

Division by 0 is not allowed. Why not? If it's allowed to introduce a symbol i and say it's the square root of -1 *which doesn't have a square root,* why not introduce some symbol, say Q, for $1/0$?

We introduce new numbers, whether negative, fractional, irrational, or complex, to preserve and extend our calculating power. We relax one rule, but preserve the others. After we bring in i, for example, we still add, subtract, multiply, and divide as before. I now show that there's no way to define $(1/0) \times 0$ that preserves the rules of arithmetic.

One basic rule is,

$0 \times (\text{any number}) = 0.$

(Formula I) So $0 \times (1/0) = 0.$

Another basic rule is

$(x) \times (1/x) = 1$, provided x isn't zero. (But if we want $1/0$ to be a number, this proviso becomes obsolete.)

(Formula II) So $0 \times (1/0) = 1$

Putting Formulas I and II together,

$1 = 0.$

Addition gives

$2 = 0$, $3 = 0$, and so on, $n = 0$

for every integer n.

Since all numbers equal zero, all numbers equal each other.

There's only one number—0.

The supposition that $1/0$ exists and satisfies the laws of arithmetic leads to collapse of the number system. Nothing is left, except—nothing.

What Is Modus Ponens?

In scholastic (medieval Aristotelian) logic, Latin names were given to the different permutations of Aristotle's syllogisms. Modus ponens is the simple argument: "If

A implies B, and A is true, then B is true." In modern formal logic the other syllogistic arguments can be eliminated. Modus ponens turns out to be sufficient.

Formalizable

A statement is "formalized" when it's translated into a formal language. Computer languages like Basic, Pascal, Lisp, C, and others—are formal languages. The notion of a formal language goes back to Peano, Frege, Russell, and Leibniz. A formal language has a vocabulary specified in advance—x_1, x_2, $+$, $\times$, $=$, etc. It has an explicit grammar, which prescribes the admissible permutations of the vocabulary. Whether a sentence in natural language is formalizable depends on the formal language under consideration. To be formalizable a sentence is supposed to be unambiguous, and to mention only objects that have names in the formal language.

How One Contradiction Makes Total Chaos

Suppose some sentence A and its negation "not-A" are both true.

We claim that (A and not-A) together implies B, no matter what B says. First, notice that:

(I) not-(A and not-a) means the same thing as (not-A or A), which is a "tautology"—it's true, no matter what A says. Also, notice that a tautology is implied by any sentence at all, because an implication is false only when the antecedent is true and the conclusion is false; if the conclusion is a tautology, it can't be false. Therefore

(II) not-B implies the tautology (not-A or A)

Now, by the definition of "implies," if a sentence P implies a sentence Q, then not-Q implies not-P. This deduction rule is called "contrapositive."

So, applying contrapositive to II,

(III) not-(not-A or A) implies not-(not-B)

(IV) But not-(not-B) is the same as B (double negative.) So, substituting (IV) into (III),

(V) not-(not-A or A) implies B.

Now applying negation ("not") to both sides of (I), we get

(VI) (A and not-a) means the same thing as not-(not-A or A)

Combining (V) and (VI), we have, as claimed,

(A and not-a) implies B, for any B.

The way (A and not-A) makes the whole logical universe collapse is rather like the way $1/0$ makes the whole number system collapse. Is there a connection?

Sets

The Natural Numbers Come Out of the Empty Set

I will describe the sequence of "constructions" by which we "create" the real number system out of "nothing." It has philosophical interest, and it's ingenious.

In practice, however, we think of numbers in terms of their behavior in calculation, not in terms of this "construction."

Start with the empty set. We define it as "the set of all objects not equal to themselves," since there are no such objects. All empty sets have the same members—no members at all! Therefore, as sets they're identical, by definition of identity of sets. In other words, there's only one empty set. This unique empty set is our building block. Next comes the set whose only member is—the empty set. This set is *not* empty. Think of a hat sitting on a table—an empty hat. An example of an empty set. Then put the hat into a box. The hat is still empty. The box containing one thing—an empty hat—is an example of a set whose single member is an empty set. We say the contents of the box has cardinality 1. We have so far two entities, the empty hat and the box containing the empty hat. Now put box and another hat together into a bigger box. The contents of the bigger box has cardinality 2. The interested reader can now construct sets with cardinality three, four, and so on. From an empty set we construct the natural number system!

How the Rational Numbers Are Dense but Countable

The natural numbers are discrete—each is separated from its two nearest neighbors by steps of size 1. On the other hand, the rational numbers (fractions) are "dense." Between any two you can find a third—the average of the two. Repeat the argument, and you see that between any two rationals there are infinitely many. (A fact intensely irritating to Ludwig Wittgenstein. He called it "a dangerous illusion." See Chapter 11.)

This seems to mean there are many more rationals than naturals. But that's not true. There are just as many!

Georg Cantor thought of a simple way to associate the rationals to the naturals. To each natural a rational, to each rational a natural.

Arrange the rational numbers in rows according to denominators. In the first row, all the fractions with denominator 1, numerators in increasing order:

$$\frac{0}{1}, \quad \frac{1}{1}, \quad \frac{2}{1}, \quad \text{and so on.}$$

In the second row, the fractions with denominator 2, numerators in increasing order:

$$\frac{0}{2}, \quad \frac{1}{2}, \quad \frac{2}{2}, \quad \text{and so on.}$$

Each row is endless, and the succession of rows is also endless.

Starting in the upper left corner at 0/1, draw a zigzag line: go down one step, then go diagonally up and to the right to the top row (with ones in the denominator). Go one step to the right, then go diagonally down and to the left to the first column (the fractions with 0 in the numerator). Go another step down, and diagonally up and right again, and so on and on. This jagged line passes exactly once through every fraction in the doubly infinite array. That means you've arranged the fractions in linear order. There's now a first, a second, a third, and so on. Every rational number appears many times in this array (only once in lowest terms), so we have a mapping of the rational numbers onto a subset of the natural numbers. We describe this relationship by saying the rationals are countable.

Yet the real numbers, obtained by filling in the gaps in the rationals, are uncountable!

How the Real Numbers Are Uncountable

The basic infinite set is N, the natural or counting numbers. Many other sets can be matched one to one with N—for example, the even or odd numbers, the squares, cubes, or any other power, the positive and negative integers, and even the rationals, as explained in the previous article. Therefore it comes as a shock that the *real* numbers can't be put in one-to-one correspondence with the naturals. Any attempt to make a list of the real numbers is bound to leave some out!

The proof is simple.

Any real number can be written as an infinite decimal, like

3.14159 . . .

From any list of real numbers written as infinite decimals, Cantor found a way to produce another number *not on the list*. It doesn't matter how the list was constructed. So all the real numbers can never be written in a list.

How does Cantor produce his unlisted number? Step by step. It is an infinite decimal, constructed one digit at a time.

Look at the *first* real number on the list. Look at its *first* digit. Choose some other number from 0 to 9—any other number. That's the first digit in your new, unlisted number. Now go to the *second* real number on the list. Look at its *second* digit. Choose any other number from 0 to 9. That's the second digit of your new, unlisted number. And so on. The n'th digit of your new unlisted number is obtained by looking at the n'th digit of the nth real number on the list, and picking some *other* number for the nth digit in your new, unlisted number.

This construction doesn't terminate. But in calculus a number is well-defined if you can approximate it with *arbitrarily high* accuracy. By going out far enough in its decimal expansion, you approximate the unlisted number as accurately as you wish.

How do you know the new number isn't on the original list? It can't be the first number on the list, because they differ in the first digit. It can't be the second number on the list, because they differ in the second digit. No matter what n you choose, your new number isn't the n'th number on the list, because they differ in the n'th digit. The new number can't be the same as any number on the list! It's not on the list!

Geometry

What's "Between"? What's "Straight"?

What is the "straightness" of the straight line? There's more in this notion than we know, more than we can state in words or formulas. Here's an instance of this "more."

a, b, c, d are points on a line. b is between a and c. c is between b and d.

What about a, b, and d? How are they arranged?

It won't take you long to see that *b has to be between a and d.*

This simple conclusion, amazingly, can't be proved from Euclid's axioms! It needs to be added, an additional axiom in Euclidean plane geometry. This oversight by Euclid wasn't noticed until 1882 (by Moritz Pasch). A gap in Euclid's proof was overlooked for 2000 years!*

Some theorems in Euclid require Pasch's axiom. Without it, the proof is incomplete. The intuitive notion of the line segment wasn't completely described by the axioms meant to describe it.

More recently, the Norwegian logician Thorolf Skolem discovered mathematical structures that satisfy the axioms of arithmetic, but are much larger and more complicated than the system of natural numbers. These nonstandard arithmetics include *infinitely large integers.* In reasoning about the natural numbers, we rely on our mental picture to exclude infinities. Skolem's discovery shows that there's more in that picture than is stated in the Dedekind-Peano axioms. In the same way, in reasoning about plane geometry, mathematicians used intuitions that were not fully captured by Euclid's axioms.

The conclusion that b is between a and d is trivial. You see it must be so by just drawing a little picture. Arrange the dots according to directions, and you see b has to be between a and d. You're using a pencil line on paper to find a property of the ideal line, the mathematical line. What could be simpler?

But there are difficulties. The mathematical line isn't quite the same as your pencil line. Your pencil line has thickness, color, weight not shared by the mathematical line. In using the pencil line to reason about the mathematical line, how can you be sure you're using *only* those properties of the pencil line that the mathematical line shares?

In the figure for Pasch's axiom, we put a, b, c, and d *somewhere* and get our picture. What if we put the dots in other positions? How can we be sure the answer would be the same, "b is between a and d"? We draw one picture, and we believe it represents all possible pictures. What makes us think so?

The answer has to do with our sharing a definite intuitive notion, about which we have reliable knowledge. But our knowledge of this intuitive notion

* H. Guggenheimer showed that another version of Pasch's axiom can be derived as a theorem using Euclid's fifth postulate.

isn't complete—not even implicitly, in the sense of a base from which we could derive complete information.

A few simple questions to ponder while shaving or when stuck in traffic:

Is "straight line" a mathematical concept?

When you walk a straight line are you doing math?

When you *think* about a straight line, are you doing math?

Appletown, Beantown, and Crabtown are situated on a north-south straight line.

Must one be between the other two? Can more than one of the three be between two others? How do you know? Can you prove it?

Dogtown, Eggtown, and Flytown are on a *circle*, center at Grubtown. On that circle, must one be between the other two? Can more than one be between two others? How do you know? Could it be proved?

Is a straight line something you know from observation? From a definition in a book? Or how?

Is it something in your head?

Is the straight line in your head the same as the one in my head? Could we find out?

Is Euclid's straight line the same as Einstein's?

Is the straight line of a great-grandma in the interior of New Guinea the same as Hillary Rodham Clinton's? If Hillary Clinton visited her and they had a common language, could she find out?

Euclid's Alternate Angle Theorem

This is the first part of theorem 29, Book 1 of Euclid. "A straight line falling on parallel straight lines makes the alternate angles equal to one another." *Proof*: Let AB and CD be parallel. Let EF cross them, intersecting AB at G and CD at H. We claim the alternate angles AGH and GHD are equal, for they are both supplementary to angle CHG (adding to two right angles). For by construction

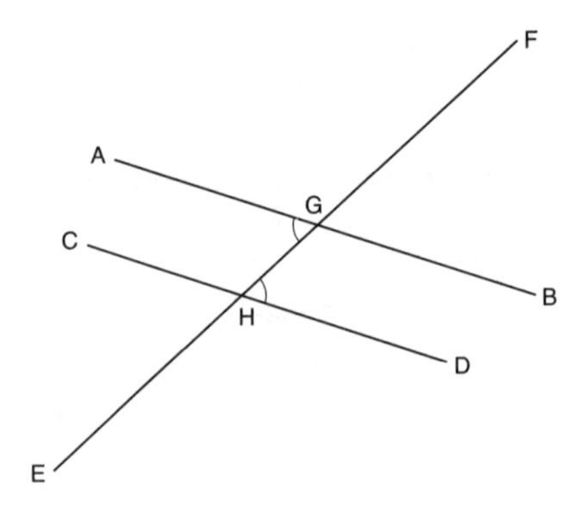

Figure 3. Alternate angles.

CHG and DHG add up to a straight angle, or two right angles. And by Euclid's fifth postulate (his definition of parallel lines) AB and CD parallel means the interior angles AGH and CHG add to two right angles.

$$\text{CHG} + \text{DHG} = 2\text{R} \qquad \text{CHG} + \text{AGH} = 2\text{R}$$
$$\text{DHG} = 2\text{R} - \text{CHG} = \text{AGH}$$

The proof is complete.

Euclid's Angle Sum Theorem

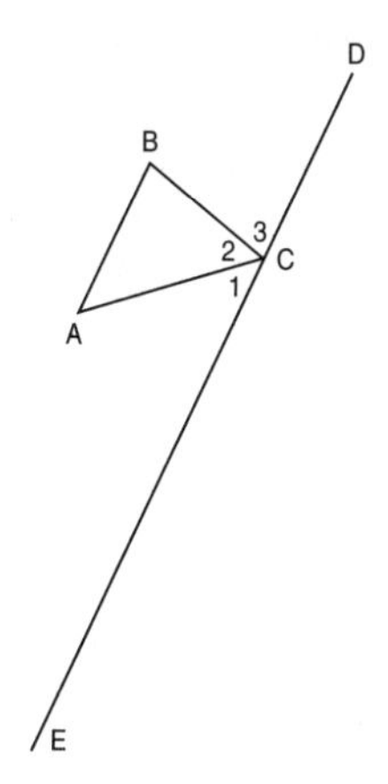

Figure 4.

In an arbitrary triangle ABC, choose a vertex, say C. Through C draw a line DCE parallel to AB. At C the three angles 1, 2, 3 add up to the sum of two right angles (180 degrees). Angle 2 is the same as angle C in triangle ABC. Angle 1 equals angle A, since they are alternate angles between two parallel lines (using Euclid's alternate angle theorem proved above). Similarly, angle 3 equals angle B. Adding,

Angle A + Angle B + Angle C =
Angle 1 + Angle 2 + Angle 3
= two right angles, q.e.d.

The Triangle Inequality

Here's an inequality valid for any six real numbers a, b, c, d, e, f:

$$\sqrt{(a-c)^2 + (b-d)^2} \leq \sqrt{(a-e)^2 + (b-f)^2} + \sqrt{(c-e)^2 + (d-f)^2}.$$

This algebraic inequality has a geometric name—the "triangle inequality."

Why?

Let the three pairs (a,b), (c,d), (e,f) be rectangular coordinates of three points P, Q, and R in the plane. Then this inequality says the distance from P to Q is less or equal the distance from P to R plus the distance from R to Q.

If P, Q, and R are vertices of a triangle, the last statement says any side of a triangle is shorter or equal to the sum of the other sides. This is the triangle inequality—than which nothing could be visually more obvious. "A straight line is the shortest distance between two points."

The complicated-looking formula is the translation into algebra of this simple geometric fact. The geometric fact "motivates" the algebraic formula. One can, with effort, give an algebraic proof of the algebraic formula, and thereby give a complicated proof of a very simple geometric fact.

But you could just as well turn the procedure around. The triangle inequality is geometrically evident. Therefore its complicated-looking algebraic statement is also true. To prove the messy algebraic inequality, use its geometric interpretation with its simple visual proof.

What's Non-Euclidean Geometry?

(See the sections on Certainty, Chapter 4, and Kant, Chapter 7.) The fifth axiom of Euclid's *Elements*, the parallel postulate, was long considered a stain on the fair cheek of geometry. This postulate says: "If a line A crossing two lines B and C makes the sum of the interior angles on one side of A less than two right angles, then B and C meet on that side."

The usual version in geometry textbooks is credited to an English mathematician named Playfair: "Through any point P not on a given line L there passes exactly one line parallel to L." This is equivalent, and easier to understand.

This parallel postulate was true, everybody agreed. Yet it wasn't as self-evident as the other axioms. Euclid's version says that something happens, but perhaps very far away, where our intuition isn't as clear as nearby. From Ptolemy to Legendre, mathematicians tried to prove the parallel postulate. No one succeeded.

Many so-called "proofs" were found. But each "proof" depended on some "obvious" principle which was only a disguised version of the parallel postulate. Posidonius and Geminus assumed there is a pair of coplanar lines everywhere equally distant from each other. Lambert and Clairaut assumed that if in a quadrilateral three angles are right angles, the fourth angle is also a right angle. Gauss assumed that there are triangles of arbitrarily large area. Each of these different-sounding hypotheses is equivalent to the fifth postulate

In the early nineteenth century Gauss, Lobachevsky, and Bolyai all had the same idea: Suppose the fifth postulate is *false*!

Euclid's axiom can be replaced in two different ways. Either "Through P pass *more than one* line parallel to L" or "Through P pass *no* lines parallel to L." The first is called "the postulate of the acute angle." The second is "the postulate of the obtuse angle." These postulates generate two different non-Euclidean geometries, called "hyperbolic" and "elliptic." The hyperbolic was studied first, and is often referred to as just "non-Euclidean geometry."

An elegant contrast between the three geometries mentioned already in Chapter 4 is the sum of the angles in a triangle. In Euclidean geometry, as we proved above, the sum equals two right angles. In elliptic geometry, the sum of the angles of every triangle is *more* than two right angles. And in hyperbolic geometry it's *less.*

Gauss was the earliest of the three discoverers. As I mentioned earlier, in the section on Kant, he didn't publish his work, to avoid "howls from the Boeotians." In classical Athens, "Boeotions" meant "ignorant hicks." To Gauss, it meant perhaps "followers of Kant." They would say non-Euclidean geometry is nonsense, since Kant proved there can be no geometry but Euclid's.

Gauss's fear was justified. When non-Euclidean geometry became public, Kantian philosophers did say it wasn't really geometry. One of them was Gottlob Frege, the founder of modern logic.

Before Gauss, deep penetrations into the problem had been made by the Italian Jesuit priest Saccheri, by Lagrange, and by Johann Heinrich Lambert (1728–1777), a leading German mathematician who was an acquaintance or friend of Kant. Decades before Gauss, Lambert wrote:

> Under the (hypothesis of the acute angle) we would have an absolute measure of length for every line, of area for every surface and of volume for every physical space. . . . There is something exquisite about this consequence, something that makes one wish that the third hypothesis be true! In spite of this gain I would not want it to be so, for this would result in countless inconveniences. Trigonometric tables would be infinitely large, similarity and proportionality of figures would be entirely absent, no figure could be imagined in any but its absolute magnitude, astronomers would have a hard time, and so on. But all these are arguments dictated by love and hate, which must have no place either in geometry or in science as a whole. . . . I should almost conclude that the third hypothesis holds on some imaginary sphere. At least there must be something that accounts for the fact that, unlike the second hypothesis (of the obtuse angle), it has for so long resisted refutation on planes.

It's astonishing that Lambert actually gives the acute angle hypothesis a fair chance. The issue is to be decided by mathematical reasoning, not by "universal intuition." He honestly contemplates the possibility that a non-Euclidean geometry may be valid. His very ability to do so refutes Kant's universal innate Euclidean intuition.

In the end, Lambert slips into the same ditch as Legendre and Saccheri. He "proves" the Euclidean postulate by getting a "contradiction" out of the acute angle postulate. Laptev exposes Lambert's fallacy.

Beltrami, Klein, and Poincaré constructed models that showed that Euclidean and non-Euclidean geometry are "equiconsistent." If either one is consistent, so is the other. Since no one doubts that Euclidean geometry is consistent, non-Euclidean is also believed to be consistent.

Kant said that only one geometry is thinkable (see Chapter 7). But the establishment of non-Euclidean geometry offers a choice between several geometries. Which works best in physics? The choice must be empirical, to be settled by observation.

It's tempting to simply declare that "obviously" or "intuitively" Euclid is correct. This was not believed by Gauss. There's a legend that he tried to settle the question by measuring angles of a gigantic triangle whose vertices were three mountain tops. (The larger the triangle, the likelier that there would be a measurable deviation from Euclideanness.) Supposedly the measurement was inconclusive. Perhaps the triangle wasn't big enough.

What's a Rotation Group?

A "group" is a closed collection of reversible actions. For instance, multiplication by the positive real numbers is a group, since the product or quotient of two positive real numbers is a positive real number. The set of rotations in 3 dimensions is a group. Motions of your arm can be thought of as rotations around your shoulder and your elbow. So awareness of how your arm moves is an intuitive acquaintance with the 3-dimensional rotation group.

The Four-Color Theorem

In political maps, countries that share a border of positive length (not just some isolated points) are required to have different colors. It turns out that four colors always suffice to meet this requirement. This was stated as a mathematical conjecture in 1852. It was first proved by Haken and Appel in 1976. They broke the problem into a great many arduous calculations, which were performed on a computer. There followed discussion and dispute on whether this way of proving was new and different in mathematics. (See article on proof, Chapter 4.)

Two Bizarre Curves

A function has a curve as its graph; and a curve (subject to mild restrictions) is the graph of a function. Today we teach the function as primary. The graph is derived from the function. Until a hundred years ago or so, it was the other way around. As a geometric object, the curve was part of the best understood branch of mathematics. Functions leaned on geometry. Mathematicians were upset when, late in the nineteenth century, they learned of functions with wild graphs impossible to visualize. Example 1 is the Riemann-Weierstrass curve. It's continuous, but at every point it has no direction! Example 2 is the Peano-Hilbert curve. It fills a two-dimensional region—actually passes through *every* point of a square.

I'll give brief sketches of these monsters.

For example 1, van der Waerden's construction is simpler than Riemann or Weierstrass. Start with two connected line segments in the x-y plane. The piece on the left has slope 1 and rises from the x-axis to height 1. There it meets the second piece, which descends back down to the x-axis with slope –1. At the corner where the two segments meet, the slope is undefined.

From this first step, define a second with two peaks, having slope twice that of the first, but height or "amplitude" only half as great. It oscillates twice as fast as the first, but rises only half as high.

In this manner define a sequence of graphs made of connected line segments, each half as high and twice as steep as the previous one, with corners half as far apart, or twice as frequent.

Then—add them up!

The sum converges, because the terms are getting smaller in a ratio of 1:2. As you add more and more terms, the corners get closer and closer, and the slope in between gets bigger and bigger. In the limit, the corners are dense, and the slope in between is infinite. There is no direction.

In example two, start with a square. Cut it into four subsquares, then 16 subsquares, then $256 = 16^2$ subsquares, and so on. At each stage, draw a broken line (polygonal curve) connecting the centers of all the subsquares. You obtain a polygonal line through the centers of many small squares that cover the whole original square. This sequence of polygonal curves converges to a limit curve, which actually passes through every point of the original square.

Square Circles

> Mad Mathesis alone was unconfin'd,
> Too mad for mere material chains to bind,
> Now to pure Space lifts her ecstatic stare,
> Now running round the circle finds it square.
>
> —Alexander Pope, *The Dunciad, Book IV*

This article is an imaginary conversation with David Hume, who rashly presumed there could be no such thing as a square circle (see Chapter 10).

We are not concerned here with the classic Greek problem, proved in modern times to be unsolvable, of constructing with ruler and compass a square with area equal to that of a given circle. By "circle" we mean, as usual, a plane figure in which every point has a fixed distance (the radius) from a fixed point (the center.)

Suppose I live in a flattened, building-less war zone. Transportation is by taxi. Taxis charge a dollar a mile. There are no buildings, so they can run anywhere, but for safety, they're required to stick to the four principal directions: east, west, north, and south.

People measure distance by taxi fare. If two points are on the same east-west or north-south line, the fare in dollars equals the straight-line distance in miles. Otherwise, the fare in dollars equals the shortest distance in miles, traveling only east-west and north-south.

The taxi company has a map showing the points where you can go for $1. These points form a square, with corners a mile north, south, east, and west of the taxi office. In the taxicab metric, *this square is a circle*—it's the set of points $1 from the center.

Yes, a square circle! Inconceivable, yet here it is!

But Hume says, "You can't just change the meaning of 'distance' that way. You know I mean the regular Euclidean distance."

"Very well, David. What's a square?"

"A quadrilateral with equal sides and equal angles."

"Fine. Take your regular Euclidean circle, and inscribe four equally spaced points on it. Then the circumference is divided into four equal sides, and they all meet at the same angle, 180 degrees. Another square circle!"

"No, no!" cries Hume in exasperation. "I mean a quadrilateral with straight sides, not curved sides!"

"O.K. Let the regular Euclidean circle be the equator of the earth. That's a great circle, as straight as a line can be, here on earth. Doesn't *that* have four equal *straight* sides and four equal angles?"

"No, no, no! The four equal angles have to be *right* angles!" shouts Hume.

"That wasn't part of your definition."

"Any way," says he, "call the equator straight if you like, but it isn't! It's a circle! No line on the surface of the earth is straight."

"What's this? You say that what the Mind can conceive is possible, and what the Mind can't conceive is impossible. Now you tell me there's no straight line on the surface of the earth! I grant your mind conceives that, but most minds can't conceive it. Either give up geometry, or give up your notion that what you can't conceive is impossible."

To carry the argument a step further, I leave David Hume behind, and introduce the equation

$$(x^p) + (y^p) = 1.$$

Here x and y are the usual rectangular "Cartesian" coordinates. To avoid irrelevant complications, take x and y in their absolute values, so the graphs are symmetric in the x- and y-axes. p is an arbitrary positive real number, a parameter.

For each different value of p we have a different equation and a different graph in the x - y plane.

If $p = 2$, $x^2 + y^2 = 1$

The graph is the familiar Euclidean circle. For any p bigger than 1 the graph is a smooth convex curve passing through four special points:

$(1, 1), (-1, 1), (1, -1), (-1, -1)$.

This curve is the unit circle in a new metric, where distance from the origin to the point (x,y) is defined as

the p'th root of $(x^p + y^p)$.

Two cases are especially simple:

$p = 1$ and p infinite.
For $p = 1$,

the graph is exactly the unit square of the "taxicab metric" defined above!

For p infinite, the graph is a larger square, with horizontal and vertical sides: the four lines

$x = +1, \quad x = -1, \quad y = +1, \quad y = -1.$

The square for $p = 1$ is inscribed in the square for p infinite. Its corners are at the midpoints of the sides of the larger square.

Call the small square inner, and the large one outer. If you let p increase, starting with $p = 1$, the graph on your computer's monitor will expand smoothly from the inner square through a family of smooth convex curves to become the outer square. A student who watched this transformation could inform Hume, "We have infinitely many unit circles of various shapes. The first and the last are square!"

Embedded Minimal Surfaces

Soap bubbles and soap films are constrained by a physical force called "surface tension." This force makes the surface area as small as possible, subject to appropriate side conditions. In a bubble, the side condition is the volume occupied by the air inside. In a soap film, the side condition is where the film is attached to something—a bubble pipe, another bubble, or the fingers of a child.

It's easy to guess or observe that for a soap bubble the minimal surface is a sphere. For a soap film, with endless different possible boundaries, the problem is more complicated.

One basic fact is clear. In order to have minimal area "in the large" (globally) the film must have minimal area "in the small" (locally). That is, if you mentally mark out any small simple closed curve in the soap film, the area of film enclosed by that curve must be the smallest area that can be enclosed by that curve.

Because of this "local" property, the soap film surface satisfies a certain complicated-looking partial differential equation discovered by Joseph-Louis Lagrange in 1760.

This suggests the famous "Plateau's problem": Given a curve in 3-space, construct a soap film (a minimal surface) having that curve as boundary. J. A. F. Plateau was a blind Belgian physicist who made the problem known to the world in 1873. Jesse Douglas, a New York mathematician, won a Fields Medal in 1936 for his solution.

In the late nineteenth-century Karl Weierstrass, Bernhard Riemann, Hermann Amandus Schwarz, and A. Enneper discovered a number of interesting new minimal surfaces. For years the corridors of university mathematics departments were lined with glass-fronted cabinets displaying plaster models of these surfaces.

It turned out that a mathematical minimal surface need not be a physical one. The mathematical conditions permit the surface to intersect itself, which soap film doesn't ordinarily do. A surface is called "embedded" if it has no self-intersection.

The use of computer graphics, starting some fifteen years ago, has revitalized the subject by making possible the visualization of complicated surfaces formerly described only by equations. The computer pictures often reveal instantly whether a surface has self-intersections, and may show other properties of the surface that can be used to provide rigorous proofs. An infinite number of new complete embedded minimal surfaces have been found in this way. David Hoffman and Jim Hoffman, whose computer-generated video is the basis of our dust-jacket picture, have been leaders in this work.

How Imaginary Becomes Reality

In these notes we have already "constructed" the natural numbers from the empty set. Now we go the rest of the way. Step by step, we construct the integers (positive and negative whole numbers); then the fractions or rational numbers; then the real numbers, rational and irrational; and at last the complex numbers. I show five different ways to construct the complex numbers! And then we go even further, to exotic creatures called quaternions. These extensions of number systems show where the axiom-theorem model is misleading. We don't just obey axioms, we modify them.

Creating the Integers

From the natural numbers we wish to construct the integers—the natural numbers *plus* zero *plus* the negative whole numbers. We can subtract one natural number from another—for instance, 3 from 7—if the former isn't bigger than the latter. But with the natural numbers, we can't subtract a larger from a smaller. "Seven from three you can't take," say the first-graders.

Mathematicians have two ways to extend a mathematical system. One is brute force—create a needed object by fiat. For instance, there's no natural number x such that

$$x + 1 = 0.$$

But we might need such a number. (For instance, to keep track of money we owe.) No problem. Just make up a symbol: -1 and state *as a definition* that

$$-1 + 1 = 0.$$

That's how it's done in school.

But can we really create what doesn't exist, just by definition? Who gave us a license for such presumption?

The fact is, it's not necessary to "create" anything. We can use a more sophisticated approach known as "equivalence classes." Since we want -1 to equal $(0 - 1)$ and $(1 - 2)$ and $(2 - 3)$ and so on, we just collect all these ordered pairs into a great big "equivalence class":

$$\{ (0 , 1), (1, 2), (2, 3) \ldots \}$$

It includes infinitely many elements; each element is an ordered pair of natural numbers. Yet it deserves and has the name you expect: -1.

My definition of equivalence class was vague. I just wrote down three members of this infinite class "to give you the idea," and then wrote three dots. . . . I can't possibly write the complete list. We need a membership rule for the class.

We find this by elementary-school arithmetic. When does

$$a - b = c - d$$

without using minus signs? Of course, when

$$a + d = b + c.$$

So we make a precise definition of equivalence:

$(a - b)$ is equivalent to $(c - d)$ if and only if
$a + d = b + c$.

To make a fresh start, not depending on a fiat, we temporarily give up the expression $a - b$, and instead write $\{a,b\}$—an ordered pair in curly brackets.

$$\{a, b\} = \{c, d\} \text{ if and only if } a + d = b + c.$$

We have to figure out how to add, subtract, and multiply the equivalence class of $\{a,b\}$ and the equivalence class of $\{c,d\}$. (Division isn't generally possible, since we don't have fractions yet.) Now, from junior high school we know the rules to add, subtract, and multiply positive and negative whole numbers. It's simple to rewrite those rules in terms of ordered pairs in curly brackets. You can do it, if you wish.

One class plays a special role:

$$\{ (0, 0), (1, 1) \ldots \}.$$

It's the additive identity, and has a special name—zero. But that name is already in use, for the whole number before 1. So we allow the same word to have two meanings.

We say two classes are "negatives" of each other, or "additive inverses," if their sum is zero. You can easily check that for any natural numbers a and b, the class of (a, b) is the negative of the class of (b, a). That is to say,

$$\{a, b\} + \{b, a\} = \{0, 0\}.$$

This means, in particular, that each equivalence class has one and only one negative or additive inverse.

In every equivalence class, there's exactly one ordered pair that includes the natural number 0. If that pair is {a, 0}, we call that equivalence class "a." (It's in a sense the "same" as the natural number a.) And if that pair is {0, a}, we call it −a.

We Have Constructed the Integers, Including the Negative Numbers!

"Constructed" out of some infinite sets, to be sure. Rather than ordered into existence "by fiat."

Why $-1 \times -1 = 1$

In extending from positive whole numbers to integers we preserve *all* the rules of arithmetic except *one*. The rule we give up is:

No number comes before 0.

But we still have:

Rule A $\quad a \times 0 = 0 \times a = 0$, for all a.

So in particular

Rule A′ $\quad -1 \times 0 = 0.$

And we still have

Rule B $\quad a \times 1 = 1 \times a = a$, for all a.

So in particular

Rule B′ $\quad -1 \times 1 = 1 \times -1 = -1$

And we still have the distributive law:

Rule C $x \times (y + z) = (x \times y) + (x \times z)$.

Consider

$$(-1) \times (-1 + 1).$$

This is -1×0, which is zero, by Rule A'. On the other hand, by Rule C,

$$(-1) \times (-1 + 1) = (-1 \times -1) + (-1 \times 1).$$

The last term on the right equals -1, by Rule B'. So

$$0 = (-1 \times -1) + (-1).$$

That is to say,

$$(-1 \times -1)$$

is the additive inverse of -1. But -1 has a unique additive inverse: 1.

So -1×-1 is 1, as we claimed.[3]

Creating the Rationals

In the course of human progress people acquired property and money. So they needed fractions. When a baker had a whole loaf and a half, he had to know he had three halves to sell. Anybody can add

$$\frac{1}{2} + \frac{1}{2} = \frac{2}{2} = 1.$$

Confusion arises with improper fractions:

$$\frac{4}{5} + \frac{4}{5} = \frac{8}{5},$$

and still worse with unequal divisors:

$$\frac{1}{3} + \frac{1}{8} = ??.$$

But people did manage to extend addition and multiplication to fractions (both positive and negative.) This enlarged system is called "the rational numbers." ("Rational" meaning, not "reasonable" or "logical," but just *ratio* of whole

[3] Thanks to Howard Gruber for suggesting this example.

numbers.) With these numbers, any problem of addition, subtraction, multiplication, or *division* (except by zero) has a solution.

We pay a penalty for this enlargement. The natural numbers are ordered "discretely"—every one has a unique follower, and all but 1 have a unique predecessor. This beautiful property makes possible a powerful method—"proof by induction." It's no longer true for the rationals.

Ordinarily we write fractions as a/b, but I will temporarily write them as ordered pairs (in *square* brackets, [a,b]).

Since

$$\frac{1}{2} = \frac{2}{4} = \frac{3}{6},$$

the rule now is,

$$[a,b] = [c,d] \text{ (or } a/b = c/d) \text{ if } ad = bc.$$

This defines "equivalence classes of pairs of integers"—what we call "equal fractions" in the fourth grade. The ones we used when we practiced reducing fractions to lowest terms.

With fractions, as with negatives, we need rules for calculating with these new ordered pairs. And again, we just take the known rules for fractions and rewrite them in terms of ordered pairs in square brackets. We have constructed the rational numbers! Jacob Klein shows that to the Greeks, number meant "positive whole number greater or equal to 2." Number 1 wasn't like other numbers. Fractions were a commercial and practical necessity, but they weren't *numbers.* Klein writes that the broadening of "number" to include positive fractions took place only in the late middle ages and early Renaissance, and with difficulty.

Why $\sqrt{2}$ *Is irrational*

The most famous theorem in Euclid is the "Pythagorean": "In any right triangle, the sum of the squares of the lengths of the two shorter sides equals the square of the length of the long side" (the "hypotenuse"). You can construct a pair of right triangles by drawing a diagonal in a square of side 1. Then Pythagoras's theorem says the diagonal has length $\sqrt{2}$. On the other hand, the Pythagoreans also discovered that *there is no ratio equal to* $\sqrt{2}$!

Since it doesn't exist, there's nothing to exhibit or construct. All the proof can do is show that the presumption such a ratio exists is absurd. This is called "indirect proof." Suppose that for some pair of numbers p and q,

$$(p/q)^2 = 2.$$

If so, p/q can be put in "lowest terms"—p and q should have no common factor. In particular, they don't both have 2 as a factor—they aren't both even. Multiplying both sides by q^2 gives

$$p^2 = 2\,q^2.$$

A factor 2 is visible on the right side, so the right side of the equation is an even number. Therefore p^2, the left side of the equation, is also even. It's easy to check that the square of an even number is always even, the square of an odd number always odd. Since p^2 is even, p is even. That means p is twice some other whole number. Let's call it r, so $p = 2\,r$. Then

$$p^2 = 4\,r^2.$$

We replace p^2 by $4\,r^2$ in the previous equation, and get

$$4\,r^2 = 2\,q^2,$$

which simplifies to

$$2\,r^2 = q^2.$$

This is just like the equation $p^2 = 2\,q^2$ we started with, but p is replaced by q and q by r. So the same argument as before proves that q is even, as p was proved to be even. But p and q aren't allowed to both be even. CONTRADICTION!

The contradiction shows that the presumption that such a fraction p/q exists is impossible, or absurd.

If we only have whole numbers and ratios, we're stuck with the conclusion that $\sqrt{2}$ doesn't exist. It exists as a line segment, the diagonal of the unit square, but not as a number. The diagonal of the unit square does not have a length! Yet, using operations of Euclidean geometry, wc can add it to other line segments, and also subtract, multiply, and divide. Line segments constitute an arithmetical system richer than the system of arithmetical numbers! This impasse suggests we go beyond the rational numbers. We need a theory of irrational numbers.

Creating the Real Numbers—Dedekind's Cut

So we want the "real numbers"—rationals and irrationals together. (The name "real" is in contrast to the imaginary and complex numbers, which we will meet shortly.) We use $\sqrt{2}$ to motivate our construction of the irrationals. No rational number when squared can equal 2 (proof is above). Yet we can approximate $\sqrt{2}$:

1
1.4
1.41
1.414

and so on, as far as our computing budget permits. This sequence converges, but what does it converge *to*? $\sqrt{2}$, naturally. But what *is* $\sqrt{2}$, if it can't be a rational number?

We want mathematics to include $\sqrt{2}$ —and many other irrational numbers, of course. We have to somehow take such "convergent" sequences of rationals, which don't have rational limits, and make them into numbers—"real numbers."

Georg Cantor, Karl Weierstrass and Richard Dedekind each found a way to do this. Dedekind's is especially easy.

Arrange the rational numbers in a row or a line in the usual way, increasing from negative to positive as you go from left to right. By a "cut" Dedekind means a separation of this row into two pieces, one on the left, one on the right. The row can be cut in infinitely many different places. Dedekind regards such a split or "cut" in the rationals as being a new kind of number! He shows in a natural way how to add, subtract, multiply, or divide any two cuts (not dividing by zero, of course). In an equally natural way, he defines the relation "less than" for cuts, and the limit of a sequence of cuts. Once these rules of calculation are laid out, the cuts are established as a number system.

Every rational number x defines an associated cut. The left piece is simply the set of rational numbers less or equal x, and the right piece is the set of rationals greater than x. By this association between cuts and rational numbers, we make the rational numbers a subsystem of the system of cuts. To identify Dedekind cuts as the sought-for "real number system," we must show that they include *all* the rationals and irrationals—all the numbers that can be approximated with arbitrary accuracy by rationals.

I'll be satisfied to show that one particular irrational is included as a Dedekind cut— $\sqrt{2}$. To do so, I must identify a left half-line and right half-line associated with $\sqrt{2}$. What rationals are less than $\sqrt{2}$? Certainly all the negative ones, and also all those whose squares are less than 2. All numbers x such that either $x < 0$ or $x^2 < 2$. That specifies the left piece of the cut, the left half-line associated to $\sqrt{2}$. Its complement is the corresponding right half-line. It's easily verified that when this cut is multiplied by itself, it produces the cut identified with the rational number 2. Among Dedekind cuts 2 does have a square root!

All that's left to prove is that no numbers are missing. Dedekind's cuts provide a limit for every convergent sequence of rationals, but we need more. We need a limit for every convergent sequence of *real* numbers—every convergent sequence of cuts. This property is called completeness. The proof is in every text on real analysis and many texts on advanced calculus. I give the essence of it. Let a_n be a convergent sequence of Dedekind cuts (real numbers.) We want to produce a cut *a* which is the limit of this sequence. We know that every cut a_n is the limit (in many ways) of a convergent sequence of rational numbers. So we replace each a_n by an approximating rational number, choosing the rational approximation more and more accurately as we go out in the a_n sequence. This is easily shown to be a convergent sequence of rationals, and it's easily shown that its limit cut is the limit of the original sequence of cuts.

These constructions are "existence proofs." If you believe Dedekind cuts exist, you have proved that the real numbers exist.

What's the Square Root of −1?

Does $\sqrt{-1}$ exist? There's no real number that yields −1 when squared. That's the reason we say $\sqrt{-1}$ doesn't exist.

Yet in our next breath we bring it into existence!

I'll show you five different ways to do it.

The simplest way is the high-school way. Just *define i* as a "quantity" that obeys the laws of arithmetic and algebra in all respects, except that

$$i^2 = -1.$$

If you wish, instead of a "quantity" you may call it a "symbol," which *by definition* satisfies

$$i^2 = -1.$$

This approach is direct. It is clear cut. i is treated algebraically like any "letter" or "indeterminate." It can be added and multiplied. These operations and their inverses obey the same commutative, distributive, and associative laws as the real numbers do. The only difference is

$$i^2 = -1.$$

Real multiples of i, like $2i$ or $-3i$, are called "imaginary" or "pure imaginary." Numbers of the form $z = x + iy$, where x and y are real numbers, are called "complex." x is called "the real part" of z, and y is "the imaginary part."

Either x or y or both can be 0, so the imaginary numbers and the real numbers are among the complex numbers! (0 is the only complex number that is both real and imaginary.) This shouldn't be a shock. The positive and negative whole numbers (the integers) are among the rational numbers (fractions.) When we enlarge a number system, we want the numbers we start with to be included among the numbers we "construct."

But since no real number satisfies $x^2 = -1$, is it legitimate to simply "introduce" the square root of −1? Isn't this cheating?

We've seen that pretending some number equals 1/0 leads to disaster. If 1/0 is fatal, how can we be sure $\sqrt{-1}$ is O.K.?

One answer might be that analysis with complex numbers is a powerful theory that has never led to a contradiction. That would be saying, "We never had trouble so far, so we never will have trouble." A dubious defense. To resolve such worries, we renounce "introducing" or "creating" the square root of −1. Instead, we'll *find* it, already there! As promised, we'll do it in five different ways.

1. A point in the x-y coordinate plane.
2. An ordered pair of real numbers.
3. A 2-by-2 matrix of real numbers.

4. An equivalence class of real polynomials.
5. In the Grand Universal Super-Structure of Sets.

1. After centuries of skepticism, mathematicians accepted complex numbers when they found them "already there," as points in the x-y plane. The complex number $3 + 4i$, for example, is associated to the point with coordinates $x = 3$, $y = 4$. In this way, every complex number gets a point in the coordinate plane, and every point in the plane gets a complex number.

Addition and multiplication of complex numbers turn out to be elementary geometric operations! Addition is just shifting. Adding $3 + 4i$, for example, shifts any complex number 3 units to the right and 4 units up.

Multiplying is stretching and turning. To see this, use polar coordinates. The "polar distance r" of a point $x + iy$ is its distance from the origin. For

$$3 + 4i,$$

the Pythagorean theorem gives $r = 5$.

The "polar angle Q" of a point is the angle between the positive x-axis and the ray from the origin to that point. Multiplying by $3 + 4i$ then turns out to be simply *multiplying* distance by $r = 5$ and *increasing* polar angle by Q.

For $i = 0 + 1i$, evidently $x = 0$ and $y = 1$. The point corresponding to i is on the (vertical) y-axis. So we call the y-axis the "imaginary axis." The "imaginary unit" i is there, one unit above the origin. The (horizontal) x-axis is the "real axis."

For the point i, polar distance r is 1, and polar angle Q is a right angle, 90 degrees.

Multiplying $i \times i$ results in *squaring* r and *doubling* Q.

Since $r = 1$, $r^2 = 1$.

Since Q is a right angle, 90 degrees, its double is two right angles—180 degrees.

This means that i^2 is on the x-axis (the real axis) one unit *left* of the origin. It has coordinates $(-1,0)$. Its complex number is $-1 + 0i$, or simply -1. We have demonstrated geometrically that

$$i^2 = -1.$$

That is, the point i or $0 + i$ is a square root of -1!

Since classical times geometry was the most venerated part of mathematics. Identifying the complex numbers with plane geometry made them respectable.

2. From a more critical viewpoint, something is still missing. The complex numbers are defined by laws of arithmetical operations. They're an independent *algebraic* system, defined prior to their geometric interpretation. We should give an *algebraic* proof of consistency. This was done by Ireland's greatest mathematician, William Rowan Hamilton (remembered also for quaternions, the Hamilton-Jacobi equations, and Hamiltonian systems of differential equations).

To construct the complex numbers, Hamilton creates from the real numbers a simple new kind of thing: an *ordered pair* of real numbers. This will look a lot like how we constructed the integers and the rationals—but historically Hamilton's construction of the complex numbers came first!

He defines equality of his ordered pairs:

$$(a, b) = (c, d) \text{ if and only if both } a = c \text{ and } b = d.$$

He defines addition in a very natural way:

$$(a, b) + (c, d) = (a + c, b + d)$$

Multiplication is more complicated:

$$(a, b) \times (c, d) = (ac - bd, ad + bc).$$

Hamilton didn't pull this multiplication rule out of thin air. He just translated the known multiplication of $(a + bi)$ times $(c + di)$ into his notation of ordered pairs. Seen this way, the whole performance looks trivial. But it gets rid of the suspicious i, and replaces it by the innocent $(0, 1)$. Please check that its square is $(-1, 0)$, which is $-1 + 0i$, which is -1.

One should verify the arithmetical laws that complex numbers share with real numbers: commutative laws of addition and multiplication, associative laws of addition and multiplication, and distributive law of multiplication over addition. These verifications are straightforward calculations that the interested reader can carry out.

Notice that ordered pairs whose second component is 0 behave just like real numbers. The zero in the second place never "gets in the way." The multiplicative identity is $(1,0)$; it's algebraically "the same" as 1, the multiplicative identity of the reals.

The pair $(-1, 0)$ is algebraically "the same" as the real number -1. The additive identity is $(0, 0)$; it's algebraically "the same" as the real number 0.

It's straightforward to define subtraction:

$$-(3, 4) = (-3, -4) \qquad -(a,b) = (-a, -b)$$

and to check that

$$(a, b) + (-(a, b)) = (0, 0).$$

Division is trickier. I'll save time by just telling you how to do it—you can check that it works. First, for the special example $(3,4)$,

$$\frac{1}{(3, 4)} = \frac{(3, -4)}{(3^2 + 4^2)} = \left(\frac{3}{25}, \frac{-4}{25}\right).$$

And in general,

$$\frac{1}{(a, b)} = \left(\frac{a}{[a^2 + b^2]} \quad , \quad \frac{-b}{[a^2 + b^2]}\right).$$

which you can check by multiplying

$$(a,b) \times \frac{1}{(a, b)}$$

using the multiplication rule given above. The answer is (1 , 0), or simply 1.

If the definitions of multiplication and division seem baffling, go back to the geometric interpretation of complex numbers to make them intuitively clear.

From a strict formal point of view, one oughtn't to write

$$a + 0i = a.$$

That's "equating apples and oranges." A single real number a just isn't the same as the pair of real numbers (a, 0). Instead of "=" one could say "is isomorphic to."

3. Another way to construct complex numbers uses 2 × 2 matrices of real numbers instead of ordered pairs. The complex number $a + bi$ corresponds to the matrix

$$\begin{pmatrix} a & b \\ -b & a \end{pmatrix}$$

If you know how to multiply 2 × 2 matrices, you can check that the usual rules of matrix algebra correspond to the usual rules of addition and multiplication of complex numbers. The number −1 corresponds to the matrix

$$\begin{pmatrix} -1 & 0 \\ 0 & -1 \end{pmatrix}$$

The matrix

$$\begin{pmatrix} 0 & 1 \\ -1 & 0 \end{pmatrix}$$

gives −1 when squared. We are entitled to call this matrix "i"!

We found a square root of −1 by interpreting −1 as a 2 × 2 matrix. What does this say about existence of $\sqrt{-1}$? It exists if you interpret −1 the right way!

4. A fourth way of finding $\sqrt{-1}$ is inspired by a branch of modern algebra called Galois theory, after Evariste Galois. He was a student, killed in a duel in 1838 at age 21, before being recognized as a precocious genius.

Instead of matrices or ordered pairs we use "polynomials with real coefficients." For instance,

$$5x^4 + 3x^3 + 7x^2 - x^1 + 5.$$

We divide all our polynomials by $x^2 + 1$. Why? Because the thing we're after, *i*, is a root of $x^2 + 1$!

As in division of numbers, so in division of these polynomials, we get a quotient and a remainder. And the remainder has *degree* less than the *degree* of the divisor. We're dividing by $x^2 + 1$—a second degree polynomial—so the remainder has degree 1 or 0. There might be *zero* remainder, or a constant remainder different from zero (a zero-degree polynomial), or a first-degree remainder—a polynomial of the simple form $ax + b$.

Two polynomials, whatever their degree, are *equivalent* if they have the *same remainder* on division by $x^2 + 1$. This equivalence splits the polynomials into equivalence classes—sets of polynomials having the same remainder on division by $x^2 + 1$.

An equivalence class is a sack. We're putting polynomials into sacks. All the polynomials in any sack have the same remainder, which is some polynomial of degree 1 or 0. The polynomials that are multiples of $x^2 + 1$, including the number 0, all have zero remainder, so that sack, or if you will that equivalence class, is the zero class, the zero of this algebra.

It's straightforward to define operations between classes or sacks—multiplication, addition, subtraction, division, additive inverse, multiplicative inverse. Everything is done

"mod $(x^2 + 1)$."

Meaning: "Whenever $x^2 + 1$ shows up, throw it away." It's equivalent to zero, because the remainder of $(x^2 + 1)$ on division by $(x^2 + 1)$ is 0.

The multiplicative inverse of the sack of polynomials with remainder $(ax + b)$ is the sack of polynomials with remainder

$$\frac{(-ax + b)}{(a^2 + b^2)}$$

Why? When multiplied together, they yield a polynomial whose remainder is 1, which, naturally, is the multiplicative unit in this algebraic structure.

And of course its additive inverse, which we denote by -1, is the sack of polynomials that leaves the remainder -1.

Now the big question. What about a square root of -1? The answer is so easy, it feels like swindle.

Since $x^2 + 1$ is equivalent to 0, or as an equation, $x^2 + 1 = 0$, then subtracting 1 from both sides,

$$x^2 = -1.$$

Hey, that's it! We've found the thing that when squared equals minus one! It's the equivalence class containing the simple special polynomial x. If you prefer, its the class with remainder $ax + b$, where $a = 1, b = 0$.

In a fussier notation, $x^2 = -1$ modulo $(x^2 + 1)$. So x^2 is "congruent" (equivalent) to -1, and x is equivalent to $\sqrt{-1}$.

I'll say it once more. x^2 is equivalent to -1 because both give the same remainder on division by $x^2 + 1$. (Or, put even more simply, adding 1 to either gives 0.) If x^2 is equivalent to -1, that means x is equivalent to the square root of -1. So x in our algebra of polynomial equivalence classes is "the same" as the complex number i! Polynomials with remainder 1 are equivalent to the real number 1. A combination of x and 1, say, $ax + b$, corresponds to the complex number $ai + b$. All our equivalence classes correspond to remainders of the form $ax + b$, so the equivalence classes and the complex numbers are in a one-to-one correspondence. They're "isomorphic." These sacks correspond precisely to the complex numbers!

Why go through all this when we can just adjoin *i*? Because adjoining something new and prescribing rules for it to follow is a leap in the dark. In using equivalence classes, on the other hand, we add nothing and risk nothing. We just notice what's there. The step from real numbers to real polynomials involves bringing in *x*, but we don't require x to satisfy any weird conditions (like $x^2 = -1$.) We just divide by the polynomial

$$x^2 + 1$$

and look at the remainder. Given two polynomials, we can find their remainders on division by

$$x^2 + 1$$

and see if they're the same or different. This relation automatically sorts the polynomials into classes. Then behold! These equivalence classes are the complex numbers!

Let's compare the three constructions—by ordered pairs, by 2×2 matrices, and by polynomials mod $(x^2 + 1)$.

The construction by ordered pairs uses an algebraic structure created specifically for constructing the complex numbers. Conceptually and computationally, it's the simplest.

The construction by matrices uses something already available—the algebra of 2×2 matrices. It isolates a special subset of them—those whose diagonal elements are equal, and whose off-diagonal elements are equal in absolute value but opposite in sign. One checks that this matrix algebra is closed—sums, products, and inverses of matrices of this type are again of this type. Then, since we know that the identity element is

$$\begin{pmatrix} 1 & 0 \\ 0 & 1 \end{pmatrix}$$

we know that -1 corresponds to

$$\begin{pmatrix} -1 & 0 \\ 0 & -1 \end{pmatrix}$$

and simply check that

$$\begin{pmatrix} 0 & 1 \\ -1 & 0 \end{pmatrix}$$

squared is

$$\begin{pmatrix} -1 & 0 \\ 0 & -1 \end{pmatrix}$$

Just call

$$\begin{pmatrix} 0 & 1 \\ -1 & 0 \end{pmatrix}$$

"*i*," and you have

$$i^2 = -1$$

You could say Hamilton "constructed" the complex numbers with his algebra of ordered pairs. In the matrix approach, you can't say anything has been *constructed*—the matrices are here already. You might say we "isolated" or "discovered" the complex numbers embedded in the algebra of 2×2 matrices.

What about the method of polynomials mod $(x^2 + 1)$? Here the objects that correspond to Hamilton's ordered pairs or to 2×2 matrices are equivalence classes of polynomials mod $(x^2 + 1)$. (See section below on "Equivalence Classes.") We take all the polynomials that have the same remainder, say $2x + 1$, and throw them into the same sack. We think of the sackful—the whole class of mutually equivalent polynomials—as a single object, which can be added to or multiplied by any other equivalence class. The multiplicative identity is the class of all polynomials with remainder 1. -1 is the class of all polynomials with remainder -1. x is a square root of -1, because the remainder when x^2 is divided by $(x^2 + 1)$ is -1.

Proof: $x^2 = [1 \times (x^2 + 1)] - 1.$

5. Finally, let's see how the complex numbers might be regarded by some anonymous set theorist.

There are two approaches to set theory. One is axiomatic. If something satisfies the 12 axioms of Zermelo and Frankel, it exists. The other way is constructive. Start with the empty set, and step by step, using axiomatically authorized set-theoretic operations, construct ever bigger uncountable sets. Everything you can get by iterating uncountably infinitely often the set-theoretic operations of enlargement is thought to have *already* existed, in advance. Modern set theory is a fascinating and difficult study.

The most famous example of constructing by means of equivalence classes was Frege's "construction" of the natural numbers as equivalence classes of sets. He ended his career resigned to the failure of his set-theoretic foundation. Yet those ideas continue to permeate philosophical logic and set theory.

Where does this put the complex numbers? In the number system they're at the top of the heap, but in the grand set-theoretic structure they're near the bottom. Whether there's a number whose square is -1 is of little set-theoretic interest. But if by chance you want a square root of -1, you have to look in the set-theoretical structure. There isn't any place else!

Recall Hamilton's ordered pairs. Forming ordered pairs is licensed by one of Zermelo's axioms, so all ordered pairs always existed, whether Hamilton knew it or not. Hamilton gave his ordered pairs an algebraic structure. How is that algebraic structure understood set-theoretically? To explain, let's go back to multiplication of natural numbers. We get a natural number as product. The formula $a \times b = c$ describes a *function*, the "times" function, which operates on the pair a, b and yields the value c. So the formula is "really" a set of ordered triples, a, b, c. Therefore, it's a subset of the set of *all* ordered triples. Since it's a subset of a set, it's a set—it exists, by another of Zermelo's axions. This "proves," in a certain strange sense, that multiplication of natural numbers exists. If we go from the natural numbers to ordered pairs (rational numbers) we get the set of all ordered triples of pairs—sextuples. A certain subset of that set represents multiplication of rational numbers. It wouldn't be essentially different to treat a *pair* of operations, like "plus" and "times." Proceeding further, we would find that the real numbers, the complex numbers, and all their operations, are already there in the grand set-theoretic super-universe. The problem is to find them. That means showing that certain sets have certain required properties. To do that requires the same checking we've been doing.

Are more representations of the complex numbers waiting to be discovered? If you look in the right math book, you'll find a theorem, "There's only one system of complex numbers." If we line up our representations carefully, they look like merely verbal variants of each other. In Hamilton`s ordered pairs, (1,0) is the multiplicative unit, and (0,1) is the imaginary unit. In the matrix representation of complex numbers,

$$\begin{pmatrix} 1 & 0 \\ 0 & 1 \end{pmatrix}$$

is the multiplicative unit, and

$$\begin{pmatrix} 0 & 1 \\ -1 & 0 \end{pmatrix}$$

is the imaginary unit i. The correspondence is obvious.

In the polynomials mod $(x^2 + 1)$, the class of polynomials having remainder $1 + 0x$ is the multiplicative unit, the class having remainder $0 + 1x$ is the imaginary unit. So the standard complex number $a + bi$, the ordered pair (a, b), the matrix

$$\begin{pmatrix} a & b \\ -b & a \end{pmatrix}$$

and the equivalence class of a + bx are four names for the same thing.

These correspondences between algebraic systems are called "isomorphisms." They are one-to-one invertible mappings, which preserve algebraic structure. In mathematics teaching an impression is often given that isomorphic systems should be regarded as the same.

The difference between (a, b), the equivalence class of a + bx and

$$\begin{pmatrix} a & b \\ -b & a \end{pmatrix}$$

is regarded as trivial or meaningless. Or their difference is mere notation, like the difference between x^2 as a function of x and t^2 as a function of t.This would be an error. The mathematician describes this situation by saying that the same structure has several representations. The structure is the abstract thing that each representation represents, in a particular language and from a particular viewpoint. An investigation often is possible only by means of some concrete representation. It can be advantageous to have several representations. Several famous theorems are representation theorems—the Riesz theorems, the Radon-Nikodym theorem, the spectral theorems. An attitude that structure is all, representations are trivial, is a serious misrepresentation.

This question comes up in use of coordinates in geometry. Geometric results should be independent of coordinates; therefore, the story goes, they should be proved without coordinates. Yet the coordinate proof may be more accessible to find and to teach.

Anything that claims to be a new representation must be *substantially new*. Somebody could report a new representation for complex numbers by using 3 × 3 matrices instead of 2 × 2. He could simply augment the 2 × 2 representation by one more row and one more column, all zeroes. This would be new formally but not substantially. Such a change from 2 × 2 to 3 × 3 would be uninteresting and obvious. What we think interesting today isn't always what Euler thought interesting, nor what geniuses in 2997 will consider interesting. This question is esthetic. Esthetic questions play a small part in deciding what's correct, a major part in deciding what's interesting. Esthetic considerations are spared little space in the journals, but they're crucial for understanding the development of mathematics.

At present our number system looks stable, although Abraham Robinson's nonstandard real numbers have proved their worth, and John Conway's "surreal numbers" may have a future.

What Are Quaternions?

Hamilton's passion was to find a number system to do for 3 dimensions what the complex numbers do for 2. To define an algebraic structure, each element of which could be identified with a point of x-y-z-space, with addition and multiplication corresponding to translation and rotation in 3-space.

This proved impossible. Hamilton came as near as anyone could.

His quaternions include three independent "imaginaries," i, j, and k. Each of them squared yields -1!

A general quaternion has the form

$$a + bi + cj + dk$$

where a, b, c, d are real numbers.

To multiply quaternions, you have to multiply i, j, k by each other. This was the hard part. Hamilton discovered the system worked if

$$ij = k \qquad jk = i \qquad ki = j$$
$$ji = -k \qquad kj = -i \qquad ik = -j$$

The commutative law has vanished! Instead, an "anticommutative law." This was the first time anyone imagined an algebraic structure without commutativity. Hamilton was so delighted that he carved

$$ij = -ji$$

on a bridge he crossed going to church.

Hamilton and his disciples tried hard to make quaternions useful in mathematical physics. But Gibbs's vector analysis accomplished similar things more conveniently.

Quaternions are hyper-complex numbers. They add, subtract, multiply, and divide. Gibbs's vectors, on the contrary, have two different multiplications, but no division.

Quaternions are four-dimensional—a 3-vector linked to a number. They don't fit in higher dimensions. Gibbs vectors generalize to any dimensions.

Crowe reports the competition between quaternions and Gibbs-Heaviside vectors for modeling electromagnetism. Both formalisms can describe electromagnetic fields. But physicists preferred the one they found more convenient for calculation—Gibbs vectors.

Do and did quaternions exist? They existed as mathematical concepts from the day Hamilton discovered them. But they weren't sitting on that Irish bridge from the beginning of time, patiently waiting to be discovered. And they didn't start to exist on the day Hamilton started trying to fit them to physics.

They're a permanent piece of algebra, and they continue to be proposed for use in physics and engineering. But from the viewpoint of Platonist set theory, the quaternions were always ready and waiting in the grand abstract universal set structure, their anticommutative multiplication merely a certain subset of the set of all sets of sets of sets of sets of empty sets.

Extension of Structures and Equivalence Classes

The extensions of number systems we have just presented are in a sense optional, but in a stronger sense not optional. Nothing in the natural numbers *logically* forces us to introduce negatives. The enlargements to integers, rational numbers, real numbers, and complex numbers were all compelled, slowly and reluctantly.

For another example of how these optional enlargements are in a deep sense compulsory, look at the Fibonacci numbers. These are the sequence

$$1, 1, 2, 3, 5, 8, 13, 21 \ldots$$

Each number after the first two is the sum of the two previous ones.

It's obvious that all the Fibonacci numbers are positive integers.

A little analysis shows that they're all combinations of the solutions of this quadratic equation:

$$x^2 = x + 1.$$

You can solve this equation with your high-school quadratic formula. You find two roots, both involving the square root of 5 (in a combination known to fame as the "golden ratio"). Both roots are irrational. But if you combine them, with the right irrational coefficients, you get the Fibonacci numbers! This is a sequence of natural numbers, yet to write a formula for them, we're forced to use an irrational number, $\sqrt{5}$!

It's enough to make you think $\sqrt{5}$ was already there when we learned to count—or even before, since, as Martin Gardner tells us,

$$2 + 2 = 4$$

was already true with the dinosaurs.

No wonder the mathematician in the street thinks $\sqrt{5}$ existed even before Fibonacci.

Another example. The infinite series

$$1 + x^2 + x^4 + x^6 + \ldots$$

converges if the absolute value of x is less than 1. It diverges if absolute x is greater than 1. To see why, notice that if absolute x is greater than 1, then the terms farther and farther out in the series get bigger and bigger. But for convergence they must get smaller and smaller.

For absolute value of x less than 1, this series sums to

$$\frac{1}{(1 - x^2)}$$

This fraction blows up when $x = 1$, because it becomes $1/0$. This is a good reason why the series can't converge for $x = 1$.

On the other hand, there's the series

$$1 - x^2 + x^4 - x^6 + \ldots$$

Like the previous one, this converges if absolute x is less than 1, and diverges for absolute x greater or equal 1. It sums to

$$\frac{1}{(1 + x^2)}$$

This denominator is always greater than or equal to 1, so this fraction doesn't blow up for any real x. Then why should the series diverge, if it's equal to a fractional algebraic expression that is well behaved for all real x?

Try replacing x by $z = x + iy$. That means, let the independent variable run around the complex x-y plane, not just the real x-axis. If you choose $z = i$ (the square root of -1) then the denominator *is* zero—the fraction blows up. There's a singularity on the *imaginary* axis at $z = i$, one unit away from the real x-axis. The singularity on the *imaginary axis* is responsible for divergence on the *real* axis! A phenomenon in real analysis, which, in a reasonable sense, can't be understood in terms of real numbers only. The complex numbers, whether or not we recognize them, are already controlling some of our real-number computations!

Finally, consider the trigonometric functions sin x and cos x and the exponential function e^{ax}, where a is some positive real number that we can choose at will. If the variable x is real, the behavior of this exponential function is completely different from that of the trigonometric functions. The exponential grows steadily from zero at $x = -\infty$, and its rate of growth is a. If a is any positive number, the exponential function grows faster and faster as x increases. The trigonometric functions, in contrast, remain bounded for all x, however large. They oscillate periodically between a minimum of -1 and a maximum of 1. By use of complex numbers, Euler made the astounding and brilliant discovery that these functions are "essentially" the same! If you do the unorthodox thing—choose a to be, not real, but imaginary—then you find that

$$e^{ix} = \text{cox } x + i \sin x.$$

Making the domain of the exponential function imaginary turns it into a combination of sine and cosine!

A gap is yawning. A unification and deeper understanding beckon, which demand going out of the given mathematical structure—allowing the existence of $\sqrt{-1}$ —*changing the axioms.*

How to change the axioms? How to change or enlarge our mathematical structure? These questions go beyond axioms and theorems. As well as working within given axiomatic structures, mathematicians tear structures down, to replace them with others more powerful.

Calculus

Newton, Leibniz, Berkeley

Berkeley's famous *Analyst* (famous in the history of mathematics, forgotten in the history of philosophy) is an attack on the differential calculus of Newton and

Leibniz. The fallacy Berkeley exposed is simple. To compute the speed of a moving body, you divide the distance traveled by the time elapsed. If the speed is variable, this fraction depends on how much time elapses. But we want the speed at one instant—a time interval of length *zero.* For a falling stone, for example, we want its speed when it hits the ground—its final or "ultimate" velocity. But that seems to require dividing by zero—which is impossible.

Newton explained: "By the ultimate velocity is meant that with which the body is moved, neither before it arrives at its last place, when the motion ceases, nor after, but at the very instant when it arrives. . . . And in like manner, by the ultimate ratio of evanescent quantities is to be understood the ratio of the quantities, not before they vanish, nor after, but that with which they vanish."

This gives us a physical intuition of ultimate velocity. But when Newton calculated he used a mathematical algorithm, not physical intuition. Starting with a time interval of positive duration (call it h), he got an average speed depending on h. He simplified the answer algebraically, and finally set $h = 0$. The resulting expression was the instantaneous speed. Newton called it the "fluxion," and the associated distance function the "fluent."

"But," wrote Berkeley, "It should seem that this reasoning is not fair or conclusive. . . . For when it is said, let the increments vanish, let the increments be nothing, or let there be no increments, the former supposition that the increments were something, or that there were increments, is destroyed, and yet a consequence of that supposition, i.e., an expression got by virtue thereof, is retained. Which is a false way of reasoning. . . . Nor will it avail to say that [the term neglected] is a quantity exceedingly small; since we are told that *in rebus mathematicis errores quan minimi non sunt contemnendi.*" ("In mathematics not even the smallest errors are ignored.")

Berkeley admitted that Newton got the right answer, and that his use of it in physics was correct. He merely showed that Newton's reasoning was obscure.

Leibniz was co-inventor, with Isaac Newton, of the infinitesimal calculus. Unlike Newton, Leibniz used "actual infinitesimals," though he couldn't explain coherently what they were. Cavalieri and others had calculated areas by dividing regions into infinitely many strips, each having infinitesimal positive area. Unfortunately, as Huygens showed, this method could give wrong answers. Problems of rates and velocities also led to infinitesimals. Think of a stone that in 2 seconds falls a distance of 4 feet, in 3 seconds a distance of 9 feet, and in general in t seconds a distance of t^2 feet. Leibniz got the stone's instantaneous speed by calculating its average speed over a time interval of infinitesimal duration. The calculation is so easy we do it right now.

Let dt be the duration of an infinitesimal time interval. At the beginning of the interval your watch reads t seconds, where t is some positive number. The distance fallen up to that time is t^2 feet. An infinitesimal time interval of duration dt elapses. Your watch then reads $(t + dt)$ seconds, and the stone has fallen $(t + dt)^2$ feet. So in the infinitesimal time dt, from instant t to instant $t + dt$, the distance

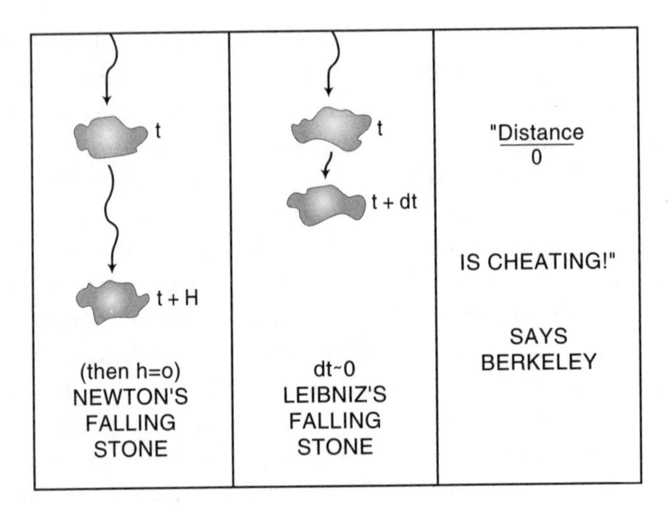

Figure 5. Falling body.

traveled is the distance from its starting point, which was t^2 feet below the stone's initial height, to its ending point, which is $(t+ dt)^2$ feet below the stone's initial height. The distance between the two points is the difference, $(t + dt)^2 - t^2$. "Average speed" is defined in general as a ratio: distance traveled divided by time elapsed. In the present case, that ratio is $[(t + dt)^2 - t^2]/(dt)$. A little algebra simplifies this expression to $(2t + dt)$. dt is infinitesimal, so we "neglect" it—throw it away—and find the instantaneous speed after t seconds: exactly 2t feet per second.

Leibniz's algebra was just like Newton's. He got the same answer as Newton after algebraic simplification, except that his formula had the infinitesimal dt where Newton had the small finite h. Then, instead of setting h equal to zero, as Newton did, Leibniz simply threw away the terms involving dt, *because they were infinitesimal*—negligible compared to the finite part of the answer.

This reasoning was also torn to shreds by Bishop Berkeley. He admitted that the answer, 2t, is right. Berkeley rightly objected to "throwing away" anything not equal to zero, no matter how small. He pointed out that, infinitesimal or not, dt has to be either zero or not zero. If it's not zero, then $2t + dt$ isn't the same as 2t. If it's zero, Leibniz had no right to divide by it. Either way, a fallacy.

Today Berkeley's objections don't disturb us. We show that the average speed *converges to a limit* as the time interval gets shorter. That limit is then *defined* as the instantaneous speed. This limit-and-continuity approach was developed by Cauchy and Weierstrass in the nineteenth century. It is adequate to demystify calculus.

Newton and Leibniz didn't have an explicit definition of limit. The careful use of limits requires explication of the real number system. This subtle task even now may not be quite finished. Still, we see today that Newton was essentially using limits.

Leibniz explained that his infinitesimal dt is "fictitious." This fiction is like an ordinary positive number, but smaller than any ordinary positive number. This is not easy to grasp. How do we decide which properties of ordinary positive numbers apply to dt because it's "just like an ordinary positive number," and which ones don't apply because "it's smaller than any ordinary positive number"? What's the square root of dt? It must be infinitesimal, yet bigger than dt. How many infinitesimals are we going to need? What about the cube root, the fourth root, the tenth root? These puzzles were solved by Abraham Robinson 200 years later, using the theory of formal languages—modern mathematical logic.

The infinitesimal has a fascinating history. At least as far back as Archimedes, it's been used by mathematicians who were perfectly aware that it didn't make sense. It surfaced from underground in the 1960s, when Robinson legitimized it with his "nonstandard analysis."

Nonstandard analysis is the fruit of a century's development of mathematical logic. The basis of it is to regard the language in which we talk about mathematics as itself a mathematical object, obeying explicit formal rules. This formal language then is subject to mathematical reasoning. (Which we carry on, as usual, in ordinary, everyday language, just as we use ordinary language to talk about Basic or C.) Then it makes mathematical sense to say that an infinitesimal is greater than zero and smaller than all the positive numbers *expressible in the formal language*. When Robinson rehabilitated infinitesimals with his nonstandard analysis, he borrowed the word "monad" from Leibniz's metaphysics. In his nonstandard analysis, a monad is an infinitesimal neighborhood (the set of points infinitely close to some given point.)

So today we have two distinct rigorous formulations of calculus. The creators of the calculus were using tools whose theories were centuries in the future.

A Calculus Refresher

Calculus is the heart of "modern mathematics"—mathematics since Newton. It's the part of mathematics most important in science and technology, the part engineers must know.

It's built around two main problems. The central discovery of calculus is that these problems are related—in fact, as we will see, they're opposites.

The first problem is speed. How fast is something changing? The second main problem is area. How big is some curved region?

First, speed. The speed is simple if it's constant:

$$\text{Speed} = \frac{\text{Distance}}{\text{Time}}$$

Divide distance traveled by time elapsed.

Figure 6. Motion at constant speed.

But speed isn't constant. When you drive you start at speed zero, gradually go to the speed limit, then finally slow down to zero. Your speed varies from instant to instant. What is your speed at some particular instant?

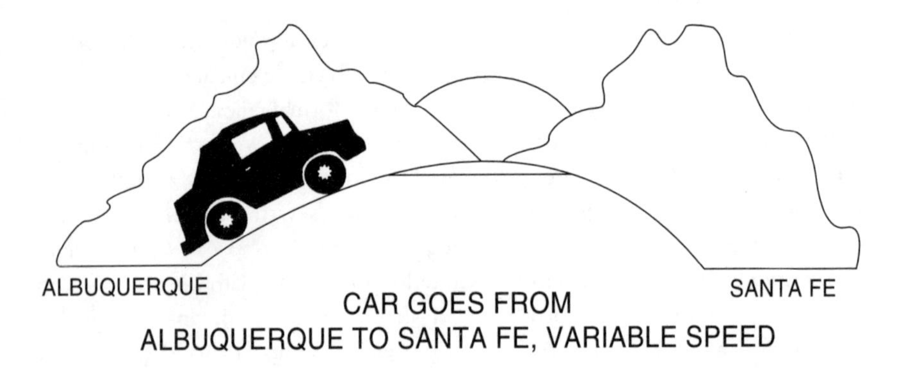

Figure 7.

Example: a body falling in vacuum near the surface of the earth travels $16\ t^2$ feet in t seconds. How fast is it falling after 2 seconds?

In the time interval between 2 seconds and 2.1 seconds—time lapse of .1 second—it falls

$$16\ (2.1)^2 - 16\ (2^2) \text{ feet} = 6.56 \text{ feet.}$$

Dividing distance by time (.1 second), its average speed was 65.6 ft/sec.

Exercise. Repeat the calculation with a time lapse of .01 second. (You'll get an average speed of 64.16 ft/sec, between time 2 seconds and time 2.01 seconds.) Do it again, with a *very small* time lapse, .001 seconds. (Its average velocity over this time period is 64.016 ft/sec.) ### (### means "end of exercise.")

But I don't want an *average* speed. I want the *exact* speed after 2 seconds! That means a time lapse of *zero*. Division by zero is impossible. The formula

$$\text{Speed} = \frac{\text{Distance}}{\text{Time}}$$

becomes meaningless.

However, without setting time lapse to 0, you've crept closer and closer to 0. You used lapses of .1, .01, .001. and found speeds of 65.6, 64.16, and 64.016.

NOW! A giant conceptual leap! If the average speeds approach a limit as the time lapse approaches zero, we declare, *as a definition*, the instantaneous speed *is* that limit! In this example, the limit is 64 ft/sec when $t = 2$. It makes sense! We agree, that's what we'll mean by instantaneous velocity.

The notion of speed as a limit took centuries to formulate. Medieval and Renaissance mathematicians calculated rates of change without defining mathematically what they wanted. The founders of the calculus, Isaac Newton and Gottlob Leibniz, fought bitterly about who had priority in the fundamental theorem of calculus (explained below.)

Exercise. Make a graph of this falling body function: distance = time squared, or $d = t^2$.

(I dropped the 16 to simplify your graphing and my calculating.) This is a quadratic function. Its graph is a parabola. Mark the points (2, 4) and (2.1, 4.41) on the parabola. The second is above and right of the first. Draw a straight line (*secant*) between the two. What's the slope of this line? ("Rise over run.")

$$\text{Rise} = 2.1^2 - 2.0^2 = .41$$
$$\text{Run} = 2.1 - 2.0 = .1$$

Slope = .41/.1 = 4.1, which we just found is the average velocity (allowing for the factor of 16 which we took out). *The average rate of change of a function*

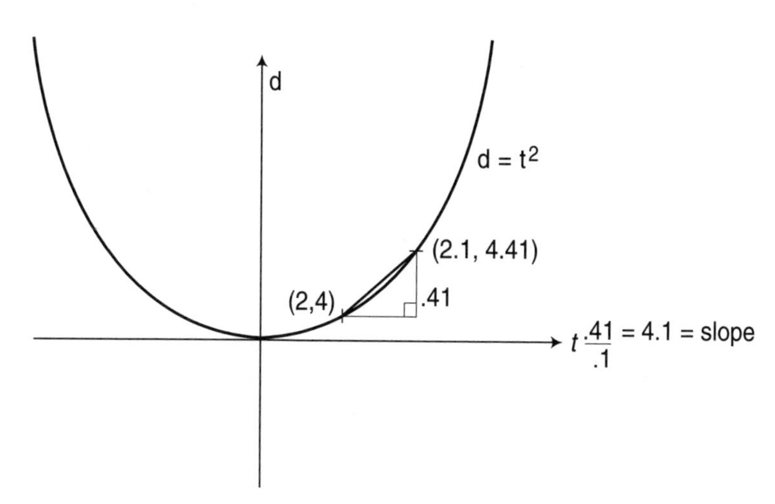

Figure 8. Differentiating x^2 = finding its slope (this graph is called a parabola).

of time is identical to the slope of its secant! Again replace .1 by .01 and .001. The corresponding marks on the graph are creeping closer and closer to (2, 4). The slopes of the secants are exactly the numbers you found to approximate the instantaneous rate of change. As the two points approach closer and closer, and the denominator approaches zero, the secant becomes a tangent, and its slope becomes the instantaneous speed, the *derivative*. ###

Calculating the derivative (the speed) is called *differentiation*. Simple functions often have simple derivatives. The derivative of t^n is n t^{n-1}. (n is any number, integer or fraction, positive or negative.) The derivative of the natural logarithm of t is 1/t. The derivative of e^t is e^t. The derivative of sine t is cosine t; of cosine t, –sine t.

Exercise. In a way similar to how you found the rate of change of $f(t) = t^2$ at $t = 2$, find the rate of change of that function at an arbitrary time t. Do the same for the cubic $f(t) = t^3$. Check with the formula in the previous paragraph for t^n. ###

Now the second main problem of calculus, area. First, a different-sounding problem. Given the velocity of a moving body, calculate the total distance traveled, at every instant of the trip. This is the opposite of the problem above, where we had the distance and found the velocity.

Start with the simplest case—constant velocity. From 2 P.M. to 3 P.M. you drive a steady 50 miles an hour. How far do you go in that hour? in half an hour? at any time t between 2 P.M. and 3 P.M.?

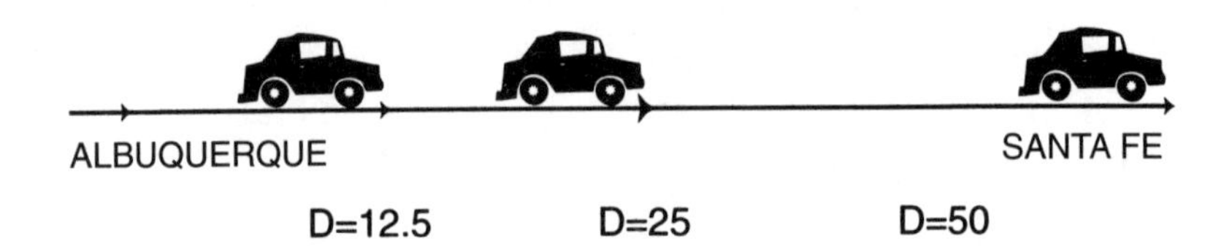

Figure 9. Driving 50 miles at a constant speed of 50 m.p.h.

In one hour you go 50 miles. In half an hour, 25 miles. 50 m.p.h. for t hours goes 50 t miles—where t can be a fraction.

The graph of the constant velocity is a horizontal line 50 units above the time axis. Time is measured on the horizontal time axis. Distance is speed times time—in the graph, height of the velocity line times length on the time axis from start to finish. *The product of these two lengths is the area of the rectangle they enclose.* Distance is graphically an area!

Now vary your speed. The graph of v(t), velocity as a function of time, becomes a curve, not a horizontal line. How can we find the distance traveled

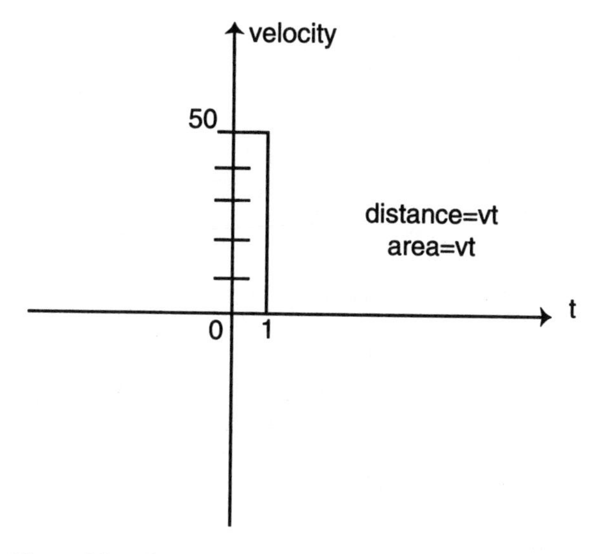

Figure 10. On a time-velocity graph, distance = area.

now? Since we know how to do it in the case of constant speed (horizontal graph), *replace the curved graph by a piecewise horizontal graph.*

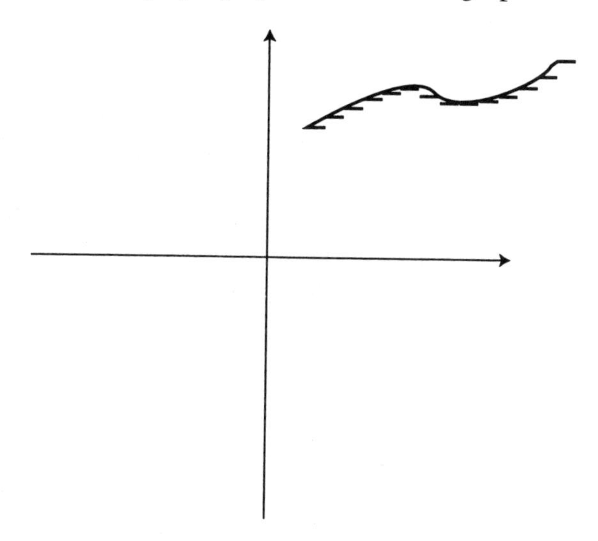

Figure 11. Variable velocity is approximated by piecewise-constant velocity.

Make the speed constant for a second, then a different constant for the next second. The sum of the distances traveled by this piecewise constant, rapidly changing velocity is close to the actual distance. The distance traveled in each second equals the speed in miles per second times the time, one second. In the graph it's

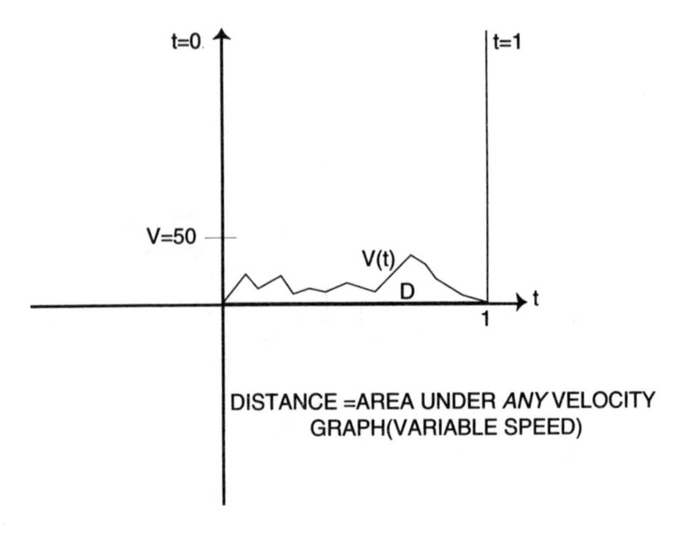

Figure 12.

the area of a skinny vertical rectangle one second wide. These skinny rectangles *nearly* fill the region under the velocity curve. The sum of their areas is very close to the total area. If we make their times shorter and shorter, we see that as in the case of constant speed, *distance traveled equals area under the velocity curve.*

To summarize: To a *distance* function d(t) is associated a velocity function v(t), the derivative of d(t). To v(t) in turn is associated an *area* function A(t), the area under the graph of v(t) up to the vertical line t. *The area A(t) is the same as the distance d(t).*

The area A(t) under the graph of v(t) is called the "integral" of v(t). The function d(t), from which v(t) was obtained by differentiation, is called the antiderivative of v(t). Finding A(t) is called "integrating" v(t). We have just found the "Fundamental Theorem of Calculus": the area function of v (the integral of v) is equal to the antiderivative of v:

$$A(t) = d(t).$$

We've been thinking of v(t) as velocity. But any function can be interpreted as a velocity!

So the Fundamental Theorem says: the integral of the derivative of any function is the function itself (except possibly for an additive constant.)

Computing the derivative directly from its definition often is easy; computing the integral directly from its definition often is hard. The Fundamental Theorem shows you how to do the hard part by doing the easy part. Make a collection of differentiation formulas. If in your collection you find a function w(t) whose derivative is v(t), then the integral of v(t) is w(t).

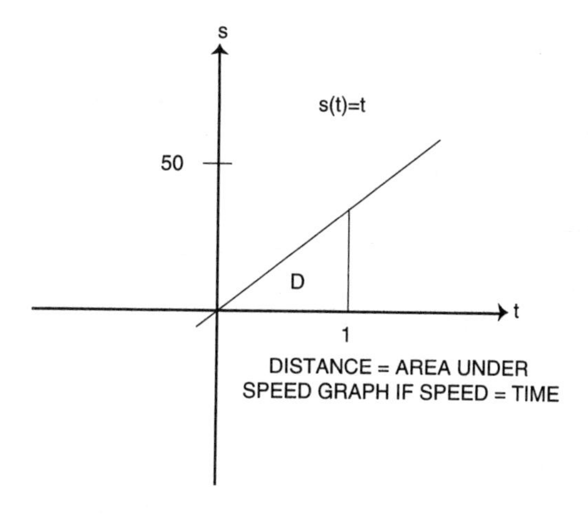

Figure 13.

Let's do a simple area problem—an isosceles right triangle with sides of length 1. By high school geometry the area is 1/2. To do it by calculus, take one side on the x axis, the other side on the vertical line x = 1.

The hypotenuse is the upper boundary of the triangle. It has slope 1 and passes through the origin, so it's a segment of the graph of y = x. The left boundary, x = 0, is a point. The right boundary at x = 1 is the vertical segment 0 < y < 1.

Cut up the triangle with vertical lines .01 apart. Each piece is long and skinny, almost a rectangle, with a tiny triangle at the top of the rectangle. Each rectangle is .01 wide. How high? The upper boundary of each rectangle is part of hypotenuse, which is on the line y = x. The point on the graph above x = .23, for example, has y-coordinate .23. That's the height of the rectangle at the 23d

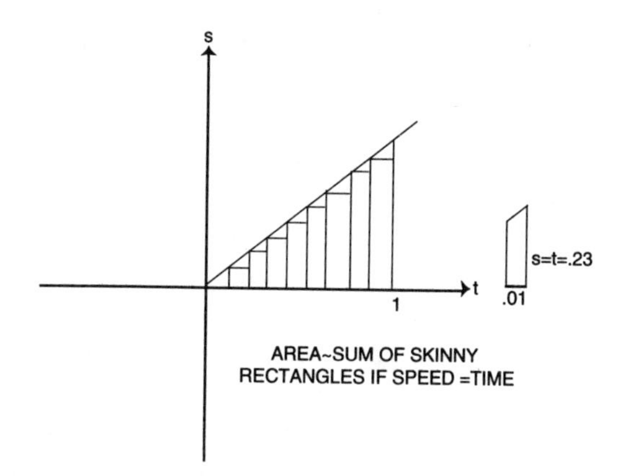

Figure 14.

piece. The area of any rectangle is height times width, so the area of this rectangle is .01 times .23 = .0023.

To approximate the whole area, add the areas of all the rectangles. As the 23d has area .0023, the 38th has area .0038, and so on. Add them all up and factor out .0001. For your approximation to the area of the whole triangle you get

.0001 times (1 + 2 + 3 + . . .).

How many rectangles are there? Your base of length 1 is in pieces .01 = 1/100 wide. So there are 100 rectangles. The last term of the sum in parentheses is 100.

This is a nice puzzle:

1 + 2 + 3 . . . + 100 = ?

A lovely trick does the job. It was discovered by the famous mathematician Karl Friedrich Gauss in school in the first grade. Karl noticed he could write the sum twice—once forward, once backward. The number in the first sum plus its neighbor below in the second sum always add to 101. He had 100 such pairs. So the two sums together equal 100 times 101. The single sum is half of that: 5,050.

In our area calculation we must multiply by .0001. We get for the approximate area .5050. Not too far off from the exact answer, .5. The error, .0050, comes from the little triangles on top of the skinny rectangles. Make the skinny rectangles skinnier and skinnier. The error gets smaller and smaller, as you see from the picture.

It wouldn't do to set the thickness of each little piece *equal* to zero. Then each little rectangle would have area *zero*, and they'd all add up to *zero*, which is wrong.

Calculating area by adding tiny rectangles is called "*integration.*" It's exactly what we did to calculate total distance from variable speed. The method works, it makes sense, so we *define* it to be correct! The area under a curve is *defined* to be the limit of the sum of areas of very skinny inscribed or circumscribed rectangles. There can't be a *proof* that the limit equals the area, because for curved regions we have no other definition of area except that limit!

What you have accomplished isn't just a roundabout way to measure triangles. It works for about any area that comes along. Problems on arc length, volume, probability, mass, electrical capacity, work, inertia, linear and angular momentum, all lead to integrations such as you just did.

What if we do our two calculus operations in the opposite order—first integrate, then differentiate? Before, we started with a distance function, differentiated to get a velocity function, then integrated that and got back our original distance function. Now start with a velocity function, integrate it to get an area and distance function, then differentiate that.

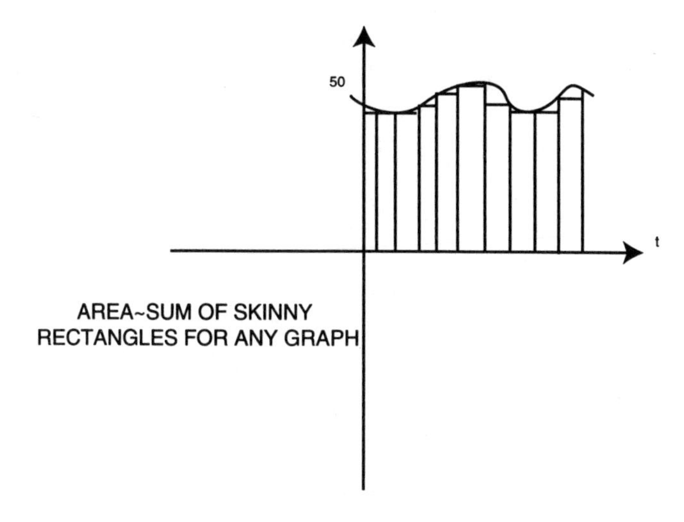

Figure 15.

We can use Figure 13, but now the right side of the triangle is movable. Let t be the distance from the origin. It is a variable. The right boundary could be a stick you slide to the right. As t increases, the stick moves to the right and the area A increases. A is a function of t, call it A(t).

You're going to calculate the rate of change of the area as the stick moves to the right. That is, you will differentiate the variable which we have named A(t).

Our argument does not require the region to be a triangle. As long as it's bounded below by the x-axis, on the left by the y-axis, and on the right by the moving vertical line $x = t$, its upper boundary can be the graph of any function v(x) you like. (Figure 15)

To differentiate—find a rate of change—you must increase the independent variable t by a little bit h, see how much your function increases from A(t) to A(t + h), and divide that increase in A(t) by h. This quotient is the average increase over the interval from t to t + h. Since h is small, the numerator and denominator are both close to zero. Their ratio is close to a limit, which limit you defined as the rate of change of A(t).

In applying the definition of rate of change to the area function A(t), you're working with *two different pictures*. The integration picture computes the area of D, the region under v(x), by cutting up D with many close vertical lines. The differentiation picture computes the rate of change of any function by drawing the secant through two nearby points on its graph. We're applying the differentiation picture to the integration picture, or, if you like, plugging the integration picture into the differentiation picture.

What happens to D and its area A if the right side moves a bit farther right? The region is enlarged by a little additional piece, which differs from a rectangle only in a very small bit at the top. Its width is h, the amount of increase of t. Its height is the height of the upper boundary of D, which is the graph of v(x). So the height of the little added rectangle is v(t), and its area is hv(t). This hv(t) is the increment of A(t). The derivative of A(t) is the increment hv(t) divided by h.

That's hv(t)/h = v(t)!

We have shown that the derivative of A is v. And A is the integral of v.

> (The derivative of the integral of v) =
> (the integral of the derivative of v) = v.

Symbolically,

D: A ——————► v
I: v ——————► A

Differentiation and integration reverse each other, in either order.

IDs=DIs=s

THE FUNDAMENTAL THEOREM OF CALCULUS!!

INTEGRAL OF SLOPE FUNCTION = HEIGHT OF GRAPH
RATE OF CHANGE OF AREA = HEIGHT OF GRAPH

Figure 16.

The Fundamental Theorem is a powerful method of computing areas. Suppose we want to know the area A under the parabola $y = 3x^2$, between x = 0 and x = 3.

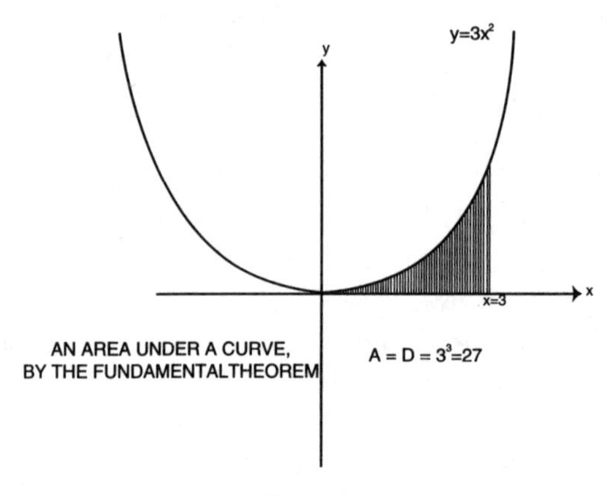

Figure 17.

According to your work a few paragraphs above, this function is the derivative of x^3. Therefore, by the Fundamental Theorem, its area function A(x) is x^3. Between 0 and 2, the area under $3x^2$ is therefore $2^3 = 8$.

Exercise. Use the Fundamental Theorem to compute the areas under the curves $y = x^2$, $y = x^3$, $y = x^4$, all for $0 < x < 1$. ###

The important scientific use of calculus is in solving differential equations. These involve an unknown function and its derivatives. In case the unknown function is a distance, the first derivative is the velocity, and the second derivative is the rate of change of the velocity, the "acceleration." The fundamental law of mechanics, Newton's third law, says

$$f = ma.$$

Force equals mass times acceleration. Often the force is given by some fundamental principle governing the motion under study. Since acceleration is a second derivative of position, Newton's law is a second-order differential equation. To find out how a body moves under the influence of a force, we try to solve this differential equation.

In the case of the planets and the sun, the force is gravity, which is directly proportional to the masses of the attracting bodies, and inversely proportional to the square of their distance. In the case of only two bodies, such as the earth and the sun, the differential equation can actually be solved. By doing so Newton proved that the three laws of Kepler (elliptic orbits; position vector covering equal areas in equal times; and length of year proportional to the 3/2 power of the radius) are equivalent to his law of gravity and his third law. This calculation requires more technique than we assume here.

This triumph of Newtonian calculus and physics ignored the mutual attractions of the planets. If we think of Mars, the earth, and the sun as a system of three bodies none of whose mutual interactions may be ignored, we have the stubbornly intractable three-body problem, which has been tempting and frustrating us for 300 years.

To glimpse how differential equations solve problems of motion, suppose I throw a ball up into the air in a room with a 16-foot ceiling. I want the ball to just barely touch the ceiling. This depends on the velocity V with which I toss the ball up. Determine the correct V.

As usual in elementary treatments, we ignore air resistance. The only force is gravity, which creates a downward acceleration of 32 feet per second per second. We measure distance in feet from ground level, so the initial height of the ball is zero. Now Newton's third law is just

$$\text{Mass} \times \text{Acceleration} = -32 \times \text{Mass}.$$

The minus sign is needed because the acceleration of gravity is ***downward***, ***decreasing*** the height h(t) as a function of time t. Newton's equation gives the

Figure 18.

acceleration, which is the second derivative of h(t). To find h(t) from a(t) we must do two integrations. (Integrating is the opposite of differentiating!)

First divide both sides by the mass. Acceleration a(t)—the second derivative of h(t)—is the derivative of velocity v(t), the first derivative of h(t). We already know that the constant -32 is the derivative of -32t + K, where K is any constant. So from Newton's law, with m divided out,

$$a(t) = -32,$$

integration gives

$$v(t) = -32t + K.$$

To find K, see what the equation says when t = 0. It reduces to

$$v(0) = 0 + K.$$

So the constant of integration K is V, the initial velocity, which we want to find.

(NOTATION! We use the upper-case, capital letter V for "initial velocity"—how many feet per second the ball moves when I release it from my hand. The lower-case v(t) means the velocity varying with time, starting with t = 0, and continuing until the ball returns to the ground. Capital V is just another name for little v(0).)

Since velocity v(t) is the derivative of height h(t), and $-32t + V$ is the derivative of

$$-16t^2 + Vt + L,$$

where L is another arbitrary constant, integrating the above equation for v(t) gives

$$h(t) = -16t^2 + Vt + L.$$

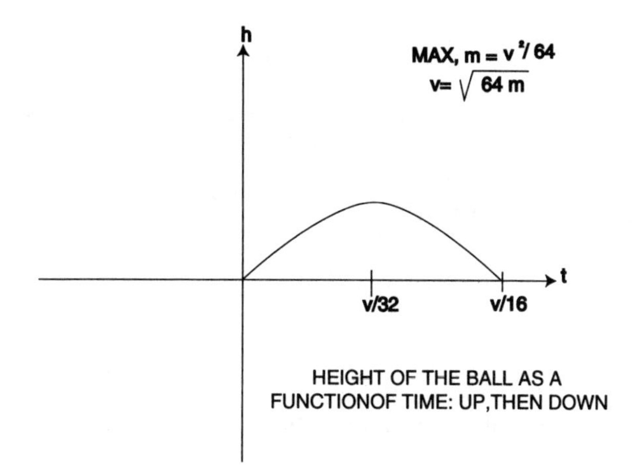

Figure 19.

Again set t = 0 to determine the arbitrary constant. Since h(0), the initial height, is 0, we get L = 0. So we have a formula for height as a function of time:

$$h(t) = -16t^2 + Vt.$$

When is h(t) = 0—when is the ball at the ground?

When h = 0 the equation for h(t) is easy to solve. We find that when h = 0, t = 0 or t = V/16. Why two answers? Because the ball is at the ground twice! First at time 0, before I throw it, and again at time V/16, when it returns to the ground.

Finally, what is M, the greatest height reached by the ball? The height is greatest when the ball stops rising and starts to fall. That is, when v(t) changes from positive to negative. That is, when v(t) = 0. We know v(t) = −32t + V. The maximum is when −32t + V = 0, or t = V/32. To find the maximum height then we need only calculate the height function h(t) when t = V/32. With a little arithmetic simplification, we get

$$h(t) = V^2/64.$$

This tells how high the ball goes, given V—how hard I threw it. But I wanted that height to be 16 feet. So

$$16 = V^2/64$$

and $V = \sqrt{(64 \times 16)} = 32$ feet per second. (In figure 19, instead of a ceiling height of 16 feet, we have allowed the ceiling height to be an arbitrary M.)

We computed this figure here on earth, where g, the acceleration due to gravity, is 32 feet per second per second. What if we visit the Moon? Or Mars? We

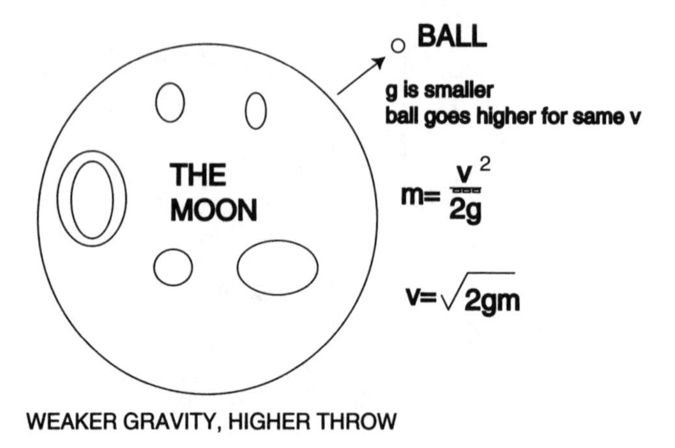

Figure 20.

must replace 32 by the gravitational constant there. On the moon g is much smaller, so we'd need a smaller initial velocity V for a given ceiling height M.

We could repeat the whole calculation, starting with an indeterminate g instead of 32. Or we can just look at our Earthly answer—

$$V = \sqrt{(64 \times 16)}$$

—and make the obvious guess for Mars or the moon. Replace 64 by 2g, 16 feet by M feet, and get $V = \sqrt{2gM}$.

Finally, let's go back to the falling body problem we started with. My friend Nicky fell off the First International Unpaid Debts Building in Miami. The Fire Department is there, life net ready to catch him. Is the net strong enough?

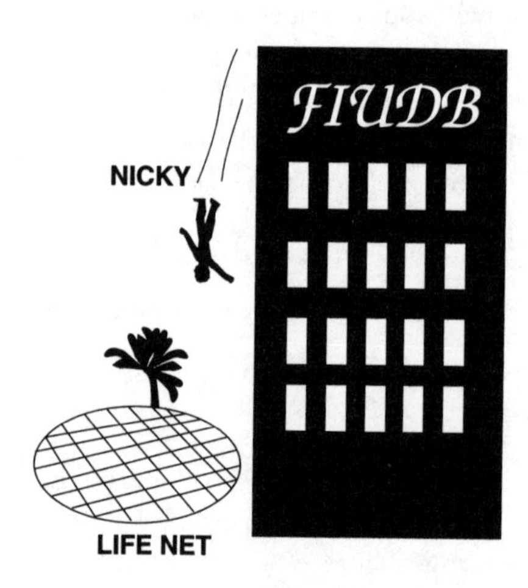

Figure 21. Will Nicky be saved?

Nicky weighs 150 pounds. The FIUDB is 1600 feet high. How fast will he reach the ground? How hard will he hit?

Like the planets and the ball, Nicky's body obeys Newton's third law, with gravitational force 32 feet per second per second. In the equation

$$\text{acceleration} = a(t) = 32$$

we again recognize that both sides are derivatives, or rates of change. Acceleration $= a(t)$ is the rate of change of

velocity $= v(t)$, and
32 is the rate of change (derivative) of
$32\,t +$ an arbitrary constant which we can call B. So
$v(t) = 32t + B$.

What's B? Set $t = 0$. The velocity equation now reads,

$$v(0) = \text{initial velocity} = B.$$

What was Nicky's initial velocity? When he slipped off the roof, he had just started to fall. His initial velocity was zero. So $B = 0$.

Our equation simplifies to

$$v(t) = 32\,t.$$

Again we recognize that both sides are rates of change. Velocity $= v(t)$ is the rate of change of distance fallen, and $32\,t$ is the rate of change (derivative) of

$16\,t^2 +$ another arbitrary constant C.
So distance fallen $= d(t) = 16\,t^2 + C$

Nicky's distance fallen at time 0 is 0, so $d(0) = 0$ and therefore $C = 0$.

We have managed to find a formula for the distance Nicky falls as a function of time:

$$d(t) = 16\,t^2.$$

When does he hit the ground? He'll hit when he has fallen a distance equal to the height of the FIUDB, 1600 feet. That is, when

$$16\,t^2 = 1600.$$

Divide both sides by 16: $t^2 = 100$.
Take square roots of both sides: $t = 10$.
So he hits the ground after falling for ten seconds.

We already have his velocity as a function of time—

$$v(t) = 32t.$$

So his velocity at time $t = 10$ is

$$v(10) = 32 \times 10 = 320 \text{ feet/second.}$$

Multiplying by his weight, the life net must sustain a momentum of 150 times 320, or 48,000 foot-pounds per second!

Only calculus can provide this life-saving information.

Should You Believe the Intermediate Value Theorem?

Here's a theorem of elementary differential calculus, the "intermediate value theorem." It seems indubitable, yet it's denied by intuitionist and constructivist mathematicians.

Theorem: If f(x) is continuous, f(0) is negative, and f(1) is positive, then at one or more points c between 0 and 1, f(c) = 0.

Just picture in your mind the graph of f. It's below the x-axis at x = 0, above the axis at x = 1, and doesn't have any jumps. (It's continuous.) It's visually obvious that the graph can't get from below the axis at x = 0 to above the axis at x = 1 without crossing the axis at least once.

But for today's calculus teaching, this visual argument isn't rigorous enough. (A rigorous proof mustn't depend on a picture.)

Here's the rigorous proof:

To start, ask "What's f(.5)?" If it's zero, we're done—the proof is finished. If it isn't zero, it's positive or negative. (This is the Law of Trichotomy.)

Suppose it's positive. Then *f* has to change from negative to positive in the closed half-interval from 0 to .5.

What's f(.25)? Again, it's 0, + , or −. If +, *f* has to change from negative to positive in the closed quarter-interval from 0 to .25. If f(.25) is negative, *f* has to change from negative to positive in the closed quarter-interval from .25 to .5.

Continuing, we generate a nested sequence of closed subintervals of width 1/2, 1/8, 1/16, and so on. In each subinterval, *f* changes from negative to positive as x goes from left to right. Such a nested sequence of closed subintervals, with width converging to zero, contains exactly one common point. Call it c. The left sides of all the nested subintervals converge to that point, and so do the right sides. What's f(c)?

In all these nested subintervals, *f* is negative on the left side and positive on the right side. The left sides and right sides both converge to c. Since *f* is continuous, the values of *f* approach some limit there.

The value of *f* at all the left sides is negative. A sequence of negative numbers can converge to a negative number or to zero, but not to a positive number. On the other side, approaching c from the right, the values of *f* are positive, and their limit can't be negative. Since f is continuous, the two limits are the same, f(c).

So f(c) equals both a nonpositive limit from the left and a nonnegative limit from the right. There's only one number that's both nonpositive and nonnegative—zero. Therefore, f(c) = 0.

End of proof.

This is a fair sample of proof in modern (since 1870) analysis or calculus. Is it constructive? Teachers who teach it and pupils who learn it think it's constructive. The point *c* seems to be constructed—approximated with any degree of accuracy—by straightforward, elementary steps. At the first step we merely check whether f(x) is positive, negative, or zero at $x = 1/2$, $x = 1/4$, and $x = 3/4$. At the second step, we look at f(x) for $x = 1/8$ and $x = 3/8$; at the third step, for $x = 5/8$ and $x = 7/8$. And so on.

The visual proof is already enough for most people. The rigorous proof is perhaps a bit of overkill. What could be left to disagree about, after firing all those bullets into a dead horse?

The trouble is that in thinking of the continuous function *f*, you had in mind something like x^2, something easily computed. However, the theorem makes an assertion about *all* continuous functions, most of which are not elementary. The guru of intuitionism, L. E. J. Brouwer, showed a constructively continuous function f(x) such that it's impossible to constructively determine whether f(0) is positive, negative, or zero. Such determinations are the heart of the proof we just presented. There are constructively continuous functions that are negative at $x = 0$, positive at $x = 1$, but for which there is no constructively defined point *c* where $f = 0$.

Whether a proof is constructive depends not only on whether it uses the law of the excluded middle (proof by contradiction), but also on whether it needs other theorems whose proof is not constructive.

In this case, we needed the law of trichotomy:

"Every real number is either positive, negative, or zero." (See the section on Brouwer in Chapter 8.) This is a still more elementary theorem in beginning calculus. Its proof is indirect—uses the law of the excluded middle. Constructivists and intuitionists don't accept indirect proof or the law of the excluded middle. They don't accept the law of trichotomy. They don't accept the intermediate value theorem.

To be fair, I add that they don't stop with a purely negative position. Bishop presents *two* constructive intermediate value theorems. In one, he strengthens the hypotheses. In the other, he weakens the conclusion.

What Is a Fourier Series?

We want to calculate the sum of $1/n^2$ as n runs from 1 to infinity through the odd numbers.[4] We'll need a few tricks.

I have to remind you about "cosine," often shortened to "cos." Imagine a circle of radius 1. Through its center, the "origin," draw two perpendicular lines, one horizontal and one vertical. Now let P be any point on the circle, and draw a "ray" connecting it to the center.

We're interested in two numbers associated to P. The first number is the magnitude of its angle, measured between the rightward horizontal axis and the ray to P from the origin. Defying custom, I will call this angle-magnitude x.

[4] See Chapter 4, "Certainty."

The second number is the distance from P to the vertical axis.

If the angle x is given, then that distance, from P to the vertical axis, is thereby determined. Mathematicians refer to such a dependence of one number on another as a "function." This function—distance to the vertical axis as a function of the angle x—is called "the cosine of x," or "cos x" for short. It's a periodic function, because every time x makes a complete revolution around the circle, the cosine returns to its original value.

It's a deep fact of mathematics that every "decent" function f(x) (for example, every continuous function with a continuous derivative) has a Fourier cosine expansion. That is to say, there are numbers a_n such that

f(x) = sum of [a_n cos(nx)], n running from 0 through all natural numbers.

This is not a triviality. You may accept it on authority, or read the proof in a text on Fourier analysis.

How do we find the coefficients a_n? This delightful trick is what makes Fourier series fun.

You multiply both sides of the equation by cos(mx), where m is an arbitrary integer.

Then integrate both sides from 0 to π.

According to a formula from first-year calculus, all the terms in the sum are zero, except one—the m'th! The infinite sum reduces to a single term!

So the left side of our formula is now the given function f(x), multiplied by cos(mx), and integrated from x = 0 to x = π. The right side is the unknown coefficient a_m, multiplied by the integral of $\cos^2(mx)$ from x = 0 to x = π.

Calculus tells us that if m isn't 0, the last mentioned integral equals π/2. If m is 0, the integral equals π.

So, dividing, we have (when m isn't 0)

$$a_m = \frac{2}{\pi} \int \cos(mx)f(x)dx$$

The limits of integration are still from 0 to π.

If m = 0, the coefficient in front of the integral is 1/π.)

This formula for a_m is the fundamental formula of Fourier series.

Now back to our problem: to sum the squares $1/m^2$ for odd m. We can do it, using our formula for Fourier cosine expansions, if we know a function whose Fourier coefficients a_m are $1/m^2$ for odd m and zero for even m.

Forgive me if I now save time and trouble by doing the usual thing: pulling a rabbit out of a hat.

Here's the rabbit: consider the function

$$f(x) = \frac{\pi^2 - 2\pi x}{8} \qquad 0 < x < \pi$$

In its Fourier cosine expansion the even terms all integrate to zero. For odd m, we can look up the integral in standard tables, and find: $a_m = 1/m^2$.

Just right!

The function is very decent, so we know it's equal to its Fourier cosine expansion for all x between 0 and π. In particular, they're equal at $x = 0$. So we set $x = 0$ on both sides of the equation.

Voilà!

Sum of $1/m^2$ for odd m equals $\frac{(\pi^2)}{8}$.

Brouwer's Fixed Point

Suppose we want to solve an equation of the form

$$AX = X.$$

It is often possible to transform a different-looking equation into this form. X is an unknown function or vector. A is a given "operator." It operates by transforming a function or vector into another function or vector. Familiar operators are multiplication, translation, rotation, differentiation, integration. "$AX = X$" says the operator A leaves the function or vector X unchanged. We call X a fixed point of A. Solving the equation $AX = X$ means finding the fixed points of A.

For instance, A could be the operation, "Rotate 3-space around the z-axis through 90 degrees," and X would be a vector in three-space. This operator leaves every vector parallel to the z axis fixed, so all those vectors are solutions of the equation. If we restrict A to operate on vectors in the x-y plane ($z = 0$), it becomes rotation around the origin ($x = 0, y = 0$). Now the zero vector is the only fixed point.

A second example is the simple differential equation,

$$\frac{d}{dt} b(t) = b(t).$$

The operator is now d/dt, differentiation with respect to t. The "fixed point" is any function that equals its own derivative. All such functions are of the form

$$f(t) = ce^t,$$

some multiple of the exponential function.

Some operators have *no* fixed points. An example is the operator of shifting the x-axis to the right. There's no fixed point. In our example above, rotating the plane through 90 degrees, suppose our plane had been "punctured"—the origin had been punched out. In such a punctured plane, rotation has no fixed point.

Brouwer proved that if A is continuous in a reasonable sense and it maps a ball (in any finite dimension) into itself, then it has at least one fixed point. That's "the Brouwer fixed point theorem." To use it in differential equations, think of the set of possible solutions as a "space." For the important special case of an "ordinary differential equation"—an equation in only one independent variable—the space of solutions is finite dimensional. We may be able to rewrite the equation in the form AX = X, where A maps some finite-dimensional space into itself. If we can show that A maps some "ball"—the set of solutions of "norm" less than or equal to 5, say—into itself, then we know a solution exists. *The theorem doesn't say how to find it or construct it.*

What Is Dirac's Delta Function?

Enlarging number systems is no longer a popular sport of mathematicians, but we do often enlarge function spaces. Laurent Schwartz of Paris received the Field Medal for his "distributions" or "generalized functions." Important work in the same direction was also done by Solomon Bochner, J. L. Sobolev, and Kurt Otto Friedrichs.

We will construct a distribution by differentiating a nondifferentiable function. Recall that differentiating a function f(x) means finding the slope of its graph. (Slope is rise/run, y-increment divided by x-increment. See Calculus Refresher above.) At a smooth point on a curve the slope is a number, depending on the position x. It's called f'(x). If $f(x) = x^3$, for example, f'(x) is $3\ x^2$. I'll call this "the classical derivative," to distinguish it from the generalized derivative.

What if the graph has jumps or breaks? The simplest jump function is the Heaviside function, H(x). (Oliver Heaviside was a great telephone engineer and applied mathematician in Victorian England.) H(x) is 0 for all *negative x*, and 1 for all *positive x*. So the graph is horizontal everywhere, except at x = 0. There it jumps instantly, from 0 on the left to 1 on the right. H'(x), the derivative of H(x), is trivial to calculate where x is not zero. For x positive or negative, H is constant, its graph horizontal, and its slope or derivative H' is 0. But what happens at x = 0? If you connect the two pieces of the graph, you draw a vertical line segment rising *instantly* from

$$(x = 0, y = 0) \text{ to } (x = 0, y = 1).$$

The y-increment is 1 and the x increment is 0. The "slope" is 1/0. But there is no number equal to 1/0! (see above.) The "slope" of this vertical segment doesn't exist as a real number.

We're in the position of the Pythagoreans when they discovered that among the rational numbers, $\sqrt{2}$ doesn't exist. They couldn't find a number equal to $\sqrt{2}$. Before distributions, mathematicians couldn't find a function equal to the derivative of H(x). The key is a different way to think about functions.

The classical definition of function is: "a rule that associates to every point in its domain some point in its range." If domain and range are the real line, it's a real-valued function of a real variable, such as cos(x) and H(x). But there's a more sophisticated way to think of a function—as something we call a "functional."

First I'll explain functionals. Then I'll explain how a function can be regarded as a functional—the key insight in Schwartz's theory.

A functional is a function of functions. Its domain is some set of functions. To each function in that domain, it associates a number. A functional operates on a function to give rise to a number. In contrast, an ordinary function maps a number into a number. A functional maps a function into a number.

The simplest functional is "evaluate at $x = 0$." This applies to any function $w(x)$ that's defined at $x = 0$. The outcome is the number, $w(0)$.

Now we have to see how Schwartz interprets an ordinary function $f(x)$ as a functional. As a functional, it must operate on suitable "test functions" $w(x)$. We restrict the test functions to be identically zero for x very large. Say $w(x) = 0$ for all $x > a$ and all $x < b$, for some numbers a and b. Consider the product, $f(x)w(x)$ and its graph. Because $w(x)$ is zero if $x < a$ or $x > b$, the graph of $f(x)w(x)$, together with the x-axis, encloses a finite area, which is the integral of $f(x)w(x)$. (See "Calculus Refresher" above.)

This area or integral is *defined* to be the number that the functional f obtains by operating on the test function $w(x)$. It's written $f(w)$ or (f, w).

A functional by definition doesn't have a value at a number, because it operates on $w(x)$ as a whole, not on a point x. So the standard definition of derivative can't be applied directly to a functional. But there's a natural way to define a "generalized" derivative, using a formula from elementary calculus: integration by parts (explained in every calculus text).

By thinking of $H(x)$ as a functional, we can define its generalized derivative *without ignoring the crucial singularity, the discontinuity at $x = 0$.* The generalized derivative of $H(x)$ is a new entity, a functional. Although it's not a function, it's called the delta function, or *Dirac's delta function $\delta(x)$.*

The physicist Paul Dirac and his followers used this illegitimate creature to good effect without license from mathematicians. Dirac defined it as a "function" whose values are zero for all x different from zero, but infinite at $x = 0$, in such a way that its area or integral is 1! Mathematicians were amused. They knew that infinity is not a number, and the value of a function at a single point can't affect the value of its integral. Since Dirac's $\delta(x)$ equaled 0 at every point except $x = 0$, its integral must be 0, not 1!

But notice that the derivative of the Heaviside function $H(x)$, like $\delta(x)$, is zero everywhere except at $x = 0$. Where H' blows up is precisely where $\delta(x)$ is infinite! Could it be that if we look at things the right way, we'll see that $\delta(x) = H'(x)$? Yes! That's what distribution theory does for us.

If $\delta(x)$ doesn't make sense, how does it work in the physicists' calculations? Now that we understand it, there's no mystery. $\delta(x)$ can be *talked about* as a function, but it's *really* the very functional we started with—evaluation at 0!

As the meaning of "number" changes with enlargement of the number system, the meaning of "function" changes with enlargement of a function space. Generalized functions (distributions) are different from functions. But we can think of them as if they were functions, *because the usual operations on functions (addition, multiplication, differentiation, integration) extend to these functionals.*

If f(x) is *smooth,* unlike the discontinuous H(x), we can prove that the functional associated with the classical derivative f'(x) is the same as the generalized derivative of the functional associated with f(x).

The result is that *every function, no matter how rough, is infinitely differentiable in the generalized sense!* Not only do we find H'(x) = δ(x), we are able to differentiate the highly singular pseudo function δ(x) as many times as we like.

Landau's Two Constants

Let f(z) be a function of a complex variable, analytic in a neighborhood of the origin. Then it has a power series expansion centered there, which has some radius of convergence, say R. If there's no point at which f = 0 and no point at which f = 1, then the radius of convergence R is not greater than a certain constant, which depends *only on f(0) and f'(0)*. For details, see Epstein and Hahn.

More Logic

Russell's Paradox

Gottlob Frege's hope of making logic a solid, secure foundation for arithmetic was shattered in one of the most poignant episodes in the history of philosophy. Bertrand Russell found a contradiction in the notion of set as he and Frege used it!

Russell's paradox is a set-theoretic pun, a tongue-twister. To follow it, you must first see how it might be possible for a set to *belong to itself.*

Consider "The class of all sets that can be defined in less than 1,000 English words." I have just defined that class, and I used only 15 words. So it belongs to itself! (This is the Berry Paradox, which Boolos exploits in the following article.)

For a simpler example, how about, "The set of all sets." It's a set, so it's a member of itself.

Let's call those two, and any others that belong to themselves, "Russell sets." Let's call the sets that don't belong to themselves "non-Russell sets." Now, asks Russell, what about the class of *all* non-Russell sets? This class ought to be very big, so I'll call it "The Monster." The Monster contains all the non-Russell sets, and nothing else. Is The Monster a Russell set, or a non-Russell set?

According to the law of the excluded middle, every meaningful statement is true or false. According to Frege's Basic Law Five, the statement "*The Monster is a Russell set*" is meaningful. It must be true or false. However, Russell discovered, it can't be true, and it can't be false!

Suppose it's false. That is, *The Monster is a non-Russell set.* Then The Monster belongs to The Monster, by definition of The Monster! It belongs to itself! *The Monster is a Russell set*! But we came to this conclusion by supposing that The

Monster is a non-Russell set. Contradiction!! The supposition that The Monster is non-Russell is impossible.

The reader should repeat the above reasoning, starting with the presumption that The Monster is a Russell set. You will easily verify that, just as in the non-Russell case, the Russell case leads in two steps to a contradiction.

If The Monster is Russell, it must be non-Russell.

If The Monster is non-Russell, it must be Russell.

The law of the excluded middle says The Monster must be either Russell or non-Russell. Both alternatives are self-contradictory. The presumption that The Monster exists leads to either one contradiction or the other contradiction.

The Monster can't exist! There is no Monster!

A New Proof the Gödel Incompleteness Theorem

BY GEORGE BOOLOS[5]

> Many theorems have many proofs. After having given the fundamental theorem of algebra its first rigorous proof, Gauss gave it three more; a number of others have since been found. The Pythagorean theorem, older and easier than the FTA, has hundreds of proofs by now. Is there a great theorem with only one proof?
>
> In this note we shall give an easy new proof[6] of the Gödel Incompleteness Theorem in the following form: *There is no algorithm whose output contains all true statements of arithmetic and no false ones.* Our proof is quite different in character from the usual ones and presupposes only a slight acquaintance with formal mathematical logic. It is perfectly complete, except for a certain technical fact whose demonstration we will outline.
>
> Our proof exploits *Berry's paradox.* In a number of writings Bertrand Russell attributed to G. G. Berry, a librarian at Oxford University, the paradox of *the least integer not nameable in fewer than nineteen syllables.* The paradox, of course, is that that integer has just been named in eighteen syllables. Of Berry's paradox, Russell once said, "It has the merit of not going outside finite numbers."[7]
>
> Before we begin, we must say a word about algorithms and "statements of arithmetic," and about what "true" and "false" mean in the present context. Let's begin with "statements of arithmetic."
>
> The *language of arithmetic* contains signs + and × for addition and multiplication, a name *0* for zero, and a sign *s* for successor

[5] George Boolos was Professor of Philosophy at MIT.

[6] Saul Kripke has informed me that he noticed a proof somewhat similar to the present one in the early 1960s.

[7] Bertrand Russell, "On Insolubilia and Their Solution by Symbolic Logic," in *Essays in Analysis*, ed. Douglas Lackey, George Braziller, New York, 1973, p. 210.

(plus-one). It also contains the equals sign =, as well as the usual logical signs ¬ (not), ∧ (and), ∨ (or), → (if . . . then . . .), ↔ (. . . if and only if . . .), ∀ (for all), and ∃ (for some), and parentheses. The variables of the language of arithmetic are the expressions x, x', x'' . . . built up from the symbols x and $'$: they are assumed to have the natural numbers (0,1,2, . . .) as their values. We'll abbreviate variables by single letters: y, z, etc.

We now understand sufficiently well what truth and falsity mean in the language of arithmetic; for example, $\forall\exists yx = sy$ is a *false* statement, because it's not the case that every natural number x is the successor of a natural number y. (Zero is a counter-example: it is not the successor of a *natural* number.) On the other hand, $\forall x\exists y(x = (y + y) \vee x = s(y + y))$ is a true statement: for every natural number x there is a natural number y such that either $x = 2y$ or $x = 2y + 1$. We also see that many notions can be expressed in the language of arithmetic, e.g., less than: $x < y$ can be defined: $\exists z(sz + x) = y$ (for some natural number z,the successor of z plus x equals y). And, you now see that $\forall x\forall y[ss0 \times (x \times x)) = (y \times y) \rightarrow x = 0]$ is—well, test yourself, is it true or false? (Big hint: $\sqrt{2}$ is irrational.)

For our purposes, it's not really necessary to be more formal than we have been about the syntax and semantics of the language of arithmetic.

By an *algorithm*, we mean a computational (automatic, effective, mechanical) procedure or routine of the usual sort, e.g., a program in a computer language like C, Basic, Lisp, . . . , a Turing machine, register machine, Markov algorithm, . . . a formal system like Peano or Robinson arithmetic, . . . , or whatever. We assume that an algorithm has an *output*, the set of things it "prints out" in the course of computation. (Of course an algorithm might have a *null* output.) If the algorithm is a formal system, then its output is just the set of statements that are provable in the system.

Although the language of arithmetic contains only the operation symbols s, $+$, and $\times$, it turns out that many statements of mathematics can be reformulated as statements in the language of arithmetic, including such famous propositions as Fermat's last theorem, Goldbach's conjecture, the Riemann hypothesis, and the widely held belief that $P \neq NP$. Thus if there were an algorithm that printed out all and only the true statements of arithmetic—as Gödel's theorem tells us there is not—we would have a way of finding out whether each of these as yet unproved propositions is true or not, and indeed a way of finding out whether or not any statement that can be formulated as a statement S of arithmetic is true: start the algorithm, and simply wait to see which of S and its negation $-S$ the algorithm prints out. (It must eventually print out exactly one of S and $-S$ if it prints out all truths and no falsehoods, for, certainly, exactly one of S and $-S$ is true.) But alas, there is no worry that the algorithm might take too long to come up with an answer to a question that interests us, for there is, as we shall now show, no algorithm to do the job, not even an infeasibly slow one.

To show that there is no algorithm whose output contains all true statements of arithmetic and no false ones, we suppose that M is an algorithm whose output contains no false statements of arithmetic. We shall show how to find a true statement of arithmetic that is not in M's output, which will prove the theorem.

For any natural number n, we let $[n]$ be the expression consisting of 0 preceded by n successor symbols s. For example, $[3]$ is $sss0$. Notice that the expression $[n]$ stands for the number n.

We need one further definition: We say that a formula $F(x)$ ***names*** the (natural) number x if the following statement is in the output of M: $\forall \times F(x) \leftrightarrow x = [n]$. (Observe that the definition of 'names' contains a reference to the algorithm M.) Thus, for example, if $\forall x(x + x = ssss0 \leftrightarrow x = ss0)$ is in the output of M, then the formula $x + x = ssss0$ names the number 2.

No formula can name two different numbers. For if both of $\forall x(F(x) \leftrightarrow x = [n])$ and $\forall x(F(x) \leftrightarrow x = [p])$ are true, then so are $\forall x(x = [n] \leftrightarrow x = [p])$ and $[n] = [p]$, and the number n must equal the number p. Moreover, for each number i, there are only finitely many different formulas that contain i symbols. (Since there are 16 primitive symbols of the language of arithmetic, there are at most 16^i formulas containing i symbols.) Thus for each i, there are only finitely many numbers named by formulas containing i symbols. For every m, then, only finitely many (indeed, $\leq 16^{m-1} + \ldots + 16^1 + 16^0$) numbers are named by formulas containing fewer than m symbols; some number is not named by any formula containing fewer than m symbols; and therefore there is a least number not named by any formula containing fewer than m symbols.

Let $C(x,z)$ be a formula of the language of arithmetic that says that x is a number that is named by some formula containing z symbols. The technical fact mentioned above that we need is that whatever sort of algorithm M may be, there is some such formula $C(x,z)$. We sketch the construction of $C(x,z)$ below, in 3).

Now let $B(x,y)$ be the formula $\exists z(z < y \wedge C(x,z))$. $B(x,y)$ says that x is named by some formula containing fewer than y symbols.

Let $A(x,y)$ be the formula $(\neg B(x,y) \wedge \forall a(a < x \rightarrow B(a,y)))$. $A(x,y)$ says that x is the least number not named by any formula containing fewer than y symbols.

Let k be the number of symbols in $A(x,y)$. $k > 3$.

Finally, let $F(x)$ be the formula $\exists y(y = ([10] \times [k] \wedge A(x,y))$. $F(x)$ says x is the least number not named by any formula containing fewer than $10k$ symbols.

How many symbols does F contain? Well, $[10]$ contains 11 symbols, $[k]$ contains $k + 1$, $A(x,y)$ contains k, and there are 12 others (since y is x^1): so $2k + 24$ in all. Since $k > 3$, $2k + 24 < 10k$, and $F(x)$ contains fewer than $10k$ symbols.

We saw above that for every m, there is a least number not named by any formula containing fewer than m symbols. Let n be the least such number for $m = 10k$. Then n is not named by $F(x)$; in other words, $\forall x(Fx) \leftrightarrow x = [n]$ is not in the output of M.

But $\forall x(F(x) \leftrightarrow x = [n])$ is a true statement, since n *is* the least number not named by any formula containing fewer than $10k$ symbols! Thus we have found a true statement that is not in the output of M, namely, $\forall x(F(x) \leftrightarrow x = [n])$. Q.E.D.

Some comments about the proof:

1. In our proof, the symbols are the "syllables," and just as "nineteen" contains $2 << 19$ syllables, so the term $([10]x[k])$ contains $k + 15 << 10k$ symbols.

2. In his memoir of Kurt Gödel,[8] Georg Kreisel reports that Gödel attributed his success not so much to mathematical invention as to attention to philosophical distinctions. Gregory Chaitin once commented that one of his own incompleteness proofs resembled Berry's paradox rather than Epimenides' paradox of the liar ("What I am now saying is not true").[9] Chaitin's proofs make use of the notion of the *complexity* of a natural number, i.e., the minimum number of instructions in the machine table of any Turing machine that prints out that number, and of various information-theoretic notions. None of these notions are found in our proof, for which the remarks of Kreisel and Chaitin, which the author read at more or less the same time, provided the impetus.

[8] Georg Kreisel, "Kurt Gödel, 28 April 1906–14 January 1978." *Biographical memoirs of Fellows of the Royal Society* 26 (1980), p. 150.

[9] Cf. Martin Davis, "What is a computation?" in *Mathematics Today*, ed. Lynn Arthur Steen, Vintage Books, New York, 1980, pp. 241–267, especially pp. 263–267, for an exposition of Chaitin's proof of incompleteness. Chaitin's observation is found in Chaitin, Gregory, "Computational complexity and Gödel's incompleteness theorem," (Abstract) *AMS Notices* 17 (1970), p. 672.

3. Let us now sketch the construction of a formula $C(x,z)$ that says that x is a number named by a formula containing z symbols. The main points are that algorithms like M can be regarded as operating on "expressions," i.e., finite sequences of symbols; that, in a matter reminiscent of ASCII codes, symbols can be assigned code numbers (logicians often call these code numbers *Gödel* numbers); that certain tricks of number theory enable one to code expressions as numbers and operations on expressions as operations on the numbers that code them; and that these numerical operations can all be defined in terms of addition, multiplication, and the notions of logic. Discussion of symbols, expressions (and finite sequences of expressions, etc.) can therefore be coded in the language of arithmetic as discussion of the natural numbers that code them. To construct a formula saying that n is named by some formula containing I symbols, one writes a formula saying that there is a sequence of operations of the algorithm M (which operates on expressions) that generates the expression consisting of $\forall$, x, $($, the i symbols of some formula $F(x)$ of the language of arithmetic, $\leftrightarrow$, x, $=$, n consecutive successor symbols s, 0, *and*$)$. Gödel numbering and tricks of number theory then allow all such talk of symbols, sequences, and the operations of M to be coded into formulas of arithmetic.

4. Both our proof and the standard one make use of Gödel numbering. Moreover, the unprovable truths in our proof and in the standard one can both be seen as obtained by the substitution of a name for a number in a certain crucial formula. There is, however, an important distinction between the two proofs. In the usual proof, the number whose name is substituted is the code for the formula into which it is substituted; in ours it is the true number of which the formula is *true*. In view of this distinction, it seems justified to say that our proof, unlike the usual one, does not involve *diagonalization*.

In a later issue of the journal where his proof appeared, Boolos made some interesting remarks about it. "Several readers of my 'New Proof of the Gödel Incompleteness Theorem,' (*Notices*, April 1989, pp. 388–90) have commented on its shortness, apparently supposing that the use it makes of Berry's paradox is responsible for that brevity. It would thus seem appropriate to remark that once syntax is arithmetized, an even briefer proof of incompleteness is at hand, essentially the one given by Gödel himself in the introduction to his famous "On Formally Undecidable Propositions. . . ."

Say that m applies to n if F[n] is in the output of M, where F(x) is the formula with Gödel number m. Let A(x,y) express "applies to" and let n be the Gödel number of −A(x,x). If n applies to n, the false statement −A([n],[n]) is in the output of M, impossible; thus n does not apply to n and −A([n],[n]) is a truth not in the output of M.

What is concealed in this argument is the large amount of work needed to construct a suitable formula A(x,y); proving the existence of the key formula C(x,z) in the "New Proof" via Berry's paradox requires at least as much effort. What strikes the author as of interest in the proof via Berry's paradox is not its brevity but that it provides ***a different sort of reason*** for the incompleteness of algorithms.

Bibliography

Aczel, P. *Non Well Founded Sets.* Stanford: Center for the Study of Language Information, 1988.

Agard, W. R. *What Democracy Meant to the Greeks.* Chapel Hill: University of North Carolina Press, 1942.

Angelelli, I. *Studies on Gottlob Frege and Traditional Philosophy.* Dordrecht: Reidel, 1967.

Anglin, W. S. *Mathematics, A Concise History and Philosophy.* New York: Springer-Verlag, 1994.

Anscombe, G. E. M. and P. Geach. *Three Philosophers.* Ithaca: Cornell University Press, 1961.

Apostle, H. G. *Aristotle's Philosophy of Mathematics.* Chicago: University of Chicago Press, 1952.

Appel, K. and W. Haken. "The Four-Color Problem." In L. A. Steen, ed., *Mathematics Today*, pp. 153–80. New York: Springer-Verlag, 1978.

Appel, K., W. Haken, and J. Koch. "Every Planar Map Is Four-Colorable." *Illinois Journal of Mathematics* 21: 429–567, 1977.

Aquinas, St. T. *The Pocket Aquinas.* Edited by Vernon J. Bourke. New York: Washington Square Press, 1960.

Archer-Hind, R. D. *The Timaeus of Plato.* London: Macmillan, 1888.

Aristotle. *Basic Works.* Edited by R. McKeon. New York: Random House, 1941.

———. *Physics, Book VI*, chapters 2 and 9. In R. M. Hutchins, ed., *Great Books of the Western World*, Vol. 8. Chicago: Encyclopedia Britannica, 1952.

———. *Poetics.* Chapel Hill: University of North Carolina Press, 1986, p. 1451

Ascher, M. *Ethnomathematics: A Multicultural View of Mathematical Ideas.* San Francisco: Brooks/Cole, 1991.

Aspray, W. and P. Kitcher, P., eds. *History and Philosophy of Modern Mathematics, Volume XI, Minnesota Studies in the Philosophy of Science.* Minneapolis: University of Minnesota Press, 1988.

Balz, A. G. A. *Cartesian Studies.* New York: Columbia University Press, 1951.

Bartley, W. W. *Wittgenstein.* Lasalle: Open Court, 1985.

Barwise, J. and J. Etchemendy. *The Liar.* New York: Oxford University Press, 1987.

Baum, R. J. *Philosophy and Mathematics from Plato to the Present.* San Francisco: Freeman, Cooper, 1973.

Beck, L. J. *The Method of Descartes.* New York: Oxford University Press, 1952.

Benacerraf, P. and H. Putnam. *Philosophy of Mathematics.* New York: Prentice-Hall, 1964.

Benacerraf, P. "Frege, The Last Logicist." In P. French, T. Ukehling, H. Wettstein, eds., *Midwest Studies in Philosophy VI.* Minneapolis: University of Minnesota Press, 1981.

Berkeley, G. *The Analyst.* London: J. Tonson, 1774.

Berlin, I., ed. *The Age of Enlightenment. The 18th Century Philosophers.* New York: New American Library, 1956.

Bernays, P. "Comments on Ludwig Wittgenstein's Remarks on the Foundations of Mathematics." In P. Benacerraf and H. Putnam.

Beth, E. W. and J. Piaget. *Mathematics, Epistemology and Psychology.* Translated by W. Mays. Dordrecht: Reidel, 1966.

Bishop, E. *Foundations of Constructive Analysis.* New York: McGraw Hill, 1967.

———. *Aspects of Constructivism.* Las Cruces, N.Mex.: Department of Mathematical Sciences, New Mexico State University, 1972.

———. "The Crisis in Contemporary Mathematics." *Historia Mathematica* 2: 507–17, 1975.

———. "Schizophrenia in Contemporary Mathematics." In Rosenblatt, M., ed., *Errett Bishop: Reflections on Him and His Research.* Providence: American Mathematical Society, 1985.

Bledsoe, W. W. and D. W. Loveland. *Automated Theorem Proving: After 25 Years.* Contemporary Mathematics 29. Providence: American Mathematical Society, 1984.

Blitz, D. *Emergent Evolution.* Boston: Kluwer, 1992.

Bloor, D. *Knowledge and Social Imagery.* Boston: Routledge and Kegan Paul, 1976.

———. *Wittgenstein: A Social Theory of Knowledge.* New York: Columbia University Press, 1983.

Bluck, R. S. *Plato's Meno.* New York: Cambridge University Press, 1964.

Boolos, G. "New Proof of the Gödel Incompleteness Theorem." *Notices of the American Mathematical Society* 36: 388–90, 1989.

———. "A Letter from George Boolos." *Notices of the American Mathematical Society* 36: 676, 1989.

Bourbaki, N. "Foundations of Mathematics for the Working Mathematician." *Journal of Symbolic Logic* 14: 1–8, 1949.

Boyer, C. "Galileo's Place in the History of Mathematics." In E. McMullin, ed., *Galileo Man of Science*. New York: Basic Books, 1968, p. 251.

———. *History of Analytic Geometry*. New York: Scripta Mathematica, 1956.

———. *The History of the Calculus and Its Conceptual Development*. New York: Dover, 1959.

———. *A History of Mathematics*, 2nd. ed., revised by Uta C. Merzbach. New York: Wiley, 1991.

Brouwer, L. E. J. *Life, Art and Mysticism*. In *Collected Works*. Amsterdam: North-Holland, 1975.

———. *Brouwer's Cambridge Lectures on Intuitionism*. Edited by D. Van Dalen. New York: Cambridge University Press, 1981.

Brown, M., ed. *Plato's Meno*. Indianapolis: Bobbs-Merrill, 1971.

Brumbaugh, R. S. *Plato's Mathematical Imagination*. Bloomington: Indiana University Press, 1954.

Bunge, M. *Intuition and Science*. Englewood Cliffs: Prentice Hall, 1962.

———. *The Mind-Body Problem*. New York: Pergamon, 1980.

Carnap, R. *Introduction to Symbolic Logic and Its Applications*. New York: Dover, 1958.

Carr, H. W. *Leibniz*. New York: Dover, 1960.

Castonguay, C. *Meaning and Existence in Mathematics*. New York: Springer-Verlag, 1972.

Chihara, C. *Ontology and the Vicious Circle Principle*. Ithaca: Cornell University Press, 1973.

———. "Wittgenstein's Discussion of the Paradoxes in his 1939 Lectures on the Foundations of Mathematics." *Philosophical Review* 86: 365–81, 1977.

Cohen, D. I. A. "The Superfluous Paradigm." In J. H. Johnson and M. J. Loomis, eds., *The Mathematical Revolution Inspired by Computing*. Oxford: Clarendon Press, 1991.

Cohen, P. "Comments on the Foundations of Set Theory." In Dana Scott, ed., *Axiomatic Set Theory*, pp. 9–15. Providence: American Mathematical Society, 1971.

———. *Set Theory and the Continuum Hypothesis*. New York: W. A. Benjamin, 1966.

Cohen, R. S. and L. Lauden. *Physics, Philosophy and Psychology: Essays in Honor of A. Grunbaum*. Dordrecht: Reidel, 1983.

Cornford, F. M. *Plato's Cosmology*. London: Routledge & Kegan Paul, 1937.

———. *From Religion to Philosophy*. New York: Harper and Row, 1957.

Corrington, R. S. *An Introduction to C. S. Peirce*. Lanham, Md.: Rowman & Littlefield, 1993.

Courant, R. and Robbins, H. *What Is Mathematics?* New York: Oxford University Press, 1948.

Crowe, M. J. *A History of Vector Analysis*. New York: Dover, 1967.

———. "Ten 'Laws' Concerning Patterns of Change in the History of Mathematics." *Historia Mathematica* 2: 161–66, 1975.

———. "Ten Misconceptions about Mathematics and Its History." In Aspray and Kitcher, 1988.
Curley, E. M. *Descartes Against the Skeptics.* Cambridge: Harvard University Press, 1978.
Currie, G. "Frege's Realism." *Inquiry* 21: 1978.
Curry, H. *Outline of a Formalist Philosophy of Mathematics.* Amsterdam: North-Holland, 1958.
D'Alembert, J. L. R. *Preliminary Discourse to the Encyclopedia of Diderot.* Translated by R. N. Schwab. Indianapolis: Bobbs-Merrill, 1963.
Damasio, A. *Descartes' Error: Emotion, Reason and the Human Mind.* New York: Putnam, 1994.
Dancy, J. and E. Sosa. *A Companion to Epistemology.* Oxford: Blackwell, 1992.
Dantzig, T. *Number, the Language of Science.* New York: Macmillan, 1959.
Davis, C. "Materialist Mathematics." *Boston Studies in the Philosophy of Science*, Vol. 15. Dordrecht: Reidel, 1974, pp. 37–66.
Davis, E. W. In Cajori, F., ed., *Teaching and History of Mathematics in the U.S.* Washington, D.C.: 1890.
Davis, M. and R. Hersh. "Nonstandard Analysis." *Scientific American* June 1972, pp. 768–84.
———. "Hilbert's Tenth Problem." *Scientific American* November 1973, pp. 84–91.
Davis, P. J. "Fidelity in Mathematical Discourse: Is 1 + 1 Really 2?" *American Mathematical Monthly* 78: 252–63, 1972.
———. "Proof, Completeness, Transcendentals, and Sampling." *Journal of the Association for Computing Machinery* 24: 298–310, 1977.
———. "Mathematics by Fiat?" *The Two-Year College Mathematics Journal* June 1980.
Davis P. J. and R. Hersh. *The Mathematical Experience.* Cambridge: Birkhauser, 1981.
Davydov, V. *Problemy Razvivayuschego Obucheniaya* (The problems of development-generated learning). Moscow: Pedagogika, 1986.
Dedekind, J. W. R. *Essays on the Theory of Numbers.* Lasalle: Open Court, 1901. Reprinted by Dover Books, 1963.
Dejnozka, J. "Zeno's Paradoxes and the Cosmological Argument." *Philosophy of Religion* 25: 65–81, 1989.
DeMillo, R. A., R. J. Lipton, and A. J. Perlis. "Social Processes and Proofs of Theorems and Programs." *Communications of the ACM* 22: 271–80, 1970.
Descartes, R. "Objectiones Septimae cum Notis Authoris sive Dissertatio de Prima Philosophia." In C. Adam and P. Tannery, eds., *Oeuvres*, 12 vols. Paris: L. Cert, 1897–1910.
———. *Correspondence.* Edited by C. Adam and G. Milhaud. Paris: Felix Alcan (Vols. 1–2). Presses Universitaires de France (Vols. 3–8), 1936–1963.
———. *Philosophical Works.* Translated by E. S. Haldane and G. R. T. Ross. New York: Dover, 1955.

———. *A Discourse on Method, Optics, Geometry and Meteorology.* Translated by Paul J. Olscamp. Indianapolis: Bobbs-Merrill, 1965.

de Villiers, M. "The Role and Function of Proof in Mathematics." *Pythagoras* 24: 17–24, 1990.

———. "Pupils' Needs for Conviction and Explanation within the Context of Geometry." *Pythagoras* 26: 18–27, 1991.

Dieudonné, J. "The Work of Nicholas Bourbaki." *American Mathematical Monthly* 77: 134–45, 1970.

Dillon, J. *The Middle Platonists.* Ithaca: Cornell University Press, 1977.

Dreben, B., P. Andrews, and S. Anderaa. "False Lemmas in Herbrand." *Proceedings of the American Mathematical Society* 69: 699–706, 1963.

Dummett, M. "Frege's Philosophy." In *Truth and Other Enigmas*, p. 89. Cambridge: Harvard University Press, 1978. Originally published as an article on Frege in P. Edwards, ed., *Encyclopedia of Philosophy.* New York: Macmillan, 1967.

———. *Frege: Philosophy of Language.* New York: Harper and Row, 1973.

———. "Frege as a Realist." *Inquiry* 19: 468, 1976.

———. *Elements of Intuitionism.* Oxford: Clarendon Press, 1977.

———. "Wittgenstein's Philosophy of Mathematics." In *Truth and Other Enigmas.* Cambridge: Harvard University Press, 1978.

———. "Frege and Wittgenstein." In Block, I., ed., *Perspectives on the Philosophy of Wittgenstein.* Cambridge: MIT Press, 1981.

Durkheim, E. *The Rules of Sociological Method.* Chicago: University of Chicago Press, 1938. Preface to 2nd edition, p. 61.

———. *Essays on Sociology and Philosophy.* New York: Harper and Row, 1964.

———. *The Elementary Forms of the Religious Life.* London: George Allen and Unwin, 1976.

Echeverria, J., A. Ibarra, and R. Mormann, eds. *The Space of Mathematics. Philosophical, Epistemological and Historical Explorations.* Berlin, New York: de Gruyter, 1992.

Eckstein, J. *The Platonic Method.* New York: Greenwood, 1968.

Edelman, G. M. *Neural Darwinism.* New York: Basic Books, 1987.

———. *Bright Air, Brilliant Fire.* New York: Basic Books, 1992.

Emerson, R. W. *Self-Reliance.* New York: Crowell, 1901.

Enderton, H. B. *A Mathematical Introduction to Logic.* New York: Academic Press, 1972.

Epstein, D. and S. Levy. "Experimentation and Proof in Mathematics." *Notices of the American Mathematical Society* 42: 670–74, 1995.

Ernest, P. *The Philosophy of Mathematics Education.* New York: Falmer, 1991.

———. *Social Constructivism in the Philosophy of Mathematics.* Albany: SUNY Press, 1997.

Euclid. *The Thirteen Books of Euclid's Elements.* Introduction and Commentary by T. L. Heath. New York: Dover, 1956.

Fann, K. T. *Wittgenstein's Conception of Philosophy.* Oxford: Blackwell, 1969.

Feferman, S. *The Logic of Mathematical Discovery vs. the Logical Structure of Mathematics.* Department of Mathematics, Stanford University, 1976.

———. "What Does Logic Have to Tell Us about Mathematical Proofs?" *Mathematical Intelligencer* 2: 20–24, 1979.

Fell, M. *Spinoza's Earliest Publication? A Loving Salutation.* In R. H. Popkin and M. A. Signer, eds. Assen, The Netherlands: Van Gorcum, 1987.

Fetisov, A. I. *Proof in Geometry.* Boston: D. C. Heath, 1963.

Feuer, L. S. *Spinoza and the Rise of Liberalism.* Boston: Beacon, 1966.

Field, H. *Science without Numbers.* Princeton: Princeton University Press, 1980.

Findlay, J. N. *Plato—The Written and Unwritten Doctrines.* London: Routledge & Kegan Paul, 1974.

Floyd, J., "Wittgenstein, Gödel and the Trisection of the Angle." To appear in J. Hintikkaa, ed., *The Foundations of Mathematics in the Early Twentieth Century.*

Floyd, J. "Wittgenstein on 2,2,2. . . : The Opening of *Remarks on the Foundations of Mathematics.*" *Synthèse* 87: 143–80, 1991.

Fogelin, R. J. *Wittgenstein.* London: Routledge, 1976.

———. "Hume and Berkeley on the Proofs of Infinite Divisibility." *The Philosophical Review* 97: 47–69, 1988.

Fowler, D. H. *The Mathematics of Plato's Academy. A New Reconstruction.* Oxford: Clarendon Press, 1987.

Frascola, P. *Wittgenstein's Philosophy of Mathematics.* London: Routledge, 1994.

Frege, G. *The Thought: A Logical Inquiry.* A translation of part of Frege, *Der Gedanke*, 1919.

———. *Translations from the Philosophical Writings of Gottlob Frege.* In P. T. Geach and M. Black, eds. Oxford: Blackwell, 1950.

———. "Begriffsschrift." In J. K. van Heijenoort, ed., *From Frege to Godel: A Source Book in Mathematical Logic.* Cambridge: Harvard University Press, 1967.

———. *On the Foundations of Geometry and Formal Theories of Arithmetic.* Edited by E.-H. W. Kluge. New Haven: Yale University Press, 1972.

———. *Conceptual Notation and Related Articles.* Edited by T. W. Bynum. New York: Oxford University Press, 1972.

———. *Logical Investigations.* Translated by P. T. Geach and R. H. Stroothoff. Oxford: Blackwell, 1977.

———. In H. Hermes et al., eds., *Posthumous Writings.* Chicago: University of Chicago Press, 1979.

———. *The Foundations of Arithmetic.* Evanston: Northwestern University Press, 1980.

———. *Philosophical and Mathematical Correspondence.* Abridged, B. McGuinness. London: Blackwell. Chicago: University of Chicago Press, 1980.

———. *Collected Papers on Mathematics, Logic and Philosophy.* New York: Blackwell, 1984.

Freudenthal, H. *Mathematics as an Educational Task.* Dordrecht: Reidel, 1973.

Friedman, M. "Kant's Theory of Geometry." *The Philosophical Review* 94: 455–506, 1985.

Furth, M. *The Basic Laws of Arithmetic.* Berkeley: University of California Press, 1964.

Gale, D. "Proof as Explanation." *The Mathematical Intelligencer* 12: 4, 1991.

Gardner, M. *Order and Surprise*. Buffalo: Prometheus Books, 1983.

Geach P. T. "On Names of Expresssions." *Mind*, 1950.

Gerrard, S. "Wittgenstein's Philosophies of Mathematics." *Synthèse* 87: 125–42, 1991.

Gleick, J. *Chaos*. New York: Penguin, 1987.

Glimm, J., J. Impagliazzo, and I. Singer, eds. *The Legacy of John von Neumann. Proceedings of Symposia in Pure Mathematics*, Vol. 50. Providence: American Mathematical Society, 1990.

Gödel, K. *The Consistency of the Axiom of Choice and of the Generalized Continuum Hypothesis with the Axioms of Set Theory*. Princeton, 1940.

Gödel, K. "What Is Cantor's Continuum Problem?" In P. Benacerraf and H. Putnam, eds., 1988.

Gödel, K. *Collected Works*, Vol. III. New York: Oxford University Press, 1995.

Goffman, E. *The Presentation of Self in Everyday Life*. Garden City, N.Y.: Doubleday, 1959.

Goldmann, L. *Immanuel Kant*. London: NLB, 1971.

Goldstine, H. H. (1972). *The Computer from Pascal to von Neumann*. Princeton: Princeton University Press, 1972.

Goodman, N. "Mathematics as an Objective Science." *American Mathematics Monthly* 81: 354–65, 1974.

——. "Worlds of Individuals." In *Problems and Projects*. Indianapolis: Bobbs-Merrill, 1972.

Grabiner, J. "Descartes and Problem-Solving." *Mathematics Magazine* 68: 83–97, 1995.

Grosholz, E. R. "Descartes' Unification of Algebra and Geometry." In S. Gaukroger, ed., *Descartes Philosophy, Mathematics and Physics*. Sussex: Harvester, 1980, pp. 156–68.

Guggenheimer, H. "The Axioms of Betweenness in Euclid." *Dialectica* 31: 187–92, 1977.

Haaparanta, L. and J. Hintikka, eds. *Frege Synthesized*, pp. 299–343. Dordrecht: Reidel, 1986.

Hadamard, J. *The Psychology of Invention in the Mathematical Field*. New York: Dover, 1945.

Hahn, H. "The Crisis in Intuition." In J. R. Newman, ed., *The World of Mathematics*, 1957–1976. New York: Simon & Schuster, 1956.

Hahn, L. E. and P. A. Schilpp, ed. *The Philosophy of W. V. Quine*. La Salle, Illinois, Open Court, 1986.

Hahn, L. S. and B. Epstein. *Classical Complex Analysis*. Sudbury, Mass.: Jones and Bartlett, 1996.

Halmos, P. Address to 75th annual summer meeting of the Mathematical Association of America. Columbus, Ohio. Tape recording, 1990.

Hanna, G. *Rigorous Proof in Mathematics Education*. Toronto: OISE Press, 1983.

———. "Some Pedagogical Aspects of Proof." *Interchange* 21: 6–13, 1990.

Hardy, G. H: "Mathematical Proof." *Mind* 38: 1–25, 1929.

———. *A Mathematician's Apology*. New York: Cambridge University Press, 1940.

Hatfield, G. *The Natural and the Normative*. Cambridge: MIT Press, 1990.

Heath, Sir T. L. *Aristarchus of Samos; the Ancient Copernicus*. Oxford: Clarendon Press, 1913.

———. *Mathematics in Aristotle*. Oxford, 1949.

———. *A History of Greek Mathematics*. New York: Dover, 1981.

Heijenoort, J. van. *From Frege to Godel*. Cambridge: Harvard University Press, 1967.

Heims, S. J. *John von Neumann and Norbert Wiener*. Cambridge: MIT Press, 1980.

Helmholtz, H. *Epistemological Writings*. Boston: Reidel, 1977.

Henrici, P. "Reflections of a Teacher of Applied Mathematics." *Quarterly of Applied Mathematics* 30: 31–39, 1972.

Hersh, R. "Introducing Imre Lakatos." *Mathematical Intelligencer* 1: 148–51, 1978.

———. "Some Proposals for Reviving the Philosophy of Mathematics." *Advances in Mathematics* 31: 31–50, 1979.

———. "Inner Vision, Outer Truth." In R. Mickens, ed., *Mathematics and Science*. Singapore: World Scientific, 1990.

———. "Proving Is Convincing and Explaining." *Educational Studies in Mathematics* 24: 389–99, 1993.

Hersh, R. and V. John-Steiner. "A Visit to Hungarian Mathematics." *Mathematical Intelligencer* 15: 13–26, 1993.

Hesse, M. "Epistemology Socialized." In E. McMillen, ed., *Construction & Constraint*. Notre Dame University Press, 1988.

Hessen, B. *The Social and Economic Roots of Newton's Principia*. New York: H. Fertig, 1971.

Hilbert, D. "On the Infinite." In P. Benacerraf and H. Putnam, 1988.

Hintikka, J. "Kant on the Mathematical Method." In L. W. Beck, ed., *Kant Studies Today*. La Salle, Ill.: Open Court, 1969.

Holton, G. *Thematic Origins of Scientific Thought*. Cambridge: Harvard University Press, 1973.

Hone, J. N. and Rossi, M. M. *Bishop Berkeley, His Life, Writings, and Philosophy*. London: Faber, 1931.

Hopkins, J. *A Concise Introduction to the Philosophy of Nicholas of Cusa*. Minneapolis: University of Minnesota Press, 1978.

———. *Nicholas of Cusa's Metaphysic of Contraction*. Minneapolis: A. J. Banning, 1983.

Hume, D. *An Abstract of a Treatise of Human Nature*. Edited by J. M. Keynes and P. Sraffa. New York: Cambridge University Press, 1938.

———. *An Inquiry Concerning Human Understanding*. Indianapolis: Bobbs-Merrill, 1955.

———. *Philosophical Works*. London: T. H. Green and T. H. Grose, 1874–1975; Scientia Verlag Aalen, 1964.

———. *A Treatise of Human Nature.* New York: Penguin, 1969.

Husserl, E. "The Origin of Geometry." In *The Crisis of European Science*, Appendix VI. Evanston: Northwestern University Press, 1970.

Iliev, L. "Mathematics as the Science of Models." *Russian Mathematical Surveys* 27: 181–89, 1972.

Irvine, A. D., ed. *Physicalism in Mathematics.* Dordrecht: Kluwer, 1990

Isaacson, D. "Mathematical Intuition and Objectivity." In A. George, ed., *Mathematics and Mind.* New York: Oxford University Press, 1994.

Janik, A. and S. Toulmin. *Wittgenstein's Vienna.* New York: Simon & Schuster, 1973.

Jesseph, D. M. *Berkeley's Philosophy of Mathematics.* Dissertation, Princeton University, January 1987.

Johnson, M. *The Body in the Mind: The Bodily Basis of Meaning, Reason and Imagination.* Chicago: University of Chicago Press, 1987.

Kac, M., G.-C. Rota, and J. Schwartz. *Discrete Thoughts.* Boston: Birkhauser, 1986.

Kant, I. *Philosophy.* Translated by J. Watson. Glasgow: James Maclehose & Sons, 1901.

———. *Critique of Pure Reason.* Chicago: Encyclopedia Britannica, 1952.

———. *Critique of Practical Reason.* New York: Liberal Arts Press, 1956.

———. *Foundations of the Metaphysics of Morals.* Indianapolis: Bobbs-Merrill, 1959.

———. *Prolegomena to Any Future Metaphysics.* La Salle, Ill.: Open Court, 1967.

———. *Metaphysical Foundations of Natural Science.* Indianapolis: Bobbs-Merrill, 1970.

———. *Logic.* Indianapolis: Bobbs-Merrill, 1974.

Kielkopof, C. F. *Strict Finitism.* Paris, Mouton, 1970.

Kitchener, R. F. *Piaget's Theory of Knowledge.* New Haven: Yale University Press.

Kitcher, P. *Kant's Transcendental Psychology.* New York: Oxford University Press, 1990.

Kitcher, P. "Frege's Epistemology." *Philosophical Review* 88: 235–62, April 1979.

———. *The Nature of Mathematical Knowledge.* New York: Oxford University Press, 1983.

———. "Mathematical Naturalism." In Aspray and Kitcher, 1988.

Klein, F. Development of Mathematics in the 19th Century. Brookline, Mass.: Math Science Press, 1979. Translation by M. Ackerman of *Vorlesungen uber die Entwichlung der Mathematik in 19 Jahrhundert.* Teil I, Berlin: Springer-Verlag, 1929.

Klein, J. *A Commentary on Plato's Meno.* Chapel Hill: University of North Carolina Press, 1965.

———. *Greek Mathematical Thought and the Origin of Algebra.* Cambridge: MIT Press, 1968.

Klenk, V. H. *Wittgenstein's Philosophy of Mathematics.* The Hague: Nijhoff, 1976.

Kline, M. *Mathematical Thought from Ancient to Modern Times.* New York: Oxford University Press, 1972.

Kneale, W. and M. Kneale. *The Development of Logic.* New York: Oxford University Press, 1962.

Knuth, D. E. *The Art of Computer Programming.* Reading, Mass.: Addison Wesley, 1968–1973.

———. "Mathematics and Computer Science: Coping with Finiteness." *Science* 194: 1235–42, 1976.

Koehler, O. "The Ability of Birds to 'Count.' " In J. R. Newman, ed., *The World of Mathematics,* Vol. 1, pp. 489–96. New York: Simon & Schuster, 1956.

Kopell, N. and G. Stolzenberg. "Commentary on Bishop's Talk." *Historia Mathematica* 2: 519–21, 1975.

Körner, S. *Kant.* New York: Penguin, 1955.

———. *The Philosophy of Mathematics.* New York: Dover, 1968.

Koyré, A. *Discovering Plato.* New York: Columbia University Press, 1945.

Kozulin, A. *Vygotsky's Psychology, A Biography of Ideas.* Cambridge: Harvard University Press, 1990.

Kreisel, G. "Wittgenstein's Remarks on the Foundations of Mathematics." *British Journal for the Philosophy of Science* 9: 158, 1958.

———. "Critical Notice: 'Lectures on the Foundations of Mathematics'." In S. G. Shanker, ed., *Ludwig Wittgenstein: Critical Assessments.* London: Croom Helm, 1986, pp. 98–110.

Kripke, S. *Wittgenstein on Rules and Private Language.* New York: Oxford University Press, 1982.

Kubitz, O. A. *Development of John Stuart Mill's System of Logic.* Urbana: University of Illinois, 1932.

Kuyk, W. *Complementarity in the Philosophy of Mathematics.* Dordrecht: Reidel, 1977.

Lakatos, I. *Proofs and Refutations: The Logic of Mathematical Discovery.* New York: Cambridge University Press, 1976.

———. *Mathematics, Science and Epistemology. Philosophical Papers, Volume 2.* New York: Cambridge University Press, 1978.

Lakoff, G. and R. E. Nuñez. "The Metaphorical Structure of Mathematics: Sketching Out Cognitive Foundations For a Mind-Based Mathematics." To appear in L. English., ed., *Mathematical Reasoning: Analogies, Metaphors, and Images.* Hillsdale, N.J.: Erlbaum, 1996.

Lambert, J. H. *Theorie der Parallelinien.* Leipzig: 1786.

Land, J. P. "Kant's Space and Modern Mathematics." *Mind* 2: 38–46, 1877.

Lanford, O. E. III. "A Computer-assisted Proof of the Feigenbaum Conjectures." *Bulletin of the American Mathematical Society* (N.S.) 6: 427–34, 1982.

———. "Computer-assisted Proofs in Analysis." *Physica* 124A: 465–70, 1984.

———. "A Shorter Proof of the Existence of the Feigenbaum Fixed Point." *Communications in Mathematical Physics* 96: 521–38, 1984.

Laptev, B. L. *Lambert as a Geometer.* Istoriko-mat. issl. 25: 248–60 (Russian), 1980.

Lazerowitz, M. and A. Ambrose. *Essays in the Unknown Wittgenstein.* Buffalo: Prometheus, 1984.

Lear, J. "Aristotle's Philosophy of Mathematics." *Philosophical Review* 91: 161–92, 1982.

Leavis, F. R. *Nor Shall My Sword*. London: Chatto & Windus, 1972.

Lehmer, D. N. *List of Prime Numbers from 1 to 10,006,721*. Washington, D.C.: Carnegie Institution of Washington, Publication No. 163, 1914.

Leibniz, G. W. F. *Philosophical Works*. New Haven: Tuttle, Morehouse and Taylor, 1908.

———. *Monadology and Other Philosophical Essays*. Indianapolis: Bobbs-Merrill, 1965.

———. *Discourse on Metaphysics*. Buffalo: Prometheus, 1992.

Leron, U. "Structuring Mathematical Proofs." *American Mathematical Monthly* 90: 174–85, 1983.

Lewis, C. I. *A Survey of Symbolic Logic*. Berkeley: University of California Press, 1918.

Lichnerowicz, A. "Rémarques sur les mathématiques et la réalité." In *Logique at connaissance scientifique*. Dijon: Encyclopédie de la Pléaiade, 1967.

Lighthill, M. J. *Fourier Analysis and Generalized Functions*. New York: Cambridge University Press, 1964.

Locke, J. "An Essay Concerning Human Understanding (1690)." In *The Empiricists*. Garden City, N.Y.: Dolphin, 1961.

MacLane, S. *Mathematics: Form and Function*. New York: Springer-Verlag, 1986.

Macrae, N. *John von Neumann*. New York: Pantheon, 1992.

Maddy, P. *Realism in Mathematics*. New York: Oxford University Press, 1992.

Malcolm, N. *Ludwig Wittgenstein: A Memoir*. New York: Oxford University Press, 1984.

Manin, Yu. I. *A Course in Mathematical Logic* New York: Springer-Verlag, 1977.

Martin, G. *Kant's Metaphysics and Theory of Science*. New York: Barnes and Noble, 1955.

Maurer, A. A. "Nicholas of Cusa." In *The Encyclopedia of Philosophy*. New York: Macmillan, 1976, pp. 496–98.

Maziarz, E. A. and T. Greenwood. *Greek Mathematical Philosophy*. New York: Ungar, 1968.

McShea, R. J. *The Political Philosophy of Spinoza*. New York: Columbia University Press, 1968.

Medawar, P. *Pluto's Republic*. New York: Oxford University Press, 1982.

Mehrtens, H. T. "T. S. Kuhn's Theories and Mathematics." *Historia Mathematica* 3: 297–320, 1976.

Menger, K. *Selected Papers in Logic etc*. Dordrecht: Reidel, 1979, chapter 18, "Square Circles" (The Taxicab Geometry), p. 217 (ref. H. C. Curtis, *Am. Math. Monthly* 60: 1953); chapter 21, p. 237, "My Memories of L. E. J. Brouwer," 1978.

Meyer, A. R. "The Inherent Computational Complexity of Theories of Ordered Sets." *Proceedings of the International Congress of Mathematicians 1972* 2: 481, 1974.

Mill, J. S. *A System of Logic, Ratiocinative and Inductive, being a connected view of the principles of evidence and the methods of scientific investigation*, 8th ed. New York: Harper & Brothers, 1874.

——. *Utilitarianism.* Indianapolis: Hackett, 1979.

Miller, G. L. "Riemann's Hypothesis and Tests for Primality." *Journal Comp. Sys. Sci.* 13: 300–17, 1976.

Miller, J. P. *Number in Presence and Absence.* The Hague: Nijhoff, 1982.

Molland, A. G. "Shifting the Foundations: Descartes' Transformation of Ancient Geometry." *Historia Mathematica* 3: 21–79, 1976.

Monk, R. *Ludwig Wittgenstein.* New York: Free Press, 1992.

Mostowski, A. "Thirty Years of Foundational Studies." *Acta Philosophica Fennica* 17: 7, 1965 (quoted by Musgrave, p. 108).

Mueller, I. *Coping with Mathematics (The Greek Way).* Chicago: Morris Fishbein Center for the Study of the History of Science and Medicine. Publication No. 2, 1980.

Musgrave, A. "Logicism Revisited." *British Journal for Philosophy of Science* 28: 99–127, 1977. Quotes Russell, *An Essay on the Foundations of Geometry*, 1897, p. 1.

Nelsen, R. B. *Proofs without Words. Exercises in Visual Thinking.* Washington, D.C.: Mathematical Association of America, 1993.

Nicholas de Cusa. *Idiota de Mente The Layman about Mind.* Translated by Clyde Lee Miller. New York: Abaris Books, 1979.

Nicomachus of Gerasa. *Introduction to Arithmetic.* Translated by Martin Luther D'Orge. Ann Arbor: University of Michigan Press, 1946.

Nidditch, P. H *The Development of Mathematical Logic.* Glencoe, Ill.: Free Press, 1962 (in basic English).

Orwell, G. *Down and Out in Paris and London.* New York: Harcourt Brace, 1961.

Parsons, C. "Quine on the Philosophy of Mathematics." In Hahn and Schilpp.

Passmore, J. *A Hundred Years of Philosophy.* Baltimore: Penguin, 1970, pp. 153–54.

Pears, D. F. *Ludwig Wittgenstein.* New York: Viking, 1970.

Peirce, C. S. "The Essence of Mathematics." In *Essays in the Philosophy of Science.* Indianapolis: Bobbs-Merrill, 1957.

——. *The New Elements of Mathematics.* The Hague: Mouton, 1976.

——. *Collected Papers.* Cambridge: Harvard University Press, 1960. Paragraph 3, p. 426. Reprinted in the *American Mathematical Monthly* 275: 1978.

Peppinghaus, B. "Some Aspects of Wittgenstein's Philosophy of Mathematics." In J. C. Bell, ed., *Proceedings of the Bertrand Russell Memorial Logic Conference.* Uldum, Denmark, 1971; Leeds, 1973.

Péter, R. *Playing with Infinity.* New York: Atheneum, 1964.

Piaget, J. *Growth of Logical Thinking from Childhood to Adolescence.* New York: Basic Books, 1958.

——. *The Child's Conception of Geometry.* With B. Inhelder and A. Szeminka. New York: Basic Books, 1960.

——. *The Child's Conception of Number.* New York: Norton, 1965.

——. *The Child's Conception of Physical Causality.* Totowa, N.J.: Littlefield, 1965.

——. *The Child's Conception of Space.* New York: Norton, 1967.

——. *Early Growth of Logic in the Child; Classification and Seriation.* New York: Norton, 1969.

——. *The Child's Conception of Movement and Speed.* New York: Basic Books, 1970.

——. *Genetic Epistemology.* New York: Columbia University Press, 1970.

——. *Insights and Illusions of Philosophy.* New York: World, 1971.

——. *Psychology and Epistemology.* New York: Grossman, 1971.

——. *Origin of the Idea of Chance in Children.* With B. Inhelder. New York: Norton, 1975.

——. *Epistemology and Psychology of Functions.* Dordrecht: Reidel, 1977.

——. *Morphisms and Categories.* With G. Henriques, E. Ascher, and T. Brown. Hillsdale, N.J.: Erlbaum, 1992.

Pistorius, P. V. *Plotinus and Neoplatonism.* Cambridge: Bowes, 1952.

Plato. *The Republic.* In *Great Dialogues.*

——. *Theaetetus.* Indianapolis: Bobbs-Merrill, 1949.

——. *Great Dialogues.* New York: New American Library, 1956.

——. *Timaeus.* Indianapolis: Bobbs-Merrill, 1959.

——. *Laws* 7: 821–22, *The Collected Dialogues.* Edited by E. Hamilton and H. Cairns. Princeton: Princeton University Press, 1961.

——. *Meno.* Indianapolis: Bobbs-Merrill, 1971.

Plutarch. *The Lives of the Noble Grecians and Romans.* Chicago: Encyclopedia Britannica, Vol. 14, 1982.

Poincaré, H. *The Foundations of Science.* New York: Science Press, 1913.

——. *Science and Method.* New York: Dover, 1952.

——. *Science and Hypothesis.* New York: Dover, 1952.

——. *Mathematics and Science; Last Essays.* New York: Dover, 1963.

——. *New Methods of Celestial Mechanics 1. Periodic and Asymptotic Solutions.* Introduction by D. L. Goroff. American Institute of Physics, 1993.

Polányi, M. *Personal Knowledge.* Chicago: University of Chicago Press, 1962.

——. *The Tacit Dimension.* New York: Doubleday, 1966.

Pole, D. *The Later Philosophy of Wittgenstein.* London: Athlone. 1958.

Pollock, F. *Spinoza, His Life and Philosophy.* London: C. Kegan Paul & Co., 1880.

Pólya, G. *How to Solve It.* Princeton: Princeton University Press, 1945.

——. *Mathematics and Plausible Reasoning.* Princeton: Princeton University Press, 1954.

Pont, J.-C. *L'Aventure des paralleles, Histoire de la géometrie non-euclidienne: précurseurs et attardés.* Berne: Lang, 1986, pp. 248–60 (Russian).

Pope, A. *Poetical Works.* New York: Thomas Y. Crowell, 1896.

Popkin, R. H. *The History of Scepticism from Erasmus to Spinoza.* Berkeley: University of California Press, 1979.

Popper, K. *The Open Society and Its Enemies.* Princeton: Princeton University Press, 1971.

——. "Epistemology without a Knowing Subject" and "On the Theory of the Objective Mind" in *Objective Knowledge.* Oxford: Clarendon Press, 1974.

Preston, R. "The Mountains of Pi." *New Yorker* 68: 36, 1992.

Pritchard, P. *Plato's Philosophy of Mathematics.* Sankt Augustin: Academia Verlag.

Putnam, H. "What Is Mathematical Truth?" In *Mathematics, Matter and Method*, Cambridge University Press; reprinted in T. Tymoczko, ed., *New Directions in the Philosophy of Mathematics.* Cambridge: Birkhauser, 1986.

———. *Representation and Reality.* Cambridge: MIT Press, 1988.

———. "Peirce the Logician." In *Realism with a Human Face.* Cambridge: Harvard University Press, 1990.

Quine, W. V. O. *Methods of Logic.* New York: Holt, 1959.

———. "On Frege's Way Out." In *Selected Logical Papers.* New York: Random House, 1966.

———. "The Scope and Language of Science." In *The Ways of Paradox.* Cambridge: Harvard University Press, 1976.

———. *Quiddities.* Cambridge: Harvard University Press, 1987.

Rabin, M. O. "Probabilistic Algorithms." In J. F. Traub, ed., *Algorithms and Complexity: New Directions and Recent Results.* New York: Academic Press, 1976.

Ratner, J., ed. *The Philosophy of Spinoza.* New York: Modern Library, 1927.

Regier, T. *The Human Semantic Potential.* Cambridge: MIT Press, 1996.

Reichenbach, H. *The Rise of Scientific Philosophy.* Berkeley: University of California Press, 1951.

Reid, C. *Hilbert.* New York: Springer-Verlag, 1970.

Rényi, A. *Dialogues on Mathematics.* San Francisco: Holden Day, 1967.

———. *Letters on Probability.* Detroit: Wayne State University Press, 1972.

———. *A Diary on Information Theory.* Budapest: Akadémiai Kiadó, 1984.

Renz, P. "Mathematical Proof: What It Is and What It Ought to Be." *The Two-Year College Mathematics Journal* 12: 83–103, 1981.

Resnik, M. D. *Frege and the Philosophy of Mathematics.* Ithaca: Cornell University Press, 1980.

Restivo, S. *The Social Relations of Physics, Mysticism, and Mathematics.* Dordrecht: Reidel, 1983.

———. *Mathematics in Society and History.* Dordrecht: Kluwer, 1992.

Restivo, S., J. P. Van Bendegem, and R. Fischer, eds. *Math Worlds.* Albany: State University of New York Press, 1993.

Robinson, A. "From a Formalist's Point of View." *Dialectica* 23: 45, 1969.

———. *Nonstandard Analysis.* Amsterdam: North-Holland, 1974.

Roche, W. J. "Measure, Number, and Weight in Saint Augustine." *New Scholasticism* XV: October 1941.

Rorty, R. *Objectivity, Relativism and Truth.* New York: Cambridge University Press, 1991.

Rosenfeld, B. A. *A History of Non-Euclidean Geometry.* New York: Springer-Verlag, 1988.

Rota, G.-C. *Indiscreet Thoughts.* Cambridge: Birkhauser, 1996.

Roth, L. *Spinoza Descartes & Maimonides.* New York: Russell, 1963.

Rotman, B. *Signifying Nothing: the semiotics of zero.* New York, St. Martin's Press, 1987.

Rotman, B. *Ad Infinitum—The Ghost in Turing's Machine.* Stanford, 1993.

Roxin, E. "A Living and Constructive View of Mathematics." Talk at Brown University, Department of Applied Mathematics, Seminar on Philosophy of Mathematics.

Russell, B. *The Principles of Mathematics.* London: Allen & Unwin, 1937.

———. *A History of Western Philosophy.* New York: Simon & Schuster, 1945.

———. "Reflections on My Eightieth Birthday." In *Portraits from Memory.* New York: Simon & Schuster, 1956.

Russell, B. and A. Whitehead. *Principia Mathematica.* Cambridge: University Press, 1925.

Ryle, G. "Plato." In *The Encyclopedia of Philosophy.* New York: Macmillan, 1967, Vol. 6, pp. 314–33.

Salmon, W. C., ed. *Zeno's Paradoxes.* Indianapolis: Bobbs-Merrill, 1970.

Saunders, J. L., ed. *Greek & Roman Philosophy after Aristotle.* New York: Free Press, 1966.

Schatz, J. A. *The Nature of Truth.* Unpublished manuscript.

Schilpp, P. A., ed. *The Philosophy of Rudoph Carnap.* La Salle, Ill.: Open Court, 1963.

Schirn, M., ed. *Studies on Frege.* Stuttgart: Bad Canstart, 1976.

Schwartz, J. T. "Fast Probabilistic Algorithms for Verification of Polynomial Identities." *Journal of the Association for Computing Machinery* 27: 701–17, 1980.

Scruton, R. *Spinoza.* New York: Oxford University Press, 1986.

Shanker, S. *Wittgenstein and the Turning Point in the Philosophy of Mathematics.* London: Croom Helm, 1987.

Shebar, W. "In Quest of Quine." *Harvard Magazine* 47–51, November–December 1987.

Shwayder, D. S. "Wittgenstein on Mathematics." In P. Winch, ed., *Studies in the Philosophy of Wittgenstein.* London: Routledge, 1969.

Skyrms, B. "Zeno's Paradox of Measure." In Cohen and Lauden.

Sluga, H. Review of *Nachgelassene Schrifte. Journal of Philosophy* 68: 265–272, 1971.

———. "Frege and the Rise of Analytic Philosophy." *Inquiry,* 18: 477, 1973.

———. *Gottlob Frege.* London: Routledge, 1980.

———. *Heidegger's Crisis.* Cambridge: Harvard University Press, 1993.

———, ed. *The Philosophy of Frege.* New York: Garland, 1993.

Smith, D. E. *History of Mathematics,* Vol. I. New York: Dover, 1958, p. 72.

Smorynski, C. "Mathematics as a Cultural System." *Mathematical Intelligencer* 5: (1), 1983.

Snapper, E. "What Is Mathematics?" *American Mathematical Monthly* 86: 551–57, 1979.

Spinoza, B. *Tractatus Theologico-Politicus* and *Tractatus Politicus.* London: Routledge, 1895.

———. *Improvement of the Understanding, Ethics and Correspondence.* Translated by R. H. M. Elwes, intro. by Frank Sewall. New York and London: M. W. Dunne, 1901.

St. Augustine of Hippo. *On the Trinity*, chapter IV. Translated by S. MacKenna. Washington, D. C.: Fathers of the Church Series No. 45, Catholic University of America.

——. *Basic Writings.* Edited by W. J. Oates. New York: Random House, 1948.

——. *The City of God.* New York: Modern Library, 1950.

——. *Introduction to the Philosophy of Saint Augustine Selected Readings and Commentaries.* Edited by J. A. Mourant. University Park, Pa.: Pennsylvania State University Press, 1964.

——. *The Confessions.* London: Collier, 1969.

——. *On Free Choice of the Will.* Indianapolis: Hackett, 1993, p. 89.

Steen, L. "The Science of Patterns." *Science* 240: 611–16, 1988.

Steiner, M. *Mathematical Knowledge.* Ithaca: Cornell University Press, 1975.

Stolzenberg, G. "Can an Inquiry into the Foundations of Mathematics Tell Us Anything Interesting about Mind?" In George Miller, ed. *Psychology and Biology of Language and Thought.* New York: Academic Press.

——. *Contemporary Mathematics.* Providence: American Mathematical Society, 1985, p. 39.

Stone, I. F. *The Trial of Socrates.* Boston: Little, Brown, 1988.

Swart, E. R. "The Philosophical Implications of the Four-color Theorem." *American Mathematical Monthly* 697–707, 1980.

Tarnas, R. *Passion of the Western Mind.* New York: Harmony, 1991.

Taylor, A. E. *Platonism and Its Influence.* New York: Cooper Square Publishers, 1963.

Thiel, C. *Sense and Reference in Frege's Logic.* Dordrecht: Reidel, 1968.

Thom, R. "Modern Mathematics: An Educational and Philosophical Error?" *American Scientist* 59: 695–99, 1971.

Thomas, J. *Musings on the Meno.* The Hague: Nijhoff, 1980.

Thomas, R. Private communication, 1996.

Tichy, P. *The Foundations of Frege's Logic.* New York: de Gruyter, 1988.

Tiles, M. *Mathematics and the Image of Reason.* London: Routledge, 1991.

Tragesser, R. S. *Husserl and Realism in Logic and Mathematics.* New York: Cambridge University Press, 1984.

Turbayne, C. M. *Introduction to Berkeley's Treatise Concerning the Principles of Human Knowledge.* Indianapolis: Bobbs Merrill, 1955–1979, pp. xviii–xix.

Tymoczko, T. "Finding a Place for the Mathematician in the Philosophy of Mathematics." *Mathematical Intelligencer*, 1981.

——. "The Four-Color Problem and Its Philosophical Significance." *Journal of Philosophy* 76: 57–83, 1979.

Ungar, P. Personal communication. 10 October 1989.

van Stift, W. P. *Brouwer's Intuitionism.* Amsterdam: North-Holland, 1990.

van der Waerden, B. L. *Science Awakening.* New York: Wiley, 1963.

van Bendegem, J. P. *Theory and Experiment.* Dordrecht: D. Reidel, 1980.

——"Zeno's Paradoxes and the Tile Argument." *Philosophy of Science* 54: 295–302, 1987.

Vartanian, A. *Diderot and Descartes*. Princeton: Princeton University Press, 1953.

Vlastos, G. "Zeno of Elea." In P. Edwards, ed., *The Encyclopedia of Philosophy*, Vol. vii, pp. 369–79. New York: Macmillan, 1967.

von Neumann, J. "The Mathematician." In R. B. Heywood, ed., *Works of the Mind*. Chicago: University of Chicago Press, 1947.

Vrooman, J. V. *René Descartes, A Biography*. New York: Putnam, 1970.

Walsh, W. H. "Immanuel Kant." In P. Edwards, ed., *The Encyclopedia of Philosophy*, pp. 305–24. New York: Macmillan, 1967.

Wang, H. "Proving Theorems by Pattern Recognition." *Communications of the Association for Computing Machinery* 3: April 1960.

——. *Popular Lectures on Mathematical Logic*. New York: van Nostrand Reinhold, 1971.

——. *From Mathematics to Philosophy*. London: Routledge & Kegan, 1972.

——. "Toward Mechanical Mathematics." International Business Machines Corporation, 1960, reprinted in J. Siekmann and G. Wrightson, eds., *Automation of Reasoning*, pp. 229–66. Berlin: Springer-Verlag, 1983.

——. *Beyond Analytic Philosophy . . . Doing Justice to What We Know*. Cambridge: MIT Press, 1988.

——. *Reflections on Kurt Gödel*. Cambridge: MIT Press, 1991.

——. "To and from Philosophy: Discussions with Gödel and Wittgenstein." *Synthèse* 88: 229–77, 1991.

——. "Computer Theorem Proving and Artificial Intelligence." In J.-L. Lassez and G. Plotkin, eds., *Computational Logic*. Cambridge: MIT Press, 1991.

——. "Imagined Discussions with Gödel and with Wittgenstein." In *Yearbook of the Kurt Gödel Society*. Vienna: Kurt Gödel Society, 1992.

——. "Quine's Logical Ideas in Historical Perspective." In Hahn and Schilpp.

Wedberg, A. *Plato's Philosophy of Mathematics*. Stockholm: Almsqvist & Wiksell, 1955.

Wheeler, J. *Magic without Magic*. San Francisco: Freeman, 1972.

Wheelwright, P. *The Presocratics*. Indianapolis: Bobbs-Merrill, 1981.

White, L. A. "The Locus of Mathematical Reality." *Philosophy of Science* 14: 289–303. Also chapter 10, *The Science of Culture: A Study of Man and Civilization*. New York: Farrar, Straus, 1949.

White, M. J. "Zeno's Arrow, Divisible Infinitesimals, and Chrysippus." *Phronesis* 27: 239–54, 1982.

——. *The Continuous and the Discrete*. New York: Oxford University Press, 1992.

Whiteside, D. T. *The Mathematical Papers of Isaac Newton*, Vol. 7. New York: Cambridge University Press, 1976.

Whitman, W. "Song of Myself," Section 51. In *Leaves of Grass*. Philadelphia: McKay, 1900.

Wiener, N. *Cybernetics*. New York: Wiley, 1948.

——. *Ex-Prodigy*. Cambridge: MIT Press, 1953.

——. *I Am a Mathematician*. Garden City, N.Y.: Doubleday, 1956.

Wilder, R. L. *Introduction to the Foundations of Mathematics.* New York: Wiley, 1968.

——. *Evolution of Mathematical Concepts An Elementary Study.* New York: Wiley, 1968.

——. *Mathematics as a Cultural System.* New York: Pergamon, 1981.

Wittgenstein, L. *Tractatus Logico-Philosophicus.* London: Routledge & Kegan Paul, 1922.

——. *Philosophical Investigations.* New York: Macmillan, 1953.

——. *The Blue and Brown Books.* London: Blackwell, 1958.

——. *Philosophical Grammar.* Berkeley: University of California Press, 1974.

——. *Lectures on the Foundations of Mathematics.* Ithaca: Cornell University Press, 1976.

——. *Remarks on Color.* Berkeley: University of California Press, 1977.

——. *Remarks on the Foundations of Mathematics.* Cambridge: MIT Press, 1983.

Wolff, C. *Preliminary Discourse on Philosophy in General.* Indianapolis: Bobbs-Merrill, 1963.

Woodger, J. *The Axiomatic Method in Biology.* New York: Cambridge University Press, 1937.

Wos, L. "The Impossibility of the Automation of Logical Reasoning." *Automated Deduction—CADE-11.* In D. Kapur, ed., pp. 1–3. New York: Springer-Verlag, 1992.

Wright, C. *Wittgenstein on the Foundations of Mathematics.* Cambridge: Harvard University Press, 1980.

Yates, F. *Giordano Bruno and the Hermetic Tradition.* Chicago: University of Chicago Press, 1964.

Zabeeh, F. "Hume's Scepticism with Regard to Deductive Reason." *Ratio* 2: 134–43, 1960.

Index